AF333979

CELL DEATH - AUTOPHAGY, APOPTOSIS AND NECROSIS

Edited by **Tobias M. Ntuli**

Cell Death - Autophagy, Apoptosis and Necrosis
http://dx.doi.org/10.5772/59648
Edited by Tobias M. Ntuli

Contributors

Daolin Tang, Rui Kang, Rubem Menna-Barreto, Solange DeCastro, Bassam Janji, Hatice Yorulmaz, Lesetja Motadi, Joanna Zarzynska, Maria Roccheri, Maria Agnello, Liana Bosco, Roberto Chiarelli, Chiara Martino, Yuzuru Imai, Taku Arano, Tatyana Volkova, Yibin Feng, Ming Hong, Ning Wang, José Antonio Sánchez-Alcázar, Manuel Oropesa Ávila, David Cotán, Mario De La Mata, Marina Villanueva Paz, Ana Delgado Pavón, Isabel De Lavera, Juan Garrido Maraver, Mario D. Cordero, Elizabet Alcocer Gómez, Alejandro Fernández Vega, Mónica Álvarez Córdoba, Maria Ines Vaccaro, Daniel Grasso, Alejandro Ropolo, Gerardo Hebert Vázquez-Nin, Ml Escobar, Echeverría Om, Tassula Proikas-Cezanne, Katharina Sporbeck, Fenja Odendall, Daotai Nie, Karin Von Schwarzenberg, Michiko Shintani, Sandra Halonen

CBS, Edition **2017**

Published by InTech
Janeza Trdine 9, 51000 Rijeka, Croatia
The moral rights of the editor(s) and the author(s) have been asserted.
All rights to the book as a whole are reserved by InTech. The book as a whole (compilation) cannot be reproduced, distributed or used for commercial or non-commercial purposes without InTech's written permission. Enquiries concerning the use of the book should be directed to InTech's rights and permissions department (permissions@intechopen.com).
Violations are liable to prosecution under the governing Copyright Law.

Notice

Statements and opinions expressed in the chapters are these of the individual contributors and not necessarily those of the editors or publisher. No responsibility is accepted for the accuracy of information contained in the published chapters. The publisher assumes no responsibility for any damage or injury to persons or property arising out of the use of any materials, instructions, methods or ideas contained in the book.

Publishing Process Manager Iva Simcic

Technical Editor InTech DTP team

Cover InTech Design team

Additional hard copies can be obtained from orders@intechopen.com

Cell Death - Autophagy, Apoptosis and Necrosis , Edited by Tobias M. Ntuli
p. cm.
ISBN 978-953-51-2236-4

Contents

Preface

Autophagy is a genetic program that secures the survival of eukaryotic cells to compensate for periods of starvation and cellular stress. Apoptosis, in contrast, is a genetically defined program leading to cell death. Both pathways are interconnected through conserved co-regulatory signaling pathways that are context-dependent. Proper co-regulation of autophagy and apoptosis, whereby autophagy exerts an anti-apoptotic function and apoptosis inhibits autophagic survival strategies, critically secures the survival of healthy cells and counteract genomic instability in eukaryotic organisms. Necrosis is a form of cell injury which results in the premature deathof cells in living tissue by autolysis.

Dr. Tobias M. Ntuli
Mangosuthu University of Technology,
Durban, KwaZulu-Natal
Republic of South Africa

Autophagy

Chapter 1

Autophagy in Cell Fate and Diseases

Daniel Grasso, Alejandro Ropolo and Maria I. Vaccaro

Additional information is available at the end of the chapter

http://dx.doi.org/10.5772/61553

Abstract

Autophagy pathway has been one of the hot topics during the last decade. From a general notion about its cellular role, autophagy becomes a more sophisticated phenomenon with significant implications in cellular homeostasis. Consequently, autophagy represents an emerging new factor in human diseases. Despite its general task, the bulk degradation of cellular constituents during starvation settings, autophagy possesses important cross talk and interrelationships with several cellular processes such as apoptosis and senescence, among others. This entire panorama gives us a complex but exciting scenario. Consequently, with the aim of encompassing the whole spectrum, in this chapter, we review three main topics: autophagy as a cellular process; autophagy in cell fate; and autophagy in disease. We discuss the emerging role of selective type of autophagy to avoid apoptosis or necrosis and the novel relationship between autophagy and senescence to understand the real extent that autophagy pathway has over cell fate. Finally, we briefly describe the current trends on autophagy in human pancreatic diseases and its role in cancer cell metabolism.

Keywords: VMP1, zymophagy, senescence, acute pancreatitis, pancreatic cancer

1. Introduction

Autophagy is a highly regulated cellular pathway for degrading long-lived proteins and is the only known pathway for clearing cytoplasmic organelles. Autophagy is a major contributor to maintain cellular homeostasis and metabolism.

Autophagy is an evolutionarily conserved and highly regulated lysosomal pathway that degrades macromolecules (e.g., proteins, glycogen, lipids, and nucleotides) and cytoplasmic

organelles [1-3]. This catabolic process is involved in the turnover of long-lived proteins and other cellular macromolecules and it might play a protective role in development, aging, cell death, and defense against intracellular pathogens [4,5]. Moreover, autophagy has been linked to a variety of pathological processes such as neurodegenerative diseases and tumorigenesis, which highlights its biological and medical importance [6,7].

Although autophagy was first identified in mammalian liver upon glucagon treatment approximately 50 years ago, its molecular understanding began only in the past decade, largely based on the discovery of the autophagy-related genes (ATGs) by genetic analyses in yeast. Since the discovery of the yeast ATG proteins, autophagosome formation has been dissected at the molecular level, but a lot of questions about this pathway remain unanswered. In mammalian cells, the sequential association of at least a subset of the ATG proteins leads to the assembly of the pre-autophagosome structure (PAS). PAS formation also requires PtdIns3P generation and it is thought that this lipid is present in specialized subdomains of membranes where the PAS is assembled and autophagosomes are generated.

One of the autophagy-related proteins is VACUOLE MEMBRANE PROTEIN 1 (VMP1), whose expression triggers autophagy in mammalian cells even under nutrient-rich conditions. Conversely, autophagy is completely blocked in the absence of VMP1. VMP1 is required for the biogenesis of autophagosomes in mammalian cells in all conditions underscoring its upstream regulatory function in autophagy. Importantly, VMP1 is also expressed early during the onset of several pathologies including diabetes mellitus, pancreatitis, and pancreatic cancer.

The presence of autophagy has been described in dying acinar cells in tissue from human diseases. Though controversy exists about whether autophagy could be at the same time a survival cell response and a programmed cell death pathway, there is no discussion that autophagy has the power to change the fate of a cell. From an energetic homeostasis point of view, autophagy is a major cellular response against starvation condition and plays key roles in other cellular stress conditions such as oxidative damage. In those cases, autophagy is not only a way to keep energy and nutrition but also a specific mechanism of cell adaptive response.

Increasing evidence suggests that autophagy may influence the pathogenesis of human diseases including cancer, neurodegenerative diseases, inflammatory diseases, and metabolic diseases. Several inherited myopathies are associated with aberrant autophagy. Recent investigations have explored the functions of autophagy in the pathogenesis of metabolic disorders such as diabetes, insulin resistance, and obesity, and the progression of aging.

It is possible to think that different types of autophagy may be involved in the initial cellular events induced by the noxa. Autophagic processes might be triggered in different cells by several diseases and probably they are more complex than we actually understand. These pathways function as adaptive responses that mostly act as protective mechanisms against cell healing. The knowledge of the molecular mechanisms of a sophisticated membrane transport system such as autophagy would provide bases for novel and more rational diagnostic and therapeutic strategies.

2. Autophagy, a complex cell event

Autophagy consists of several sequential steps: induction, autophagosome formation, autophagosome-lysosome fusion, and degradation. Depending on the delivery route of the cytoplasmic material to the lysosome, there are three major types of autophagy in eukaryotes: 1) chaperone-mediated autophagy (CMA), 2) microautophagy, and 3) macroautophagy, hereafter referred to as autophagy [8]. CMA allows the direct lysosomal import of unfolded soluble proteins that contain a particular pentapeptide motif. In microautophagy, cytoplasmic material is directly engulfed into the lysosome at the surface of the lysosome by membrane rearrangement. Finally, autophagy involves the sequestration of cytoplasm into a double-membrane cytosolic vesicle referred to as an autophagosome that subsequently fuses with a lysosome to form an autolysosome for degradation by lysosomal hydrolases [9].

2.1. The autophagic process

Autophagy is characterized by sequestration of bulk cytoplasm and organelles in double-membrane vesicles called autophagosomes that eventually acquire lysosomal-like features [9,10]. The autophagic process is described in Figure 1.

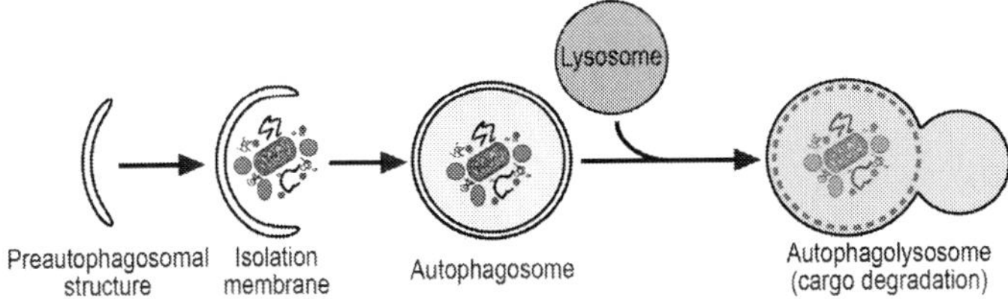

Figure 1. General scheme of autophagic process. During autophagy an isolation membrane forms, invaginates and sequesters cytoplasmic constituents. The edges of the membrane fuse to form the autophagosome. The outer membrane of the autophagosome fuses with the lysosome to form the autolysosome, where the cargo is degraded.

Briefly, sequestration of cytoplasm into a double-membrane cytosolic vesicle is followed by the fusion of the vesicle with a late endosome or lysosome to form an autophagolysosome (or autolysosome). Then, inner membrane of the autophagosome and autophagosome-containing cytoplasm-derived materials are degraded by lysosomal/vacuolar hydrolases inside the autophagosome. The molecular mechanisms underlying the transport and fusion of autophagosomes are just beginning to be understood and through active investigations, several major events involved in the process have recently been clarified including the recycling of lysosomes [11]. In mammalian cells, autophagosome maturation is a prior step for the fusion between autophagosomes and lysosomes. The degradation products, including macromolecules, are then exported to the cytosol for reuse by the cell.

2.2. The autophagosome at a molecular level

Since the discovery of yeast ATG proteins, autophagosome formation has been dissected at the molecular level but a lot of questions about the molecular mechanism underlying this process remain unanswered. Autophagosomes can be considered unique organelles because they do not contain marker proteins of other subcellular compartments [12]. In mammalian cells, the sequential association of at least a subset of the ATG proteins leads to the assembly of the PAS that is believed to be the site where the precursor structure of the autophagosomes, the phagophores, are generated [13]. The PAS and phagophore formation also requires phosphatidylinositol 3-phosphate (PI3P) [14] and it is believed to be associated to specific subdomains of the endoplasmic reticulum (ER), termed omegasomes [15,16]. Among the key mediators initiating autophagosome formation, there is a set of evolutionarily conserved ATG gene products: the kinase-containing Ulk1/2 complex (ATG1 in yeast), the Class III phosphatidylinositol 3-kinase (PI3K) complex (composed by BECN1/ATG6-hVps34, hVps15 and ATG14L), the ubiquitin-like conjugation systems leading to the formation of the ATG5–ATG12–ATG16L1 complex, and the LC3/ATG8 phosphatidylethanolamine-conjugate (e.g., LC3-II) [17]. A second group of ATG proteins, which does not have orthologous group in yeast, has also recently emerged and appears to play a key role in regulating autophagy in high eukaryotes. One of these proteins is the transmembrane protein VMP1, whose expression triggers autophagy in mammalian cells even under nutrient-rich conditions [18,19]. Conversely, autophagy is completely blocked in the absence of VMP1 [18].

The autophagosome formation process is composed of isolation membrane nucleation, elongation, and completion steps. In mammals, the Class III PI3K plays an essential role in isolation membrane nucleation during autophagy [20]. The Class III PI3K is associated with BECN1/ATG6 and p150, the homolog of Vps15 (phosphoinositide-3- kinase regulatory subunit 4), to form the PI3K complex. This kinase catalyzes the generation of PI3P on the autophagosomal membrane, favoring the localization of other ATG proteins to the PAS during autophagosomal formation. The autophagosome nucleation system is ATG12-ATG5-ATG16L, which is a ubiquitin-like protein conjugation system essential for the formation of the PAS. ATG12 is conjugated to ATG5 [21]. E1-like ATG7 activates the carboxyl-terminal glycine residue of ATG12 through a high-energy thioester bond in an ATP-dependent manner. The ATG12-ATG5 conjugate further interacts with ATG16L1 to form a ~350 kDa multimeric ATG12-ATG5-ATG16L protein complex through the homo-oligomerization of ATG16L [22]. Another ubiquitin-like protein conjugation system is the modification of LC3 (a mammalian homolog of ATG8) by the phospholipid phosphatidylethanolamine (PE) [22], an essential process for the formation of autophagosomes. The cytosolic form of LC3 (LC3-I) is cleaved by cysteine protease ATG4 and then conjugated with PE by ATG7 and ATG3. This lipidated LC3 (LC3-II) then associates with newly forming autophagosome membranes. LC3-II remains on mature autophagosomes until its fusion with lysosomes [23, 24]. The conversion of LC3-I to LC3-II is thus well known as a marker of autophagy (Figure 2). However, the increase of LC3-II alone is not enough to show autophagy activation because the inhibition of LC3-II degradation in the lysosome by the impaired autophagy flux can also cause its accumulation.

While the origin of autophagic vacuoles remains disputable, several hypotheses have been proposed for the source of autophagosomal membrane during autophagosome formation. The first hypothesis is "de novo" formation of autophagosome by ATG9 reservoirs. In the second hypothesis, various organelles such as ER, mitochondria, and plasma membrane are used as

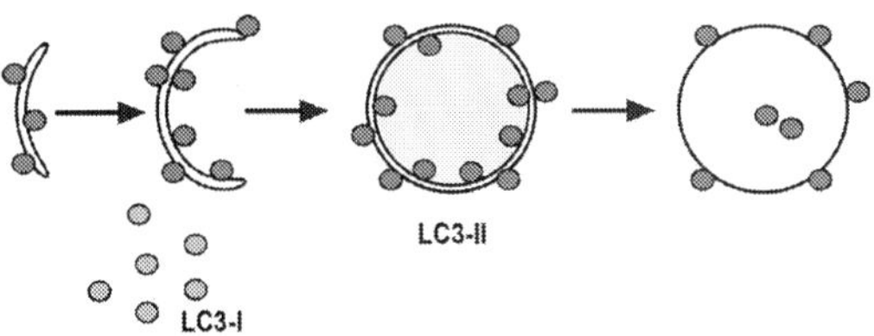

Figure 2. During autophagy the cytosolic form of LC3 (LC3-I) undergoes C-terminal proteolysis and lipidation (LC3-II) and translocates to the autophagosomal membrane. LC3 is currently used as a specific marker of autophagy.

an origin for the formation of the phagophore. Recently, cup-shaped structures called omegasome, a discrete region of the ER, were identified as a platform for autophagosome formation [25]. The ATG5 complex, LC3, and ULK1 have been shown to recruit into the omegasome after starvation, and ATG5- and LC3-positive membranes seem to emerge from the omegasome. It was also observed that omegasomes form in close proximity to the PI3K-containing vesicles which may synthesize the PI3P. This hypothesis is also supported by a notion of a physical association between the ER and early autophagic membranes [26].

2.3. Autophagy induction

Basal autophagy in unstressed cells is kept down by the action of the mammalian target of rapamycin complex 1 (mTORC1). Key upstream regulators of mTORC1 include the class I phosphoinositide 3-kinase-Akt pathway which keeps mTORC1 active in cells with sufficient growth factors and the AMP-activated protein kinase (AMPK) pathway that inhibits mTORC1 upon starvation and calcium signals [27,28].

Under stress conditions such as amino acid starvation, autophagy is strongly induced in many types of cultured cells. The effects of individual amino acids differ in their abilities to regulate autophagy. Amino acids including Leu, Tyr, Phe, Gln, Pro, His, Trp, Met, and Ala suppress autophagy in ex vivo perfused liver [29]. However, such profiles depend on cell types showing their different amino acid metabolisms in tissues. The questions on how cells sense amino acid concentration and physiological significance of autophagy regulation by amino acid starvation are not fully understood yet. It has been demonstrated that amino acid signaling pathways exist, which involve activation of mTORC1 and the subsequent regulation of the Class III PI3K. mTORC1 is involved in the control of multiple cell processes in response to changes in nutrient conditions [30]. Especially, mTORC1 requires Rag GTPase, Rheb, and Vps34 for its activation and subsequent inhibition of autophagy in response to amino acids [31, 32]. Energy levels are primarily sensed by AMPK, a key factor for cellular energy homeostasis. In low energy states, AMPK is activated and the activated AMPK then inactivates mTORC1 through TSC1/TSC2 and Rheb protein [33].

Thus, inactivation of mTORC1 is essential for the induction of autophagy and plays a central role in autophagy. In addition to amino acid signaling, hormones, growth factors, and many other factors including Bcl-2 [34] have also been reported to regulate autophagy. But, not all autophagy signals are transduced through mTOR signaling. A recent study showed that small-molecule enhancers of the cytostatic effects of rapamycin (called SMERs) induce autophagy independently of mTOR [35]. Activities of the ULK1 are regulated by mTOR depending on nutrient conditions. Under growing and high-nutrient conditions, active mTORC1 interacts with ULK1 and phosphorylates ULK1 and mATG13 and, thus, inhibits the membrane targeting of the ULK1. During starvation condition, on the other hand, inactivated mTORC1 dissociates from ULK1 and results in the ULK1 complex formation (ULK1-mATG13-FIP200-ATG101) leading to autophagy induction [36].

2.4. The autophagy-related protein Beclin 1 (BECN1)

BECN1 (former Beclin 1), the mammalian ATG6, is a haploinsufficient tumor suppressor and an important effector of autophagy. BECN1 is a subunit of the PI3K complex, the action of which is antagonized by Bcl-2 [36,37]. BECN1 contains a BH3 domain that mediates its interaction with Bcl-2 [38,39]. The interaction between Bcl-2 and BECN1 leads to inhibition of autophagy by interfering with the formation and activity of the autophagy promoter complex, BECN1- PI3K [40] (Figure 3).

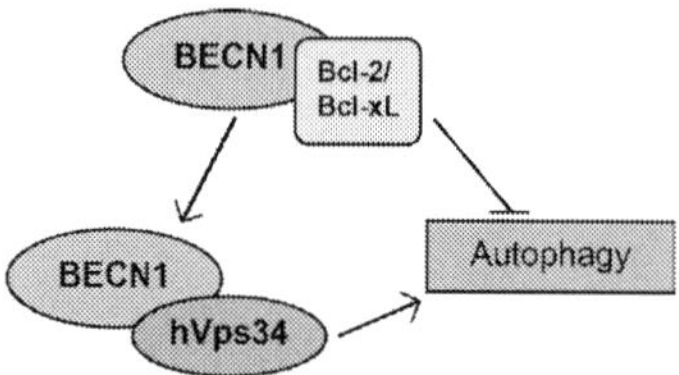

Figure 3. BECN1 is an effector of autophagy whose interaction with Bcl-2 inhibits its role as autophagy promoter. BECN1 and the PI3K complex, where hVP34 is the active subunit, are necessary for PI3P production over the autophagic structures and the consequent recruitment of additional ATG proteins.

2.5. The Vacuole Membrane Protein 1 (VMP1)

The pancreatitis-associated protein named VMP1 is a transmembrane protein with no known homologs in yeast. VMP1 was found searching for new molecules that were differentially expressed during acute pancreatitis [41]. VMP1 expression is induced by mutated K-Ras in pancreatic cancer cells [42] and by hyperstimulation of Gq-coupled cholecystokinin receptor (CCK-R) in pancreatic acinar cells during acute pancreatitis [43]. In the adult normal pancreas, VMP1 expression is not detectable but it is highly induced early during experimental acute pancreatitis and its expression levels correlate with morphological features resembling autophagy [44]. Moreover, VMP1 expression can be found in pancreatic acinar cells from rats developing spontaneous chronic pancreatitis (WBN/Kob rats) [45] and it is rapidly and highly expressed in experimental diabetes [45-47]. Finally, gemcitabine (2,2-difluorodeoxycytidine), the standard chemotherapy for the treatment of advanced pancreatic cancer, induces VMP1 expression in human pancreatic cancer cells [48-50]. We demonstrated that VMP1 expression triggers autophagosome formation in mammalian cells even under nutrient-replete conditions [18,19]. Remarkably, VMP1 pancreas-specific transgenic expression in mice promotes autophagosome formation in acinar cells. Therefore, VMP1 expression may be involved in autophagy induction during acute pancreatitis, a disease defined as pancreas self-digestion. Furthermore, VMP1 is the only human-disease-inducible ATG-protein described so far.

VMP1 interacts with BECN1 through its hydrophilic C-terminal region (VMP1-ATGD) that is necessary for early steps of autophagosome formation. Hierarchical analyses in mammalian cells shows that VMP1 along with ULK1 localizes in the autophagosome formation site [13]. VMP1-BECN1 interaction is required for the formation of the PI3K complex acting in mammalian autophagy. This complex, which is composed of BECN1-hVps34-ATG14, promotes PI3P generation on autophagosomal membrane, favoring the localization of ATG16L1 and LC3 to the autophagosomal membrane during autophagosome formation [51,52]. The interaction between VMP1 and BECN1 requires the BECN1 domain that binds to Bcl-2 (BECN1-BH3 domain) [51]. During normal growth conditions, Bcl-2 binding to BECN1 is maximal, and when autophagy is induced, this interaction is strongly reduced [53]. VMP1 expression leads to the dissolution of the BECN1-Bcl-2 complex, indicating that VMP1 is involved in driving BECN1 into the autophagic process [38]. Thus, VMP1-BECN1 interaction through the VMP1-ATGD is required for the proper localization of PI3K activity on the autophagosomal membrane during mammalian autophagy, positioning VMP1 as a key regulator of the early steps of autophagosome formation possibly acting as a platform in the autophagosomal membrane.

2.6. Autophagy, a selective process

Early studies suggested that autophagy was a nonselective process in which cytoplasmic structures were randomly sequestered into autophagosomes before being delivered to the mammalian lysosome or the plant and yeast vacuole for degradation. Now, there is growing evidence that unwanted cellular structures can be selectively recognized and exclusively eliminated within cells. This is achieved through the action of specific autophagy receptors such as Nbr1 and p62, which is a ubiquitin-binding protein that interacts with LC3 [10,54]. Thus, excess or damaged organelles, including mitochondria, peroxisomes, lipid droplets, endoplasmic reticulum, and ribosomes, can be specifically sequestered by autophagosomes and targeted to the lysosome for degradation. Importantly, there is growing evidence that selective autophagy subtypes also have a wide range of physiological functions. Selective autophagic pathways target distinct cargoes to autophagosomes including mechanisms for the clearance of aggregated protein and for the removal of dysfunctional mitochondria (mitophagy). In pancreatic cells, autophagy has recently been shown to specifically turn over secretory granules damaged by acute pancreatitis as a protective cellular response [43].

2.7. Zymophagy, a novel selective type of autophagy

The pancreatic acinar cell activates VMP1-mediated autophagy early during acute pancreatitis [44]. Relevant data about the role of autophagy in pancreas were obtained using ElaI-VMP1 mice in which the pancreaatic acinar-cell-specific elastase promoter drives VMP1 expression in pancreas. Pancreases of these transgenic mice show numerous vesicles that stain for endogenous LC3, indicating that VMP1 induces the autophagosome formation and, therefore, autophagy [18]. Interestingly, ElaI-VMP1 mice do not develop pancreatitis in normal conditions, confirming that autophagosome formation does not induce acute pancreatitis [18]. The immunomagnetic isolation of VMP1-autophagosomes containing zymogen granules from the ElaI-VMP1 transgenic mouse pancreas with acute pancreatitis allowed the discovery of a new type of selective autophagy named zymophagy that functions as an inducible cellular process that recognizes and degrades activated zymogen granules [43].

Zymophagy is characterized by the formation of autophagosomes containing zymogen granules. These organelles mediate the sequestration and degradation of pancreatitis-activated zymogen granules. CCK-R hyperstimulation with cerulein in wild-type animals, a classical model of acute pancreatitis, induced a markedly altered distribution pattern of the secretory granules such as fusion among zymogen granules as well as their fusion with condensing vacuoles. In addition, acinar cells lose their polarity, which results in the relocation of zymogen granules to the basolateral membrane. Surprisingly, ElaI-VMP1 mice subjected to CCK-R hyperstimulation reveal that acinar cells preserve their structure and polarity with negligible or no alteration in vesicular transport. Instead, pancreases from cerulein-treated ElaI-VMP1 mice presented autophagosomes containing zymogen granules displaying a distinct localization to the apical area of the acinar cell. This observation is confirmed using isolated mouse pancreas acini revealing that 15 min after cerulein treatment, zymophagy is detected [43]. The finding of different maturation levels of selective autophagic vesicles as well as the degradation of p62 provide evidence that autophagic flow remains primarily unchanged under CCK-R hyperstimulation [43].

Regarding the pathophysiological relevance of zymophagy during acute pancreatitis, it was demonstrated that zymophagy protects acinar cells from intracellular trypsinogen activation triggered in vivo by experimental pancreatitis induced by cerulein. Upon CCK-R hyperstimulation, wild-type mice developed acute pancreatitis with high amylase and lipase serum levels. On the contrary, enzymatic levels in cerulein-treated ElaI-VMP1 mice were significantly lower compared with wild-type mice. Consistently, ElaI-VMP1 mouse pancreas showed remarkably less macroscopic evidence of acute pancreatitis compared with wild-type animals that showed marked edema and hemorrhage. Histological analyses displayed a high degree of necrosis as well as inflammation in wild-type pancreas with acute pancreatitis. In contrast, neither necrosis nor significant inflammation was seen in cerulein-treated ElaI-VMP1 mice [41,43]. Thus, results obtained in the transgenic animal model showed that zymophagy functions as a protective pathophysiological mechanism against pancreatitis-associated injury.

Upon CCK-R hyperstimulation, acinar cells from wild-type mice showed early cytoplasmic trypsinogen activation, which is a hallmark of pancreatitis pathophysiology. Surprisingly, in acinar cells from ElaI-VMP1 mice, CCK hyperstimulation caused almost no activation of trypsinogen. Microscopic examinations using BZiPAR (rhodamine 110 bis-[CBZ-L-isoleucyl-L-prolyl-L-arginine amide] dihydrochloride), a cell permeable substrate that becomes fluorescent after the cleavage by the protease revealed only few activated granules that highly colocalize with VMP1, showing that zymophagy selectively sequesters the activated zymogen granules. Zymogen activation is an enzymatic chain reaction where initial zymogen granule alterations trigger rapid spread of active trypsin within the acinar cell. We think that the degradation of early-activated zymogen granules by zymophagy prevents this deleterious event. Interestingly, the inhibition of autophagic flow markedly increased trypsin activity within acinar cells in ElaI-VMP1 mouse pancreases under CCK-R hyperstimulation confirming that zymophagy specifically degrades those zymogen granules that are initially activated by acute pancreatitis [43]. This function is confirmed in the in vivo animal model of acute pancreatitis where the ability of the ElaI-VMP1 mouse developing zymophagy clearly prevents the increment of enzymatic markers of pancreatic damage and morphological changes characteristic of acute pancreatitis.

Analysis of autophagosomes containing zymogen granules magnetically immunopurified from the pancreas of ElaI-VMP1 mice treated with cerulein revealed that, apart from zymogen granules, isolated vesicles contained LC3-II and, notably, strong signals of p62. Moreover, GFP-ubiquitin-transfected acinar cells subjected to CCK-R hyperstimulation showed colocalization between activated granules and ubiquitin aggregates but do not show colocalization between unaffected or normal zymogen granules and ubiquitin, indicating that the ubiquitin system serves as a targeting signal for activated zymogen granules during zymophagy. Therefore, activated zymogen granules are directly or indirectly ubiquitinated for their recognition by autophagic membranes in which ubiquitin acts as a label for selective engulfment. Nevertheless, activated zymogen granules were ubiquitinated upon acute pancreatitis and the VMP1-mediated selective autophagic pathway sequestered these ubiquitinated granules [43]. p62 may function as a cargo receptor during zymophagy. These data demonstrate for the first time that ubiquitin modifications may possess an additional function in acinar cells by promoting the degradation of highly harmful activated zymogen granules [55].

Zymophagy prevents pancreatic acinar cell death induced by CCK-R hyperstimulation [43]. Autophagosome formation inhibition with 3-methyladenine as well as autophagy flux interruption with vinblastine significantly reduced acinar cell survival in a cell model of acute pancreatitis. Moreover, VMP1 downregulation (shVMP1) also significantly decreases acinar cell survival under CCK-R hyperstimulation, showing that VMP1 expression and autophagy is required to prevent acinar cell death in acute pancreatitis. Therefore, VMP1 expression is activated in acinar cells to mediate zymophagy as a protective cellular response against cell death [55].

Furthermore, VMP1 expression and zymophagy are present in human pancreas affected by acute pancreatitis [43,55]. VMP1 is not detectable in human normal pancreas tissue but its expression is activated in human pancreatitis pancreatic specimens and highly colocalized with LC3 in autophagosomes. Moreover, autophagosomes markedly colocalized with zymogen granules. Remarkably, the finding of large autolysosomes without trypsin signal in pancreas of human pancreatitis suggests that affected zymogen granules are eventually degraded by zymophagy during human pancreatitis.

3. Autophagy in cell fate

Though there was controversy about whether autophagy could be at the same time a survival cell response and a programmed cell death pathway, there is none that autophagy has the power to change the fate of a cell. From an energetic homeostasis point of view, autophagy is a major cellular response against starvation condition. Nevertheless, we have to keep in mind that autophagy plays key roles also in other cellular stress conditions such as oxidative damage [57,58], damaged organelles elimination [59-61], depletion of toxic proteins aggregates [62], host response to microorganisms [61], etc. In those cases, autophagy is not just a way to keep energy and cellular material but a specific cellular stress response. Therefore, it is not difficult to imagine that autophagy could be observed in several cellular life-threatening situations and, then, its pro-survival or pro-death role becomes diffuse. Solid and ineligible is the fact that, independently, its situation-specific role, a determined autophagy process, has a profound impact in the cellular fate.

3.1. Autophagy as nutritional stress response

We must begin our analysis from the most basic and evolutionarily conserved autophagy duty as nutrient stress response. In the starvation context, autophagy plays the first explored, and may be more obvious, task of autophagy, that is, the energy cellular support during nutrient-limiting conditions. Though the bulk degradation and recycling of cytoplasmic portions seem to be a simple event in cellular life, it has major consequences in cell fate and organismal adaptation. As it could be imagined, a proper autophagic flux during a challenging cellular nutritional status might be determinant for the cell survival. Furthermore, the importance of starvation-induced autophagy for cell survival can have beneficial or detrimental consequences to tissue in context dependence. For instance, autophagy could give a life opportunity to cells under urgent energy requirement such as organ starvation, ischemia, hypoxia, etc., being beneficial for a determinate tissue. On the other hand, the same mechanism gives to pancreatic cancer cells enough adaptation to survive in a highly hipoperfunded environment of the pancreatic tumor [63]. This dual behavior could also be observed with AMPK (5' AMP-activated protein kinase) which is one of the master regulators of the cellular energetic homeostasis and a direct autophagy trigger through ULK1 phosphorylation [64]. Similar to what is observed with autophagy, AMPK could be a tumor suppressor, stopping all anabolic process and activating all the catabolic ones including autophagy (probably contributing to the oncogene-induced senescence – see below) but, in other circumstance, it gives tumor high resistance to stress [65].

3.2. Autophagy and cell death

Connections between autophagy and disease have attracted an increasing amount of attention. By morphological studies, autophagy has been linked to a variety of pathological processes and autophagy was associated with cell death. Taking cell death as one major topic in cell fate, there are a large amount of data concerning the complex relationship between autophagy and apoptosis [65,66]. However, this relationship is far from being fully understood and, in many cases, it seems to be context-dependent. This autophagy–apoptosis relationship could be observed from the beginning since they share several inducing factors such as ROS [57], increased levels of cytosolic calcium concentrations [67], oncogenes, and p53 [68], among others (Figure 4A). Among those, the BH3-only proteins have a prominent role. These are pro-apoptotic and different from the Bcl-2 family proteins which contain only one BH3 (Bcl-2 homology 3) domain [69]. BECN1 is a fundamental protein in autophagic mechanism (as mentioned above). This protein was one of the first described mammalian autophagy-related proteins [70]. BECN1 interacts with proteins of Bcl-2 family (Bcl-2, Bcl-XL and Mcl-1) [71]. Through interaction with the BECN1 BH3 domain, these proteins are able to inhibit BECN1-mediated autophagy [72, 73]. What is more, this interaction makes BECN1 not only a highly relevant autophagy actor but it is indeed also a haploinsufficient tumor suppressor [74]. On the other hand, BH3-only proteins are capable of disruption of Bcl-2-BECN1 interaction inducing apoptosis by Bcl-2 blockade and autophagy by releasing BECN1 [75].

Experimentally, impairment of autophagy in starving cells is able to induce apoptosis or at least a kind of cell death. This autophagy shortage impedes cell proper management of a metabolic stress or, in some cases, such as neurons, clearance of toxic cellular metabolic products. Moreover, in some way, this effect seems to be in both sides since inhibition of apoptosis could also be a potent trigger of the autophagy process. Transgenic depletion of pro-apoptotic molecules such as Bax or Bak or the use in vitro of the pan-caspase inhibitor Z-VAD-FMK were able to strongly induce autophagy and a cell death which is inhibited by autophagy inhibitors [76,77]. This seems to be the rule in most cases, a highly intricate autophagy-apotosis cross talk, where, despite their respective main duties, they act by inhibiting each other (Figure 4A). For instance, in one way, autophagy could reduce the cytosolic concentrations of pro-apoptotic proteins [78]. In the other way, activated caspases are able to cleave some autophagic proteins in which the fragmented products indeed acquired now pro-apoptotic properties [79-83]. Such is the case of ATG5, BECN1, and ATG4D [79-83].

Beyond the central role of ATG5 in autophagy machinery, this protein, in some situations, is able to induce apoptosis by two different ways. As mentioned above, a calpain-mediated cleaved ATG5 translocated to mitochondria, favoring its depolarization and triggering of apoptosis [84]. Moreover, ATG5 can also associate to FADD (Fas-associated death domain) enhancing apoptosis cell death [85].

The autophagic cell death is a programmed alternative to other sorts of cell demise. All of the comments above about the relationship of autophagy with cell death do not imply that autophagy could be by itself a cell death mechanism. Many researchers have suggested a number of times the existence of a really programmed autophagic cell death. Nevertheless, since autophagy is a cell survival pathway, there was concern about whether autophagy was a last cellular attempt to avoid apoptosis or if autophagy-related cell death only occurs in in vitro settings. These controversies were clarified with study of the degeneration of salivary glands during the D. melanogaster embryogenesis [86]. These glands suffered a cell death that is independent of caspases and completely dependent on autophagy [87].

All this intricate cross talk among apoptosis, autophagy, and autophagic cell death makes one rethink the definition of cell death types. Hence, the 2015 Nomenclature Committee on Cell Death made a switch from morphological- to biochemical-based definitions of cell death types [87]. They stated that autophagic cell death will be defined as processes of cell death that were prevented by the use of pharmacological or genetic tools targeting at least two different components of autophagy machinery [87].

Beyond the molecular mechanism, autophagy is able to modulate the cell death response and, hence, cell fate indirectly. Mitophagy, the most studied form of selective autophagy, is aimed to eliminate damaged mitochondria contributing to its renewing and ROS homeostasis [88]. As mitochondrial impairment is one of most important apoptosis triggering events, elimination of damaged mitochondria could avoid the intrinsic apoptosis program (Figure 4B). Damaged mitochondria and subsequent decrease in the inner mitochondrial transmembrane potential ($\Delta \Psi$) led to leakage of several harmful substances such as ROS and intrinsic apoptotic-pathway-triggered leakage of others such as cytochrome c [88]. In response, damaged mitochondria may be fragmented and ubiquitinated by a PINK-Parkin-dependent mechanism [89]. Then, these damaged organelles are recognized and selectively eliminated by autophagy in order to allow the cell survival [64,83].

It is surprising that autophagy can also interfere with such an uncontrolled process as necrosis. This example was demonstrated to occur during acute pancreatitis with zymophagy [55]. In the cellular basis of this disease, there was a premature intracytoplasmic activation of digestive enzymes [41]. This last activity pushes pancreatic acinar cell to an inevitable necrosis and the catastrophe of the tissue in a sort of chain cascade [41]. Zymophagy, as selective type of autophagy, eliminates the granules where the dangerous zymogens were activated by a ubiquitin recognition mechanism [55]. This protective mechanism avoids the enzymatic autodegradation and subsequent necrosis of the acinar cell [55] (Figure 4B). This event, in fact, has consequences beyond cellular biology since it reduces gland inflammation and contributes to autolimitation of the disease.

3.3. Autophagy and cellular senescence

There exists another unexpected process related to cell fate which is influenced by autophagy: the cellular senescence [90]. This term is used to describe an irreversible (which differentiates it from quiescence) deep arrested status of the cell cycle. The senescence is a cellular alternative mechanism to apoptosis in response to certain stressors including oncogene overexpression. Then, the so-called oncogene-induced senescence is a potent tumor suppressor mechanism against cellular transformation [91-94] and it is the first option that cells have against oncogene activation. Autophagy participated in the transition phase of senescence establishment being part of the TOR-autophagy spatial coupling compartment (TASCC). The TASCC is a cytoplasmic sub-compartmentalization where mTOR is closely associated with lysosomes/autolysosomes fuelled by a constant autophagy flow outside of this area [95]. This activity resulted in an efficient synthesis-degradation coupling that seems to be crucial for senescence. Inhibition of autophagy impairs the senescent progression and it seems to be necessary to reach an intermediate state before the final setting of senescent phenotype. Finally, it was suggested that autophagy during senescence is triggered by the isoform ULK3 instead of ULK1 or ULK2; therefore, it is tempted to hypothethize that a specific type of autophagy may be related to senescence (Figure 4B).

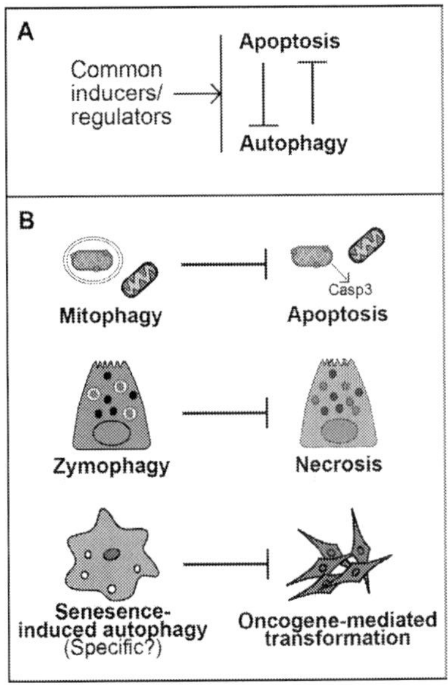

Figure 4. The implication of autophagy in cell fate. A) There is a deep crosstalk between autophagy and apoptosis. They share several common inducers and posses regulatory properties on each other. B) Autophagy is able to modify cell fate and some examples are depicted. Mitophagy eliminates the damaged mitochondria avoiding the apoptosis triggered by intrinsic pathway (upper panel). By the specific elimination of activated zymogen granules, zymophagy prevents the acinar cell damage and the necrosis in pancreas tissue (middle panel). Autophagy participates in the oncogene-induced senescence, a process capable of repress the oncogenic transformation (lower panel).

4. Autophagy in pancreatic diseases

Pathological processes such as pancreatitis and diabetes mellitus as well as cancer cell transformation and also cancer chemotherapy activate autophagy in human tissues and human tumor cells. While human normal pancreatic acinar cells do not have detectable autophagy levels, pancreatic diseases activate autophagy, confirming the relevant role of autophagy in human pancreatic disease. In addition, selective autophagy of pancreatic zymogen granules, zymophagy, has been discovered and characterized as a cell-protective process activated by the disease.

Pancreatic ductal adenocarcinoma (PDAC) is one of the most aggressive human malignancies with 2-3% five-year survival rate. It remains a devastating and poorly understood malignancy. Its poor prognosis has been attributed to the inability to make a diagnosis while the tumor is still resectable and a propensity toward early vascular dissemination and spread to regional lymph nodes. Up to 60% of patients have advanced pancreatic cancer at the time of diagnosis and their median survival time is a dismal 36 months. This is due to both the aggressive nature of the disease, the lack of specific symptoms and early detection tools, and its relatively refractory response to traditional cytotoxic agents and radiotherapy. Moreover, pancreatic cancer cells become more malignant or survive with an extremely poor blood supply. So far, little and contradictory data are available regarding the activity of autophagy and its regulation in pancreatic cancer cells. Experimental evidence pointed at autophagy as a pancreatic cancer cell mechanism to survive under adverse environmental conditions or as a defective programmed cell death mechanism that favors pancreatic cancer cell resistance to treatment.

4.1. Autophagy in cancer cell

Both downregulated and excessive autophagy have been implicated in the pathogenesis of diverse diseases such as certain type of neuronal degeneration, diabetes and its complications, and cancer [96]. Autophagy has also been implicated in cell death called autophagic or type II programmed cell death, which was originally described on the basis of morphological studies detecting autophagic vesicles during tissue involution [97].

Cancer cells in general tend to undergo less autophagy than their normal counterparts at least for some tumors [98,99]. The BECN1 autophagy gene is monoallelically deleted in 40-74% of cases of human sporadic breast, ovarian, and prostate cancer [99]. Heterozygous disruption of BECN1 increases the frequency of spontaneous malignancies and accelerates the development of virus-induced premalignant lesions [99] suggesting that defective regulation of autophagy promotes tumorigenesis. It has been proposed that autophagy suppresses carcinogenesis by a cell-autonomous mechanism involving protection of genome integrity and stability, and a nonautonomous mechanism involving suppression of inflammation and necrosis. On the other hand, autophagy may support the survival of rapidly growing cancer cells that have outgrown their vascular supply and are exposed to an inadequate oxygen supply or metabolic stress. By contrast, excessive levels of autophagy promote cell death [100]. Accordingly, it has been proposed that autophagy plays an important role both in tumor progression and in promotion of cancer cell death [101], although themolecular mechanisms responsible for this dual action of autophagy in cancer have not been elucidated.

It has been suggested that autophagy may be a cancer cell survival response to tumor-associated hypoxia. Tumor hypoxia has been used as a marker of poor prognosis [102]; however, how cancer cells become more malignant or survive with an extremely poor blood supply is poorly understood. When cancer cells are exposed to hypoxia, anaerobic glycolysis increases and provides energy for cell survival but as the glucose supply is also insufficient because of the poor blood supply, there must be an alternative metabolic pathway that provides energy when both oxygen and glucose are depleted [103,104]. Hypoxia in pancreatic cancer has been reported to increase its malignant potential [102]. Proliferating cancer cells require more nutrients than surrounding noncancerous cells do, though nutrition is supplied via functionally and structurally immature neovessels. Because autophagy-specific genes promote the survival of normal cells during nutrient starvation in all eukaryotic organisms, autophagy may react to the cancer microenvironment to favor the survival of rapidly growing cancer cells. LC3 expression in surgically resected pancreatic cancer tissue showed activated autophagy in the peripheral area, which included the invasive border, and concomitantly shows enhanced expression of carbonic anhydrase [105]. This observation suggests that autophagy may promote cell viability in hypovascularized cancer tissue.

It has also been proposed that autophagy is a cancer cell survival response to tumor-associated inflammation [106]. Cancer-associated inflammation results in promotion of carcinogenesis and resistance to therapy. Several phenotypic alterations observed in cancer cells are a result of inflammatory signals found within the tumor microenvironment [106]. The receptor for advanced glycation end products (RAGE) is an induced inflammatory receptor constitutively expressed on many murine and human epithelial tumor cell lines [107,108] and the highest levels of RAGE expression were observed in murine and human pancreatic adenocarcinoma tumors. Genotoxic and/or metabolic stress lead to modest but reproducible increases in overall expression of RAGE on epithelial cell lines. RAGE expression correlates directly with the ability of both murine and human pancreatic tumor cell lines to survive cytotoxic insult. Targeted knockdown of RAGE significantly increased cell death, whereas forced overexpression promotes survival. Recently, it was reported that the enhanced sensitivity to cell death in the setting of RAGE knockdown is associated with increased apoptosis and decreased autophagy. In contrast, overexpression of RAGE is associated with enhanced autophagy, diminished apoptosis, and enhanced cancer cell viability. Knockdown of RAGE enhances mTOR phosphorylation in response to chemotherapy, thus preventing induction of a survival response. Inhibition of autophagy by means of silencing BECN1 expression in pancreatic cancer cells enhanced apoptosis and cell death [109]. These observations suggest that RAGE expression in cancer cells has a role in tumor cell response to environmental stress through the enhancement of autophagy. However, increased sensitivity to chemotherapeutic agents in RAGE-knockdown pancreatic cancer cells is dependent on ATG5 expression but independent of BECN1 expression [109]. These last findings suggested that the role of autophagy in the resistance to microenvironment insult or in the sensitivity to chemotherapeutic agent is the result of complex molecular pathways in the tumor cell.

On the other hand, repression of autophagy has been suggested as a cancer cell response to prolonged hypoxic conditions. Pancreatic cancer cell response to prolonged hypoxia may consist of inhibition of autophagic cell death. The short isoform of single-minded 2 (SIM2s) is a member of the basic helix-loop-helix family of transcriptional regulators [110] and is upregulated in pancreatic cancer. Microarray studies identified the pro-cell death gene BNIP3 as a target of SIM2s repression. Prolonged hypoxia induces cell death via an autophagic pathway involving the hypoxia-inducible factor 1 (HIF1)-mediated upregulation of BNIP3 [30,111]. Deregulation of both SIM2s and BNIP3 were associated with poor prognostic outcomes [112]. Decreased BNIP3 levels and poor prognosis clearly correlate with elevated SIM2s expression in pancreatic cancer. The loss of BNIP3, either by hypermethylation or by transcriptional repression, was correlated with inhibition of cell death [113, 114], whereas upregulation of BNIP3 sensitized pancreatic carcinoma cells to hypoxia-induced cell death [115]. SIM2s expression, concomitant with its repression of BNIP3, enhanced tumor cell survival under prolonged hypoxic conditions. Recent data linked increased SIM2s expression with enhanced cell survival during hypoxia-stress concomitantly with BNIP3 repression and the attenuation of hypoxia-induced autophagic processes.

Thus, inhibition of autophagic cell death by BNIP3 repression enhances tumor cell survival under prolonged hypoxic conditions [115].

Decreased autophagy in some cancer cells has been related to malignant stages of the disease. Cancer cells in general tend to undergo less autophagy than their normal counterparts supporting the contention that defective autophagic cell death plays a role in tumor progression. Studies of carcinogen-induced pancreatic cancer in animal models have shown that pancreatic adenocarcinoma cells have lower autophagic capacity than premalignant cells [116]. The WIPI protein family, which includes ATG18, the WIPI-1 homolog in S. cerevisiae, was genetically identified as a gene contributing to autophagy [116]. Human WIPI-1a is a member of a highly conserved WD-repeats protein family. hWIPI-1 is linked to starvation-induced autophagy in the mammalian system. Amino acid deprivation triggered an accumulation of endogenous hWIPI-1 protein to large vesicular and cup-shaped structures where it colocalizes with LC3. Starvation-induced hWIPI-1 formation was blocked by wortmannin, a principal inhibitor of PI-3 kinase-induced autophagosome formation [117]. Interestingly, WIPI proteins are linked pathologically to cellular transformation because all human WIPI genes are reportedly expressed aberrantly in a variety of matched human cancer samples. Strikingly, hWIPI-2 and hWIPI-4 mRNA expression is substantially decreased in 70% of matched kidney (10 patients) and 100% of pancreatic (seven patients) tumor samples. The majority of these samples were derived from advanced-stage tumors such as pancreatic adenocarcinomas stages I–IV. Hence, cancer-associated downregulation of hWIPI-2 and hWIPI-4 supports the possibility that decreased autophagic activity is necessary for the malignant stages of pancreatic cancer.

5. Perspectives

There is ample evidence supporting an active role for autophagy in cell physiology and disease. During the last decade, autophagy has turned from a morphological finding to a cellular process involving a membrane transport system and complex molecular machinery. Moreover, since the discovery of ATG genes, there have been many studies on the physiological and pathological roles of autophagy in a variety of autophagy knockout models. However, direct evidence of the connections between ATG gene dysfunction and human diseases has emerged only recently [56]. Here we have overviewed the physiological bases and molecular mechanisms of the autophagic process. We have introduced the reader to a novel autophagy-related transmembrane protein – VMP1 – whose expression triggers autophagy and its role in the cell response to disease. Elucidation of the specific extracellular and intracellular conditions that stimulate autophagy and the linkage of these conditions to either cell survival or cell injury and death in different cell types and during different pathological processes is a rapidly evolving and fruitful field of research. The development of therapies to take advantage of the potential cytoprotective effect of autophagy in several pathologies such as cancer or neurodegenerative diseases is a potentially promising avenue of investigation.

Author details

Daniel Grasso, Alejandro Ropolo and Maria I. Vaccaro*

*Address all correspondence to: mvaccaro@ffyb.uba.ar

Institute of Biochemistry and Molecular Medicine, Department of Pathophysiology, National Council for Scientific and Technical Research. School of Pharmacy and Biochemistry, University of Buenos Aires, Argentina

References

[1] Cuervo AM. Autophagy: in sickness and in health. Trends Cell Biol 2004; 14: 70-77.

[2] Levine B, Klionsky DJ. Development by self-digestion: molecular mechanisms and biological functions of autophagy. Dev Cell 2004; 6: 463-477.

[3] Yang Z, Klionsky DJ. An overview of the molecular mechanism of autophagy. Curr Top Microbiol Immunol 2009; 335: 1-32.

[4] Hara T, Nakamura K, Matsui M et al. Suppression of basal autophagy in neural cells causes neurodegenerative disease in mice. Nature 2006; 441: 885-889.

[5] Qu X, Zou Z, Sun Q et al. Autophagy gene-dependent clearance of apoptotic cells during embryonic development. Cell 2007; 128: 931-946.

[6] Pattingre S, Levine B. Bcl-2 inhibition of autophagy: a new route to cancer? Cancer Res 2006; 66: 2885-2888.

[7] Shintani T, Klionsky DJ. Autophagy in health and disease: a double-edged sword. Science 2004; 306: 990-995.

[8] Klionsky DJ. Autophagy. Curr Biol 2005; 15: R282-283.

[9] Klionsky DJ, Emr SD. Autophagy as a regulated pathway of cellular degradation. Science 2000; 290: 1717-1721.

[10] Klionsky DJ, Abdalla FC, Abeliovich H et al. Guidelines for the use and interpretation of assays for monitoring autophagy. Autophagy 2012; 8: 445-544.

[11] Sridhar S, Patel B, Aphkhazava D et al. The lipid kinase PI4KIIIbeta preserves lysosomal identity. EMBO J 2013; 32: 324-339.

[12] Mari M, Tooze SA, Reggiori F. The puzzling origin of the autophagosomal membrane. F1000 Biol Rep 2011; 3: 25.

[13] Itakura E, Mizushima N. Characterization of autophagosome formation site by a hierarchical analysis of mammalian Atg proteins. Autophagy 2010; 6: 764-776.

[14] Xie Z, Klionsky DJ. Autophagosome formation: core machinery and adaptations. Nat Cell Biol 2007; 9: 1102-1109.

[15] Axe EL, Walker SA, Manifava M et al. Autophagosome formation from membrane compartments enriched in phosphatidylinositol 3-phosphate and dynamically connected to the endoplasmic reticulum. J Cell Biol 2008; 182: 685-701.

[16] Hayashi-Nishino M, Fujita N, Noda T et al. A subdomain of the endoplasmic reticulum forms a cradle for autophagosome formation. Nat Cell Biol 2009; 11: 1433-1437.

[17] Madeo F, Tavernarakis N, Kroemer G. Can autophagy promote longevity? Nat Cell Biol 2010; 12: 842-846.

[18] Ropolo A, Grasso D, Pardo R et al. The pancreatitis-induced vacuole membrane protein 1 triggers autophagy in mammalian cells. J Biol Chem 2007; 282: 37124-37133.

[19] Vaccaro MI, Ropolo A, Grasso D, Iovanna JL. A novel mammalian trans-membrane protein reveals an alternative initiation pathway for autophagy. Autophagy 2008; 4: 388-390.

[20] Kihara A, Noda T, Ishihara N, Ohsumi Y. Two distinct Vps34 phosphatidylinositol 3-kinase complexes function in autophagy and carboxypeptidase Y sorting in Saccharomyces cerevisiae. J Cell Biol 2001; 152: 519-530.

[21] He C, Klionsky DJ. Regulation mechanisms and signaling pathways of autophagy. Annu Rev Genet 2009; 43: 67-93.

[22] Karanasios E, Stapleton E, Manifava M et al. Dynamic association of the ULK1 complex with omegasomes during autophagy induction. J Cell Sci 2013; 126: 5224-5238.

[23] Ichimura Y, Kirisako T, Takao T et al. A ubiquitin-like system mediates protein lipidation. Nature 2000; 408: 488-492.

[24] Burman C, Ktistakis NT. Regulation of autophagy by phosphatidylinositol 3-phosphate. FEBS Lett 2010; 584: 1302-1312.

[25] Tooze SA, Yoshimori T. The origin of the autophagosomal membrane. Nat Cell Biol 2010; 12: 831-835.

[26] Musiwaro P, Smith M, Manifava M et al. Characteristics and requirements of basal autophagy in HEK 293 cells. Autophagy 2013; 9: 1407-1417.

[27] Hoyer-Hansen M, Jaattela M. AMP-activated protein kinase: a universal regulator of autophagy? Autophagy 2007; 3: 381-383.

[28] Zheng M, Wang YH, Wu XN et al. Inactivation of Rheb by PRAK-mediated phosphorylation is essential for energy-depletion-induced suppression of mTORC1. Nat Cell Biol 2011; 13: 263-272.

[29] Mortimore GE, Poso AR. Intracellular protein catabolism and its control during nutrient deprivation and supply. Annu Rev Nutr 1987; 7: 539-564.

[30] Nobukuni T, Joaquin M, Roccio M et al. Amino acids mediate mTOR/raptor signaling through activation of class 3 phosphatidylinositol 3OH-kinase. Proc Natl Acad Sci U S A 2005; 102: 14238-14243.

[31] Wullschleger S, Loewith R, Hall MN. TOR signaling in growth and metabolism. Cell 2006; 124: 471-484.

[32] Sancak Y, Bar-Peled L, Zoncu R et al. Ragulator-Rag complex targets mTORC1 to the lysosomal surface and is necessary for its activation by amino acids. Cell 2010; 141: 290-303.

[33] Gwinn DM, Shackelford DB, Egan DF et al. AMPK phosphorylation of raptor mediates a metabolic checkpoint. Mol Cell 2008; 30: 214-226.

[34] Levine B, Sinha S, Kroemer G. Bcl-2 family members: dual regulators of apoptosis and autophagy. Autophagy 2008; 4: 600-606.

[35] Sarkar S, Perlstein EO, Imarisio S et al. Small molecules enhance autophagy and reduce toxicity in Huntington's disease models. Nat Chem Biol 2007; 3: 331-338.

[36] Mizushima N. The role of the Atg1/ULK1 complex in autophagy regulation. Curr Opin Cell Biol 2010; 22: 132-139.

[37] Liang XH, Jackson S, Seaman M et al. Induction of autophagy and inhibition of tumorigenesis by beclin 1. Nature 1999; 402: 672-676.

[38] Pattingre S, Tassa A, Qu X et al. Bcl-2 antiapoptotic proteins inhibit Beclin 1-dependent autophagy. Cell 2005; 122: 927-939.

[39] Feng W, Huang S, Wu H, Zhang M. Molecular basis of Bcl-xL's target recognition versatility revealed by the structure of Bcl-xL in complex with the BH3 domain of Beclin-1. J Mol Biol 2007; 372: 223-235.

[40] Oberstein A, Jeffrey PD, Shi Y. Crystal structure of the Bcl-XL-Beclin 1 peptide complex: Beclin 1 is a novel BH3-only protein. J Biol Chem 2007; 282: 13123-13132.

[41] Vaccaro MI. Autophagy and pancreas disease. Pancreatology 2008; 8: 425-429.

[42] Lo Re AE, Fernandez-Barrena MG, Almada LL et al. Novel AKT1-GLI3-VMP1 pathway mediates KRAS oncogene-induced autophagy in cancer cells. J Biol Chem 2012; 287: 25325-25334.

[43] Grasso D, Ropolo A, Lo Re A et al. Zymophagy, a novel selective autophagy pathway mediated by VMP1-USP9x-p62, prevents pancreatic cell death. J Biol Chem 2011; 286: 8308-8324.

[44] Vaccaro MI, Grasso D, Ropolo A et al. VMP1 expression correlates with acinar cell cytoplasmic vacuolization in arginine-induced acute pancreatitis. Pancreatology 2003; 3: 69-74.

[45] Jiang PH, Motoo Y, Vaccaro MI et al. Expression of vacuole membrane protein 1 (VMP1) in spontaneous chronic pancreatitis in the WBN/Kob rat. Pancreas 2004; 29: 225-230.

[46] Grasso D, Sacchetti ML, Bruno L et al. Autophagy and VMP1 expression are early cellular events in experimental diabetes. Pancreatology 2009; 9: 81-88.

[47] Gonzalez CD, Lee MS, Marchetti P et al. The emerging role of autophagy in the pathophysiology of diabetes mellitus. Autophagy 2011; 7: 2-11.

[48] Pardo R, Lo Re A, Archange C et al. Gemcitabine induces the VMP1-mediated autophagy pathway to promote apoptotic death in human pancreatic cancer cells. Pancreatology 2010; 10: 19-26.

[49] Ropolo A, Bagnes CI, Molejon MI et al. Chemotherapy and autophagy-mediated cell death in pancreatic cancer cells. Pancreatology 2012; 12: 1-7.

[50] Gilabert M, Vaccaro MI, Fernandez-Zapico ME et al. Novel role of VMP1 as modifier of the pancreatic tumor cell response to chemotherapeutic drugs. J Cell Physiol 2013; 228: 1834-1843.

[51] Molejon MI, Ropolo A, Re AL et al. The VMP1-Beclin 1 interaction regulates autophagy induction. Sci Rep 2013; 3: 1055.

[52] Molejon MI, Ropolo A, Vaccaro MI. VMP1 is a new player in the regulation of the autophagy-specific phosphatidylinositol 3-kinase complex activation. Autophagy 2013; 9: 933-935.

[53] He C, Bassik MC, Moresi V et al. Exercise-induced BCL2-regulated autophagy is required for muscle glucose homeostasis. Nature 2012; 481: 511-515.

[54] Reggiori F, Komatsu M, Finley K, Simonsen A. Selective types of autophagy. Int J Cell Biol 2012; 2012: 156272.

[55] Vaccaro MI. Zymophagy: selective autophagy of secretory granules. Int J Cell Biol 2012; 2012: 396705.

[56] Jiang P, Mizushima N. Autophagy and human diseases. Cell Res 2014; 24: 69-79.

[57] Filomeni G, De Zio D, Cecconi F. Oxidative stress and autophagy: the clash between damage and metabolic needs. Cell Death Differ 2015; 22: 377-388.

[58] Navarro-Yepes J, Burns M, Anandhan A et al. Oxidative stress, redox signaling, and autophagy: cell death versus survival. Antioxid Redox Signal 2014; 21: 66-85.

[59] Okamoto K. Organellophagy: eliminating cellular building blocks via selective autophagy. J Cell Biol 2014; 205: 435-445.

[60] Stolz A, Ernst A, Dikic I. Cargo recognition and trafficking in selective autophagy. Nat Cell Biol 2014; 16: 495-501.

[61] Levine B, Mizushima N, Virgin HW. Autophagy in immunity and inflammation. Nature 2011; 469: 323-335.

[62] Nikoletopoulou V, Papandreou ME, Tavernarakis N. Autophagy in the physiology and pathology of the central nervous system. Cell Death Differ 2015; 22: 398-407.

[63] Grasso D, Garcia MN, Iovanna JL. Autophagy in pancreatic cancer. Int J Cell Biol 2012; 2012: 760498.

[64] Kim J, Kundu M, Viollet B, Guan KL. AMPK and mTOR regulate autophagy through direct phosphorylation of Ulk1. Nat Cell Biol 2011; 13: 132-141.

[65] Zadra G, Batista JL, Loda M. Dissecting the Dual Role of AMPK in Cancer: From Experimental to Human Studies. Mol Cancer Res 2015.

[66] Marino G, Niso-Santano M, Baehrecke EH, Kroemer G. Self-consumption: the interplay of autophagy and apoptosis. Nat Rev Mol Cell Biol 2014; 15: 81-94.

[67] Grotemeier A, Alers S, Pfisterer SG et al. AMPK-independent induction of autophagy by cytosolic Ca2+ increase. Cell Signal 2010; 22: 914-925.

[68] Maiuri MC, Tasdemir E, Criollo A et al. Control of autophagy by oncogenes and tumor suppressor genes. Cell Death Differ 2009; 16: 87-93.

[69] Doerflinger M, Glab JA, Puthalakath H. BH3-only proteins: a 20-year stock-take. FEBS J 2015; 282: 1006-1016.

[70] Aita VM, Liang XH, Murty VV et al. Cloning and genomic organization of beclin 1, a candidate tumor suppressor gene on chromosome 17q21. Genomics 1999; 59: 59-65.

[71] Maiuri MC, Le Toumelin G, Criollo A et al. Functional and physical interaction between Bcl-X(L) and a BH3-like domain in Beclin-1. EMBO J 2007; 26: 2527-2539.

[72] Malik SA, Orhon I, Morselli E et al. BH3 mimetics activate multiple pro-autophagic pathways. Oncogene 2011; 30: 3918-3929.

[73] Malik SA, Shen S, Marino G et al. BH3 mimetics reveal the network properties of autophagy-regulatory signaling cascades. Autophagy 2011; 7: 914-916.

[74] Elgendy M, Ciro M, Abdel-Aziz AK et al. Beclin 1 restrains tumorigenesis through Mcl-1 destabilization in an autophagy-independent reciprocal manner. Nat Commun 2014; 5: 5637.

[75] Andreu-Fernandez V, Genoves A, Messeguer A et al. BH3-mimetics- and cisplatin-induced cell death proceeds through different pathways depending on the availability of death-related cellular components. PLoS One 2013; 8: e56881.

[76] Shimizu S, Kanaseki T, Mizushima N et al. Role of Bcl-2 family proteins in a non-apoptotic programmed cell death dependent on autophagy genes. Nat Cell Biol 2004; 6: 1221-1228.

[77] Wu YT, Tan HL, Huang Q et al. Autophagy plays a protective role during zVAD-induced necrotic cell death. Autophagy 2008; 4: 457-466.

[78] Shaid S, Brandts CH, Serve H, Dikic I. Ubiquitination and selective autophagy. Cell Death Differ 2013; 20: 21-30.

[79] Shalini S, Dorstyn L, Dawar S, Kumar S. Old, new and emerging functions of caspases. Cell Death Differ 2015; 22: 526-539.

[80] Wirawan E, Vande Walle L, Kersse K et al. Caspase-mediated cleavage of Beclin-1 inactivates Beclin-1-induced autophagy and enhances apoptosis by promoting the release of proapoptotic factors from mitochondria. Cell Death Dis 2010; 1: e18.

[81] Betin VM, Lane JD. Caspase cleavage of Atg4D stimulates GABARAP-L1 processing and triggers mitochondrial targeting and apoptosis. J Cell Sci 2009; 122: 2554-2566.

[82] Cho DH, Jo YK, Hwang JJ et al. Caspase-mediated cleavage of ATG6/Beclin-1 links apoptosis to autophagy in HeLa cells. Cancer Lett 2009; 274: 95-100.

[83] Zhu Y, Zhao L, Liu L et al. Beclin 1 cleavage by caspase-3 inactivates autophagy and promotes apoptosis. Protein Cell 2010; 1: 468-477.

[84] Yousefi S, Perozzo R, Schmid I et al. Calpain-mediated cleavage of Atg5 switches autophagy to apoptosis. Nat Cell Biol 2006; 8: 1124-1132.

[85] Young MM, Takahashi Y, Khan O et al. Autophagosomal membrane serves as platform for intracellular death-inducing signaling complex (iDISC)-mediated caspase-8 activation and apoptosis. J Biol Chem 2012; 287: 12455-12468.

[86] Berry DL, Baehrecke EH. Growth arrest and autophagy are required for salivary gland cell degradation in Drosophila. Cell 2007; 131: 1137-1148.

[87] Galluzzi L, Bravo-San Pedro JM, Vitale I et al. Essential versus accessory aspects of cell death: recommendations of the NCCD 2015. Cell Death Differ 2015; 22: 58-73.

[88] Youle RJ, Narendra DP. Mechanisms of mitophagy. Nat Rev Mol Cell Biol 2011; 12: 9-14.

[89] Kim I, Lemasters JJ. Mitophagy selectively degrades individual damaged mitochondria after photoirradiation. Antioxid Redox Signal 2011; 14: 1919-1928.

[90] Grasso D, Vaccaro MI. Macroautophagy and the oncogene-induced senescence. Front Endocrinol (Lausanne) 2014; 5: 157.

[91] Collado M, Serrano M. Senescence in tumours: evidence from mice and humans. Nat Rev Cancer 2010; 10: 51-57.

[92] Nardella C, Clohessy JG, Alimonti A, Pandolfi PP. Pro-senescence therapy for cancer treatment. Nat Rev Cancer 2011; 11: 503-511.

[93] Serrano M, Lin AW, McCurrach ME et al. Oncogenic ras provokes premature cell senescence associated with accumulation of p53 and p16INK4a. Cell 1997; 88: 593-602.

[94] Gorgoulis VG, Halazonetis TD. Oncogene-induced senescence: the bright and dark side of the response. Curr Opin Cell Biol 2010; 22: 816-827.

[95] Narita M, Young AR, Arakawa S et al. Spatial coupling of mTOR and autophagy augments secretory phenotypes. Science 2011; 332: 966-970.

[96] Mizushima N, Levine B, Cuervo AM, Klionsky DJ. Autophagy fights disease through cellular self-digestion. Nature 2008; 451: 1069-1075.

[97] Hoyer-Hansen M, Jaattela M. Autophagy: an emerging target for cancer therapy. Autophagy 2008; 4: 574-580.

[98] Toth S, Nagy K, Palfia Z, Rez G. Cellular autophagic capacity changes during azaserine-induced tumour progression in the rat pancreas. Up-regulation in all premalignant stages and down-regulation with loss of cycloheximide sensitivity of segregation along with malignant transformation. Cell Tissue Res 2002; 309: 409-416.

[99] Qu X, Yu J, Bhagat G et al. Promotion of tumorigenesis by heterozygous disruption of the beclin 1 autophagy gene. J Clin Invest 2003; 112: 1809-1820.

[100] Levine B. Cell biology: autophagy and cancer. Nature 2007; 446: 745-747.

[101] Mathew R, Karantza-Wadsworth V, White E. Role of autophagy in cancer. Nat Rev Cancer 2007; 7: 961-967.

[102] Buchler P, Reber HA, Lavey RS et al. Tumor hypoxia correlates with metastatic tumor growth of pancreatic cancer in an orthotopic murine model. J Surg Res 2004; 120: 295-303.

[103] Izuishi K, Kato K, Ogura T et al. Remarkable tolerance of tumor cells to nutrient deprivation: possible new biochemical target for cancer therapy. Cancer Res 2000; 60: 6201-6207.

[104] Esumi H, Izuishi K, Kato K et al. Hypoxia and nitric oxide treatment confer tolerance to glucose starvation in a 5'-AMP-activated protein kinase-dependent manner. J Biol Chem 2002; 277: 32791-32798.

[105] Fujii S, Mitsunaga S, Yamazaki M et al. Autophagy is activated in pancreatic cancer cells and correlates with poor patient outcome. Cancer Sci 2008; 99: 1813-1819.

[106] DeNardo DG, Johansson M, Coussens LM. Inflaming gastrointestinal oncogenic programming. Cancer Cell 2008; 14: 7-9.

[107] Abe R, Yamagishi S. AGE-RAGE system and carcinogenesis. Curr Pharm Des 2008; 14: 940-945.

[108] Arumugam T, Simeone DM, Van Golen K, Logsdon CD. S100P promotes pancreatic cancer growth, survival, and invasion. Clin Cancer Res 2005; 11: 5356-5364.

[109] Kang R, Tang D, Schapiro NE et al. The receptor for advanced glycation end products (RAGE) sustains autophagy and limits apoptosis, promoting pancreatic tumor cell survival. Cell Death Differ 2010; 17: 666-676.

[110] Kewley RJ, Whitelaw ML, Chapman-Smith A. The mammalian basic helix-loop-helix/PAS family of transcriptional regulators. Int J Biochem Cell Biol 2004; 36: 189-204.

[111] Azad MB, Chen Y, Henson ES et al. Hypoxia induces autophagic cell death in apoptosis-competent cells through a mechanism involving BNIP3. Autophagy 2008; 4: 195-204.

[112] Burton TR, Gibson SB. The role of Bcl-2 family member BNIP3 in cell death and disease: NIPping at the heels of cell death. Cell Death Differ 2009; 16: 515-523.

[113] Okami J, Simeone DM, Logsdon CD. Silencing of the hypoxia-inducible cell death protein BNIP3 in pancreatic cancer. Cancer Res 2004; 64: 5338-5346.

[114] Mahon PC, Baril P, Bhakta V et al. S100A4 contributes to the suppression of BNIP3 expression, chemoresistance, and inhibition of apoptosis in pancreatic cancer. Cancer Res 2007; 67: 6786-6795.

[115] Abe T, Toyota M, Suzuki H et al. Upregulation of BNIP3 by 5-aza-2'-deoxycytidine sensitizes pancreatic cancer cells to hypoxia-mediated cell death. J Gastroenterol 2005; 40: 504-510.

[116] Guan J, Stromhaug PE, George MD et al. Cvt18/Gsa12 is required for cytoplasm-to-vacuole transport, pexophagy, and autophagy in Saccharomyces cerevisiae and Pichia pastoris. Mol Biol Cell 2001; 12: 3821-3838.

[117] Proikas-Cezanne T, Waddell S, Gaugel A et al. WIPI-1alpha (WIPI49), a member of the novel 7-bladed WIPI protein family, is aberrantly expressed in human cancer and is linked to starvation-induced autophagy. Oncogene 2004; 23: 9314-9325.

Chapter 2

Interconnected Regulation of Apoptosis and WIPI-Mediated Autophagy

Katharina Sporbeck, Fenja Odendall and Tassula Proikas-Cezanne

Additional information is available at the end of the chapter

http://dx.doi.org/10.5772/62056

Abstract

Autophagy is a genetic program that secures the survival of eukaryotic cells to compensate for periods of starvation and cellular stress. Apoptosis, in contrast, is a genetically defined program leading to cell death. Both pathways are interconnected through conserved co-regulatory signaling pathways that are context-dependent. Proper co-regulation of autophagy and apoptosis, whereby autophagy exerts an antiapoptotic function and apoptosis inhibits autophagic survival strategies, critically secures the survival of healthy cells and counteracts genomic instability in eukaryotic organisms. Cancer cells often become resistant to apoptosis and addicted to autophagy, making this scenario highly relevant to define novel therapeutic strategies. The co-regulatory crosstalk between apoptosis and autophagy converges on the production of phosphatidylinositol 3-phosphate (PI3P) essential for the onset of autophagy. WD-repeat protein interacting with phosphoinositides (WIPI) members function as essential PI3P effectors in autophagy and fulfill an important role in health and disease. Here, we summarize details on the regulation of WIPI-mediated autophagy in the context of co-regulatory signals for both apoptosis and autophagy.

Keywords: WIPI, PI3P, autophagy, autophagosomal cell death, p53

1. Introduction

The process of autophagy was conceptualized in 1963 by Christian de Duve, who coined the term based on the Greek words *auto* for self and *phagy* for eat, hence self-eating. de Duve suggested that intracellular vesicles, derived from the endoplasmic reticulum, harbor cytoplasmic material for lysosomal degradation. This novel concept was followed up by morphological studies combined with biochemical analysis, providing compelling evidence that autophagy represents an evolutionarily conserved catabolic machinery in eukaryotic cells for

On the other hand, p53 inhibits autophagy in several distinct ways (Figure 2). In exhibiting its nonnuclear function, cytoplasmic p53 competes with ULK1 for binding to FIP200 under certain circumstances [29]. As the ULK1/FIP200 association is required for the activation of PI3KC3 and subsequent PI3P production followed by WIPI-mediated autophagy (Figure 1), cytoplasmic p53 inhibits autophagy. This scenario represents an interesting connection to the induction of apoptosis, which can be initiated through the association of cytoplasmic p53 to mitochondria [30].

In general, the site-specific transcriptional transactivator factor activity of p53 in the nucleus is induced upon DNA damage and p53 target genes subsequently permit DNA repair or apoptosis to secure genomic stability. Among the p53 target genes, FBXL20 (F-box and leucine-rich repeat protein 20), known to mediate ubiquitination and proteasome-mediated degradation, is expressed and targets PI3KC3 [31]. Hence, activation of the nuclear function of p53 can also lead to the inhibition of autophagy through p53-promoted PI3KC3 degradation (Figure 2).

Further, common factors regulating autophagy and apoptosis include Bcl-2 (B-cell lymphoma 2) with antiapoptotic function and Bim (Bcl-2-interacting mediator of cell death), a BH3 (Bcl-2 homology domain 2)-only protein with pro-apoptotic function (Figure 2). Bim is bound to the dynein motor protein DYNLL1/LC8 (dynein light chain 1) at microtubules. In nutrient-rich conditions, Bim binds to Beclin 1, preventing Beclin 1 binding to complex with PI3KC3 and initiate WIPI-mediated autophagy. Bim, therefore, inhibits autophagy through Beclin 1 mislocalization to the dynein motor protein [32]. In starvation conditions, Bim releases Beclin 1 and allows Beclin 1/PI3KC3 association and subsequently the induction of autophagy [32].

Bcl-2, a well-known interacting partner of Beclin 1, also prevents Beclin 1 from binding to PI3KC3. The Bcl-2/Beclin 1 association is inhibited upon starvation, leading to JNK (c-Jun N-terminal kinase)-mediated Bcl-2 phosphorylation at multiple sites subsequently releasing Beclin 1 and permitting autophagy activation (Figure 2) [33].

The complex relationship between apoptosis and autophagy regulation is further highlighted by the notion that autophagy-related proteins fulfill pro-apoptotic functions under certain circumstances. During apoptosis, active caspases target AMBRA1, a component of the PI3KC3 complex, for degradation and hence prevent WIPI-mediated autophagy upon the apoptotic point-of-no-return route to cell death [34]. Further, it was demonstrated that calpain 1 and calpain 2, classified as cysteine proteases [35], cleave ATG5 at threonine 193 generating an 24 kDa N-terminal fragment of ATG5 [36]. This ATG5 fragment subsequently translocates to the mitochondrial membrane and initiates cytochrome c release, ultimately promoting the cleavage of pro-caspase-3 into its active form. The effect of fragmented ATG5 is inhibited by high Bcl-2 levels [36]. Moreover, ATG12, the conjugation partner of ATG5, was shown to interact with Bcl-2 via an amino acid sequence in ATG12 resembling a BH3 domain generally known to bind Bcl-2. By binding to Bcl-2, ATG12 inhibits the antiapoptotic function of Bcl-2, leading to an increase of apoptotic cell death [37].

To provide a final example for the complex regulatory crosstalk between apoptosis and WIPI-mediated autophagy, the function of DAPK (death-associated protein kinase 1), a Ca^{2+}/

calmodulin-sensitive serine/threonine kinase involved in the induction of apoptosis [38], is highlighted in the following section. DAPK phosphorylates Beclin 1 or PKD (protein kinase D) under certain conditions. When reactive oxygen species activate DAPK, DAPK phosphorylates PKD, resulting in the phosphorylation of PI3KC3 by PDK. In turn, PI3KC3 produces PI3P and initiates autophagy [39]. As mentioned, DAPK can also phosphorylate Beclin 1, which occurs at threonine 119 in the BH3-like domain of Beclin 1 responsible for binding to Bcl-2. As Bcl-2-bound Beclin 1 inhibits autophagy, threonine 119 phosphorylation releases Beclin 1 from Bcl2 and enables Beclin1 to bind to PI3KC3 and to stimulate autophagy [40].

4. Outlook

In general, apoptosis and autophagy represent mutually exclusive genetic programs co-regulated by key factors with opposing roles in both pathways. This crosstalk contributes to the survival of healthy cells and secures genomic stability more than one of the pathways alone. Interestingly, key regulatory factors for both apoptosis and autophagy converge on the regulation of PI3P production. WIPI proteins function as essential PI3P effectors at the onset of autophagy; hence, PI3P-dependent localization of WIPI proteins at autophagosomal membranes should reflect both induction of autophagy and inhibition of apoptosis.

The genetic program of apoptosis is executed when the point-of-no-return is reached. In contrast, autophagy was shown to be initiated by the same stimuli that are additionally capable of blocking final autophagic cargo destruction in the lysosomal compartment, or even superstimulate autophagic degradation leading to autophagosomal cell death. Hence, the autophagic pathway not only represents a new rational therapeutic target mechanism for future treatment of human pathologies but also a target mechanism with the benefit to be modulated in many different ways according to the context-dependent needs.

Author details

Katharina Sporbeck, Fenja Odendall and Tassula Proikas-Cezanne*

*Address all correspondence to: tassula.proikas-cezanne@uni-tuebingen.de

Autophagy Laboratory, Department of Molecular Biology, Interfaculty Institute of Cell Biology, Eberhard Karls University Tuebingen, Tuebingen, Germany

Katharina Sporbeck and Fenja Odendall conducted their bachelor thesis in biochemistry under the supervision of Tassula Proikas-Cezanne at the Eberhard Karls University Tuebingen, Germany. Sporbeck and Odendall drafted this manuscript and Proikas-Cezanne wrote the final version of the chapter.

References

[1]   Yang Z and Klionsky D. Eaten alive: a history of macroautophagy. *Nat Cell Biol*. 2010, 12(9): 814-822.

[2]   Lockshin RA and Zakeri Z. Programmed cell death and apoptosis: origins of the history. *Nat Rev Mol Cell Biol*. 2001, 2: 545-550.

[3]   Hanahan D and Weinberg RA. Hallmarks of cancer: the next generation. *Cell* 2011, 144: 646-674.

[4]   Mizushima N and Komatsu M. Autophagy: renovation of cells and tissues. *Cell* 2011, 147: 728-741.

[5]   Mariño G, Niso-Santano M, Baehrecke EH, Kroemer G. Self-consumption: the inter-play of autophagy and apoptosis. *Nat Rev Mol Cell Biol*. 2014, 15(2): 81-94.

[6]   Codogno P, Mehrpour M, Proikas-Cezanne T. Canonical and non-canonical autopha-gy: variations on a common theme of self-eating? *Nat Rev Mol Cell Biol*. 2011, 13(1):7-12.

[7]   Proikas-Cezanne T, Takacs Z, Dönnes P, Kohlbacher O. WIPI proteins: essential PtdIns3P effectors at the nascent autophagosome. *J Cell Sci*. 2015, 128(2): 207-217.

[8]   Hardie, DG. AMPK and autophagy get connected. *EMBO J*. 2011. 30: 634-635.

[9]   Høyer-Hansen M, Bastholm L, Szyniarowski P, Campanella M, Szabadkai G, Farkas T, Bianchi K; Fehrenbacher N, Elling F, Rizzuto R, Mathiasen IS, Jäättelä M. Control of macroautophagy by calcium, calmodulin-dependent kinase kinase-beta, and Bcl-2. *Mol Cell*. 2007, 25(2):193-205.

[10]   Alessi, DR, K. Sakamoto, JR Bayascas. LKB1-dependent signaling pathways. *Annu Rev Biochem*. 2006, 75:137-163.

[11]   Pfisterer SG, Mauthe M, Codogno P, Proikas-Cezanne T. Ca2+/calmodulin-dependent kinase (CaMK) signaling via CaMKI and AMP-activated protein kinase contributes to the regulation of WIPI-1 at the onset of autophagy. *Mol Pharmacol*. 2011, 80(6):1066-1075.

[12]   O Farrell F, Rusten TE, Stenmark H. Phosphoinositide 3-kinases as accelerators and brakes of autophagy. *FEBS J*. 2013, 280(24):6322-6337.

[13]   Matsunaga K, Morita E, Saitoh T, Akira S, Ktistakis NT, Izumi T, Noda T, Yoshimori T. Autophagy requires endoplasmic reticulum targeting of the PI3-kinase complex via Atg14L. *J Cell Biol*. 2010, 190(4):511-521.

[14]   Proikas-Cezanne T, Waddell S, Gaugel A, Frickey T, Lupas A, Nordheim A. WIPI-1α (WIPI49), a member of the novel 7-bladed WIPI protein family, is aberrantly expressed in human cancer and is linked to starvation-induced autophagy. *Oncogene* 2004. 23: p. 9314-9325.

[15]   Jeffries TR, Dove SK, Michell RH, Parker PJ. PtdIns-specific MPR pathway associa-tion of a novel WD40 repeat protein, WIPI49. *Mol Biol Cell* 2004. 15(6):2652-2663.

[16] Mauthe M, Jacob A, Freiberger S, Hentschel K, Stierhof YD, Codogno P, Proikas-Cezanne T. Resveratrol-mediated autophagy requires WIPI-1-regulated LC3 lipidation in the absence of induced phagophore formation. *Autophagy* 2011, 7(12):1448-1461.

[17] Proikas-Cezanne T, Ruckerbauer S, Stierhof YD, Berg C, Nordheim A. Human WIPI-1 puncta-formation: a novel assay to assess mammalian autophagy. *FEBS Lett.* 2007, 581(18):3396-3404.

[18] Proikas-Cezanne T, Robenek H. Freeze-fracture replica immunolabelling reveals human WIPI-1 and WIPI-2 as membrane proteins of autophagosomes. *J Cell Mol Med.* 2011, 15(9):2007-2010.

[19] Thost AK, Dönnes P, Kohlbacher O, Proikas-Cezanne T. Fluorescence-based imaging of autophagy progression by human WIPI protein detection. *Methods.* 2015, 75:69-78.

[20] Polson HE, de Lartigue J, Rigden DJ, Reedijk M, Urbé S, Clague MJ, Tooze SA. Mammalian Atg18 (WIPI2) localizes to omegasome-anchored phagophores and positively regulates LC3 lipidation. *Autophagy.* 2010, 6(4):506-522.

[21] Dooley HC, Razi M, Polson HE, Girardin SE, Wilson MI, Tooze SA. WIPI2 links LC3 conjugation with PI3P, autophagosome formation, and pathogen clearance by recruiting Atg12-5-16L1. *Mol Cell.* 2014, 55(2):238-252.

[22] Lu Q, Yang P, Huang X, Hu W, Guo B, Wu F, Lin L, Kovács AL, Yu L, Zhang H. The WD40 repeat PtdIns(3)P-binding protein EPG-6 regulates progression of omegasomes to autophagosomes. *Dev Cell.* 2011, 21(2):343-57.

[23] Saitsu H, Nishimura T, Muramatsu K, Kodera H, Kumada S, Sugai K, Kasai-Yoshida E, Sawaura N, Nishida H, Hoshino A, Ryujin F, Yoshioka S, Nishiyama K, Kondo Y, Tsurusaki Y, Nakashima M, Miyake N, Arakawa H, Kato M, Mizushima N, Matsumoto N. De novo mutations in the autophagy gene WDR45 cause static encephalopathy of childhood with neurodegeneration in adulthood. *Nat Genet.* 2013, 45(4): 445-449, 449e1.

[24] Zoncu R, Efeyan A, Sabatini DM. mTOR: from growth signal integration to cancer, diabetes and ageing. *Nat Rev Mol Cell Biol.* 2010, 12:21-35.

[25] Hou W, Han J, Lu C, Goldstein LA, Rabinowich H. Autophagic degradation of active caspase-8: a crosstalk mechanism between autophagy and apoptosis. *Autophagy.* 2010, 6(7):891-900.

[26] Sandilands E, Serrels B, McEwan DG, Morton JP, Macagno JP, McLeod K, Stevens C, Brunton VG, Langdon WY, Vidal M, Sansom OJ, Dikic I, Wilkinson S, Frame MC. Autophagic targeting of Src promotes cancer cell survival following reduced FAK signalling. *Nat Cell Biol.* 2011, 14(1):51-60.

[27] Youle, RJ and Narendra DP. Mechanisms of mitophagy. *Nat Rev Mol Cell Biol.* 2011, 12:9-14.

[28] Jones RG, Plas DR, Kubek S, Buzzai M, Mu J, Xu Y, Birnbaum MJ, Thompson CB. AMP-activated protein kinase induces a p53-dependent metabolic checkpoint. *Mol Cell*. 2005, 18:283-293.

[29] Yu X, Muñoz-Alarcón A, Ajayi A, Webling KE, Steinhof A, Langel Ü, Ström AL. Inhibition of autophagy via p53-mediated disruption of ULK1 in a SCA7 polyglutamine disease model. *J Mol Neurosci*. 2013, 50(3):586-599.

[30] Comel A, Sorrentino G, Capaci V, Del Sal G. The cytoplasmic side of p53's oncosuppressive activities. *FEBS Lett*. 2014, 588(16):2600-2609.

[31] Xiao J, Zhang T, Xu D, Wang H, Cai Y, Jin T, Liu M, Jin M, Wu K, Yuan J. FBXL20-mediated Vps34 ubiquitination as a p53 controlled checkpoint in regulating autophagy and receptor degradation. *Genes Dev*. 2015, 29(2):184-196.

[32] Luo S, Garcia-Arencibia M, Zhao R, Puri C, Toh PP, Sadiq O, Rubinsztein DC. Bim inhibits autophagy by recruiting Beclin 1 to microtubules. *Mol Cell*. 2012, 47(3):359-370.

[33] Wei Y, Pattingre S, Sinha S, Bassik M, Levine B. JNK1-mediated phosphorylation of Bcl-2 regulates starvation-induced autophagy. *Mol Cell*. 2008, 30(6):678-688.

[34] Pagliarini V, Wirawan E, Romagnoli A, Ciccosanti F, Lisi G, Lippens S, Cecconi F, Fimia GM, Vandenabeele P, Corazzari M, Piacentini M. Proteolysis of Ambra1 during apoptosis has a role in the inhibition of the autophagic pro-survival response. *Cell Death Differ*. 2012, 19(9):1495-1504.

[35] Dear TN, Boehm T. Identification and characterization of two novel calpain large subunit genes. *Gene* 2001, 274(1-2):245-252.

[36] Yousefi S, Perozzo R, Schmid I, Ziemiecki A, Schaffner T, Scapozza L, Brunner T, Simon HU. Calpain-mediated cleavage of Atg5 switches autophagy to apoptosis. *Nat Cell Biol*. 2006, 8(10):1124-1132.

[37] Rubinstein AD, Eisenstein M, Ber Y, Bialik S, Kimchi A. The autophagy protein Atg12 associates with antiapoptotic Bcl-2 family members to promote mitochondrial apoptosis. *Mol Cell*. 2011, 44(5):698-709.

[38] Deiss LP, Feinstein E, Berissi H, Cohen O, Kimchi A. Identification of a novel serine/threonine kinase and a novel 15-kD protein as potential mediators of the gamma interferon-induced cell death. *Genes Dev*. 1995, 9(1):15-30.

[39] Eisenberg-Lerner A, Kimchi A. PKD is a kinase of Vps34 that mediates ROS-induced autophagy downstream of DAPk. *Cell Death Differ*. 2012, 19(5):788-797.

[40] Zalckvar E, Berissi H, Mizrachy L, Idelchuk Y, Koren I, Eisenstein M, Sabanay H, Pinkas-Kramarski R, Kimchi A. DAP-kinase-mediated phosphorylation on the BH3 domain of beclin 1 promotes dissociation of beclin 1 from Bcl-XL and induction of autophagy. *EMBO Rep*. 2009, 10(3):285-292.

Chapter 3

Autophagy Regulation of the Tumor Immunity – An Old Machinery for a New Function

Tsolere Arakelian, Takouhie Mgrditchian, Elodie Viry, Kris Van Moer, Guy Berchem, Salem Chouaib and Bassam Janji

Additional information is available at the end of the chapter

http://dx.doi.org/10.5772/61474

Abstract

Cancer was initially thought to be just a disease of cells with deregulated gene expression. It may be more accurate to consider cancer as a disease of the microenvironment. Despite the remarkable and fairly rapid progress over the past two decades regarding the role of the microenvironment in cancer biology and treatment, our understanding of its actual contribution to cancer resistance is still poor and fragmented. Nevertheless, the microenvironment is now considered to be of critical importance during the initiation and progression of carcinogenesis since it is involved in shaping and remodeling stroma reactivity and in reprogramming phenotypic and functional plasticity. Therefore, the tumor microenvironment represents an important hallmark of cancer, and the challenge now is to better understand how the tumor microenvironment participates in the emergence of immune-resistant tumor cell variants, which appears to be the greatest impediment to successful immunotherapy. In this context, autophagy has recently emerged as a new player in regulating the antitumor immune response under hostile tumor microenvironment. In this review, we will summarize recent data describing how autophagy activation under hypoxic stress impairs the antitumor immune response. In addition, we will discuss how tumor manages to hide from the immune attack and either mounts a "counter-attack" or develops resistance to immune cells. In particular, we will focus on the effect of hypoxia-induced autophagy in allowing tumor cells to outmaneuver an effective immune response and escape from immunosurveillance. It is our belief that autophagy may represent a conceptual realm for new immunotherapeutic strategies aiming to block immune escape and therefore providing rational approach to future tumor immunotherapy design.

Keywords: Autophagy, hypoxia, tumor immunity, immunotherapy, immune surveillance

1. Introduction

In addition to malignant cells, tumors contain cells of the immune system, the tumor vasculature, and lymphatics as well as fibroblasts, pericytes, and adipocytes. Cells of the immune system can identify and destroy tumor cells in a process termed cancer immunosurveillance.

Several types of immune cells are involved in tumor immune surveillance. Briefly, key cells of the adaptive immune system identifying cancer cells are cytotoxic T lymphocytes (CTL), which are able to recognize tumor antigens via the T-cell receptor (TCR) [1]. Some of these antigens are expressed exclusively by tumors and thus are called tumor-specific antigens [1]. Natural killer (NK) cells of the innate immune system also play an important role in tumor immune surveillance [1] by mechanisms called "missing-self" and "induced-self" recognitions [2]. In addition to CTL and NK cells, macrophages and neutrophil granulocytes are also involved in antitumor immunity [3]. Macrophages are antigen presenting cells (APCs) that display tumor antigens and stimulate other immune cells such as CTL, NK cells, and other APCs [4]. While the molecular mechanism by which CTL and NK cells recognize their target tumor cells is fundamentally different, both immune cells kill their target following the establishment of immunological synapse (IS) [5]. The formation of IS requires cell polarization and extensive remodeling of the actin cytoskeleton at various stages [6]. It is now well established that CTL and NK cells recognize and kill target cells by two major pathways: either through the release of cytotoxic granules containing perforin and granzymes to the cytosol of target cells [7] or through tumor necrosis factor (TNF) super family-dependent killing [8].

Although various immune effector cells are recruited to the tumor site, their antitumor functions are largely downregulated in response to several microenvironmental factors. Indeed, hypoxic stress in the tumor microenvironment, which is the result of an inadequate oxygen supply to the cells and tissues, is a characteristic feature of locally advanced solid tumors and is considered as the major mechanism responsible for tumor resistance to therapies [9].

Experimental and clinical evidence indicates that the majority of mechanisms suppressing the antitumor immune functions are directly evolved in the hypoxic tumor microenvironment (reviewed in [10]). Thus, NK cells and natural killer T (NKT) cells infiltrate the tumor microenvironment but are not found in contact with tumor cells [11]. It has been reported that in colorectal, gastric, lung, renal, and liver cancer NK cells appear to predict a good prognosis [12]. However, although they are present in the tumor microenvironment, NK cells may not be able to exert their tumor-killing function. A number of studies reported that NK cells in the tumor stroma have an anergic phenotype that is induced by malignant cell-derived transforming growth factor beta (TGF-β) [13]. Furthermore, immune cells in the tumor microenvironment not only fail to exercise antitumor effector functions but also co-opted to promote tumor growth [14]. In addition, it has become clear that the immune system not only protects the host against tumor development but also sculpts the immunogenic phenotype of a developing tumor and can favor the emergence of resistant tumor cell variants [15]. Thus, it has become obvious that the evasion of immunosurveillance by tumor cells is under the control of the tumor microenvironment complexity and plasticity. Reactivating the immune system for therapeutic benefit in cancer has therefore long been a goal in cancer immunotherapy.

After decades of disappointment, cancer immunotherapy has recently emerged as a promising treatment of several cancers for which conventional therapies have failed [16]. Notably, the success of the recent proof-of-concept clinical trials of anticytotoxic T-lymphocyte-associated protein 4 (anti-CTLA4) and anti-programmed cell death 1 (anti-PD-1) based on reactivating the adaptive immune response claims that the tide has finally changed. This success is mainly attributed to an increase in our understanding of the mechanisms regulating tumor cell cytotoxicity mediated by immune cells. For several years, our group has been able to participate in this understanding by studying the mechanism responsible for the tumor escape from the immune surveillance [17, 18], which still represents the major obstacle for defining efficient cancer immunotherapeutic approaches. Therefore, it remains important to better understand how tumor cells manage to outmaneuver the immune system and evade effective immunosurveillance.

2. Hypoxic stress in the tumor microenvironment

Hypoxia in the tumor microenvironment commonly refers to a condition in tumors where the pressure of oxygen is lower than 5–10 mm Hg. The adaption of tumors to hypoxic stress is regulated by hypoxia inducible factor family of transcription factors (HIFs). It has been demonstrated in a large number of human cancer cases and/or incidents that HIFs were overexpressed and such overexpression is associated with poor response to treatment [19]. Moreover, evidence showed a clear positive correlation between enhanced hypoxic expression of HIFs and mortality [20].

2.1. Hypoxia-inducible factors

Three isoforms of HIF have been identified: HIF-1, HIF-2, and HIF-3. HIF-1 and HIF-2 (also known as EPAS1) have the same structure and are well characterized. However, HIF-3 acts as a negative regulator of HIF-1 and HIF-2 [21]. HIF-1 is ubiquitously expressed in all mammalian cells, whereas HIF-2 and HIF-3 are selectively expressed in certain tissues such as vascular endothelial cells, type II pneumocytes, renal interstitial cells, liver parenchymal cells, and cells of the myeloid lineage [22].

HIF-1 is a heterodimer composed of a constitutively expressed subunit and an O_2-regulated subunit, HIF-1β, and HIF-1α, respectively [23]. In the presence of oxygen, HIF-1α is hydroxylated on a proline residue by prolyl hydroxylase domain protein 2 (PHD2), which leads to an interaction with the von Hippel–Lindau (VHL) protein [24]. This allows the recruitment of an E3 ubiquitin ligase that catalyzes the polyubiquitination of HIF-1α and its subsequent degradation by ubiquitin proteasome system (UPS) [24]. Under hypoxia, the hydroxylation of HIF-1α is inhibited, and HIF-1α is accumulated and translocated to the nucleus where it forms a dimer with HIF-1β and activates the transcription of several genes involved in many biological processes [24]. Similar to HIF-1α, HIF-2α is also regulated by oxygen-dependent hydroxylation. While the effect of hypoxia on suppressing the activity of immune cells is relatively well defined, the mechanisms by which hypoxia educates tumor cells to escape an effective immune cell mediated killing are still largely elusive.

2.2. Hypoxia-induced autophagy

Although autophagy can be activated in response to different stimuli, including nutrient starvation and/or growth factors withdrawal, hypoxic stress is the major activator of autophagy in the tumor microenvironment [25]. Indeed, emerging recent data have showed that hypoxia-induced autophagy is an important regulator of the innate and adaptive tumor immunity mediated by NK cells and CTL, respectively. In particular, hypoxia has been described to play a central role in activating multiple overlapping adaptive mechanisms involving autophagy and leading to the emergence of resistant tumor cells able to outmaneuver an effective immune response and escape from immune cell killing. In this context, we have recently showed that the activation of autophagy in tumor cells under hypoxia dramatically decreases tumor cell susceptibility to NK- and CTL-mediated lysis [17, 26]. Therefore, autophagy activation is considered to be an important adaptive and resistance mechanism operating in tumor cells to escape the immune system. In accordance with such a role of autophagy, Lotze *et al.* showed that NK cells along with human peripheral blood lymphocytes are primary mediators in inducing autophagy in several human tumors promoting cancer cell survival [27]. Other studies showed that autophagy also plays an important role in regulating CTL-mediated antitumor immune response. While the molecular mechanisms by which autophagy impairs tumor susceptibility to NK and CTL are different, experimental evidence claims that blocking autophagy may improve tumor immunity.

Autophagy is a lysosomal degradation pathway that allows the cell to self-digest its own components, getting rid of excessive or damaged organelles and misfolded proteins in the cell. Such degradation process provides nutrients to maintain crucial cellular functions under nutrient deprivation, thus allowing the survival of cancer cells [28]. It has been reported that the activation of autophagy under hypoxic stress in tumor cells occurs either by HIF-1-dependent or -independent manner. In this section, we will briefly describe the different mechanisms involved in the activation of autophagy.

2.2.1. HIF-1-dependent activation of autophagy

Under hypoxia, HIF-1α is stabilized, and its heterodimerization with HIF-1β allows the binding of the transcription factor to hypoxia response elements (HREs) in target genes [9]. HRE is a *cis*-acting hypoxia response element (5'-TACGTGCT-3'), which can be located in either the 5' or the 3' regions of the genes [29] to confer oxygen regulation of genes expression [29–32]. The activation of HIF-1α-dependent autophagy occurs via the induction of the Bcl-2 (B-cell lymphoma 2)/adenovirus E1B 19 kDa protein-interacting protein 3 (BNIP3), which contains two HRE sites in its promoter region: HRE1 and HRE2. It has been demonstrated that HIF-1 directly binds to HRE2 site in order to induce the expression of BNIP3 [33], thus leading to the disruption of the autophagy inhibitory complex BECN1/Bcl-2 and the subsequent release of Beclin1 (BECN1) to promote the activation of autophagy.

2.2.2. HIF-1-independent activation of autophagy

Despite the role of HIF-1 in the regulation of autophagy under hypoxic conditions, it remains important to note the existence of other pathways that may regulate autophagy under hypoxia independently of HIF-1. Recently, two pathways that influence gene expression and tumor cell behavior have been described to be O_2 sensitive [34]. One of them occurs through the regulation of the mammalian target of rapamycin (mTOR) and its downstream effectors that orchestrate several biological processes, including autophagy. Indeed, mTOR signaling consists of two major pathways mediated by the specific mTOR complexes mTORC1 and mTORC2. It has been reported that mTORC1 negatively controls autophagy by the inhibition of protein kinase ATG1 involved in the formation of autophagosomes [35]. Hypoxia inhibits mTORC1 through multiple pathways; one of them is mediated through hypoxic activation of the tuberous sclerosis protein (TSC) complex, which is a heterodimeric complex formed by TSC1 and TSC2 [36]. The 5′ AMP-activated protein kinase (AMPK) is a heterotrimeric complex encoded by several genes and is the primary energy sensor in cells. Upon its activation through the increase in the AMP/ATP ratio, AMPK phosphorylates many downstream targets, including TSC2. TSC2 phosphorylation on serine residues 1270 and 1388 enhances the activity of the TSC1/TSC2 complex and thereby blocks the Ras homolog enriched in brain (RHEB)-dependent activation of mTOR [37]. Another pathway that activates autophagy in an HIF-1-independent manner is through the activation of the unfolded protein response (UPR), a program of transcriptional and translational changes that occur as a consequence of endoplasmic reticulum (ER) stress [38]. The UPR is mediated by three ER stress sensors: PKR-like ER kinase (PERK), ER to nucleus signaling 1 (ERN1), and activating transcription factor (ATF) 6 [38]. In some conditions, autophagy appears to be mediated by PERK, whereas in others, it occurs downstream of IRE1. For example, by inducing PERK-dependent phosphorylation of the eukaryotic translation initiation factor 2 alpha (elF2α), hypoxia activates autophagy by the transcriptional induction of LC3 through the expression of ATF4 [39].

2.2.3. Autophagy activation under nutrient starvation

Under starvation condition, autophagy is activated by several mechanisms. One of the well described mechanisms is the regulation of autophagy by Reactive Oxygen Species (ROS) which mainly comprise superoxide (O_2), hydrogen peroxide (H_2O_2) and hydroxyl radical (OH). Several studies have demonstrated that ROS-dependent activation of AMPK leads to the inhibition of mTOR pathway under starvation conditions, thereby activating autophagy [40].

Furthermore, recent studies showed that the upregulation of NS5ATP9 previously identified as p15PAF [proliferating cell nuclear antigen (PCNA)-associated factor] in starved HepG2 cells plays a functional role in starvation-induced autophagy and contributes to tumor cell growth. NS5ATP9 promotes autophagy by Beclin1-dependent manner under starvation condition. Indeed NS5ATP9 upregulates Beclin-1 expression at the transcriptional level, thereby inducing autophagy [41]. Another mechanism of autophagy activation under starvation conditions is MK2/MK3-dependent Beclin1 phosphorylation. MK2/MK3 are two related stress-responsive kinases members of p38 mitogen-activated protein kinase (MAPK) signaling pathway. Yobgjie Wei *et al.* showed that MK2/MK3 phosphorylate Beclin1 at serine 90, and this phosphorylation is essential for autophagy induction in response to nutrient starvation [42].

It has been reported in several studies that some microRNAs (miRNAs) are able to regulate starvation-induced autophagy. Mature miRNAs are a class of noncoding RNAs that play key roles in the regulation of gene expression by acting at the posttranscriptional level. miRNAs are short, single-stranded RNA molecules ~22 nucleotides in length. They are partially complementary to one or more messenger RNA (mRNA) molecules. By base pairing with

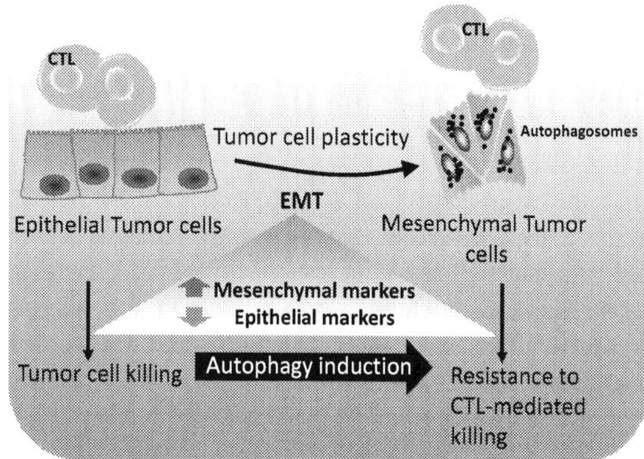

Figure 4. The acquisition of an EMT phenotype of tumor cells through loss of epithelial and gain of mesenchymal markers confers resistance to CTL-mediated lysis through autophagy induction.

that the transcription factors SNAI1 and ZEB1 bind to E-boxes in the MIR34A/B/C promoters, thereby repressing MIR34A and MIR34B/C expression. While much remains to be learned mechanistically, this result extended the role of SNAI1 as a regulator of autophagy and paved the way to an interesting topic of research. Although targeting BECN1 in mesenchymal cells is sufficient to restore CTL-mediated tumor cell lysis, it has no effect on cell morphology and the expression of EMT markers. This finding suggests that autophagy is a downstream target of the EMT program in breast cancer cells [57, 58]

4. Autophagy as a target for improving anticancer therapies

Intrinsic resistance mechanisms evolved by cancer cells are a key limitation to improve response rates and survival of patients treated with anticancer therapies. It is now clearly established that autophagy activation in cancer cells under stress conditions allows resistance to chemotherapy [59], radiotherapy [60, 61], and immunotherapy [17, 44, 62, 63]. On the basis of these observations, it is not surprising that autophagy has emerged as a potential therapeutic target, and research efforts have intensified to develop autophagy inhibitors that could be used in combination with anticancer therapies.

Pharmacological inhibitors of autophagy identified so far can be classified in two main groups depending on which stage of the autophagy process is targeted. Sequestration inhibitors such as 3-methyladenine (3-MA), LY294002, and wortmannin act at the early stage of the autophagy pathway by inhibiting the class III phosphatidylinositol-3 kinase (PI3K). Recently, a potent small molecule inhibitor of autophagy, called spautin-1, was identified which causes the degradation of the class III PI3K complex by targeting BECN1 [64]. Most of other inhibitory compounds act as later stage. Microtubule poisons such as vinca alkaloids, taxanes, nocodazole, and colchicine cause blockade of autophagosome and lysosome fusion. Inhibitors of lysosomal enzymes (e.g., leupeptin, pepstatin A, and E-64d) or compounds that elevate lysosomal pH (e.g., bafilomycin A1, chloroquine) impair autophagy through the inhibition of cargo degradation by lysosomal hydrolases (reviewed in [28]). Chloroquine (CQ) and its derivative hydroxychloroquine (HCQ) have long been used as antimalarial and antirheumatic drug, and they were the only autophagy inhibitors approved by the US Food and Drug Administration. As single agent, CQ has shown anticancer activity in lymphoma [65], pancreatic [66], and breast cancers [67]. In addition, the use of CQ or HCQ in combination with conventional therapies has provided convincing results in preclinical models [68]. Indeed, autophagy blockade enhances anticancer effects of apoptosis-inducing agents [69] and Src family kinase inhibitors [70]. Moreover, this strategy has shown promising results on patient's survival in the first phase III clinical trial using CQ as adjuvant treatment to conventional anticancer therapy for glioblastoma [71]. Currently, more than 30 clinical trials are registered with the National Cancer Institute to evaluate the effects of autophagy inhibition in a variety of human cancers (http://clinicaltrials.gov). The Table 1 summarize the clinical trials involving CQ or HCQ in combinational treatment of refractory malignancies.

Cancer type	Drug intervention	Phase status	Clinical trial ID	Title of the clinical trial
Pancreatic cancer	HCQ +Gemcitabine +Abraxane	I/II Active	NCT01506973	A Phase I/II/Pharmacodynamic Study of Hydroxychloroquine in Combination with Gemcitabine/ Abraxane to Inhibit Autophagy in Pancreatic Cancer
Breast cancer	HCQ	II Active	NCT01292408	Autophagy Inhibition Using Hydrochloroquine in Breast Cancer Patients
NSCL cancer	HCQ +Paclitaxel +Carboplatin +Bevacizumab	II Active	NCT01649947	Modulation of Autophagy in Patients with Advanced/Recurrent Non-small Cell Lung Cancer - Phase II
Renal cancer	HCQ+RAD001	I/II Active	NCT01510119	Autophagy Inhibition to Augment mTOR Inhibition: A Phase I/II Trial of RAD001 and Hydroxychloroquine in Patients with Previously Treated Renal Cell Carcinoma
SCLC	CQ +Chemotherapy +Radiotherapy	I Active	NCT00969306	Chloroquine as an Anti-Autophagy Drug in Stage IV Small Cell Lung Cancer (SCLC) Patients
Colorectal cancer	HCQ+FOLFOX +Bevacizumab	I/II Active	NCT01206530	FOLFOX/Bevacizumab/ Hydroxychloroquine (HCQ) in Colorectal Cancer

Cancer type	Drug intervention	Phase status	Clinical trial ID	Title of the clinical trial
Solid tumors	HCQ+Sorafenib	I Active	NCT01634893	Oral Hydroxychloroquine Plus Oral Sorafenib to Treat Patients with Refractory or Relapsed Solid Tumors
Solid tumors	CQ+Carboplatin +Gemcitabine	I Active	NCT02071537	Chloroquine in Combination with Carboplatin/Gemcitabine in Advanced Solid Tumors
Solid tumors	HCQ +Temsirolimus	I Active	NCT00909831	Hydroxychloroquine and Temsirolimus in Treating Patients with Metastatic Solid Tumors that Have not Responded to Treatment
Chronic myeloid leukemia	HCQ +Imatinib mesylate	II Active	NCT01227135	Imatinib Mesylate with or without Hydroxychloroquine in Treating Patients with Chronic Myeloid Leukemia
Pancreatic cancer	HCQ +Gemcitabine +Nab-Paclitaxel	II Active	NCT01978184	Randomized Phase II Trial of Pre-operative Gemcitabine and Nab Paclitacel with or without Hydroxychloroquine
Melanoma, prostate or kidney cancers	HCQ+MK2206	I Active	NCT01480154	Akt Inhibitor MK2206 and Hydroxychloroquine in Treating Patients with Advanced Solid Tumors, Melanoma, Prostate or Kidney Cancer
Multiple myeloma	HCQ +Cyclophosphamide +Dexamethasone +Sirolimus	I Active	NCT01689987	Hydroxychloroquine, Cyclophosphamide, Dexamethasone, and Sirolimus in Treating Patients with Relapsed or Refractory Multiple Myeloma
Solid tumors	HCQ+Vorinostat	I Active	NCT01023737	Hydroxychloroquine + Vorinostat in Advanced Solid Tumors
Pancreatic cancer	HCQ +Radiotherapy	II Active	NCT01494155	Short Course Radiation Therapy with Proton or Photon Beam Capecitabine and Hydroxychloroquine for Resectable Pancreatic Cancer
Glioma	HCQ +Radiotherapy	II Active	NCT01602588	A Randomised Trial Investigating the Additional Benefit of Hydroxychloroquine (HCQ) to Short Course Radiotherapy (SCRT) in Patients Aged 70 Years and Older with High Grade Gliomas (HGG)

Cancer type	Drug intervention	Phase status	Clinical trial ID	Title of the clinical trial
Soft tissue sarcoma	HCQ+Sirolimus	II Active	NCT01842594	A Phase II Trial of Combined Hydroxychloroquine and Sirolimus in Soft Tissue Sarcoma
Melanoma	HCQ +Vemurafenib	I Active	NCT01897116	A Phase I Trial of Vemurafenib and Hydroxychloroquine in Patients with Advanced BRAF Mutant Melanoma
Renal cancer	HCQ+Aldesleukin	I/II Active	NCT01550367	Study of Hydroxychloroquine and Aldesleukin in Renal Cell Carcinoma Patients (RCC)
Colorectal cancer	HCQ+Vorinostat	II Approved	NCT02316340	Vorinostat Plus Hydroxychloroquine Versus Regorafenib in Colorectal Cancer
Advanced cancer	HCQ+Vorinostat or Sirolimus	I Active	NCT01266057	Sirolimus or Vorinostat and Hydroxychloroquine in Advanced Cancer
Prostate cancer	HCQ +Navitoclax +Abiraterone acetate	II Active	NCT01828476	Navitoclax and Abiraterone Acetate with or without Hydroxychloroquine in Treating Patients with Progressive Metastatic Castrate Refractory Prostate Cancer
Melanoma	HCQ +Dabrafenib +Trametinib	I/II Active	NCT02257424	The BAMM Trial: BRAF, Autophagy and MEK Inhibition in Metastatic Melanoma: A Phase I/2 Trial of Dabrafenib, Trametinib and Hydroxychloroquine in Patients with Advanced BRAF Mutant Melanoma
Solid tumors	HCQ +Temozolomide	I Active	NCT00714181	Hydroxychloroquine and Temozolomide in Treating Patients with Metastatic or Unresectable Solid Tumors
Multiple myeloma	HCQ+Bortezomib	I/II Active	NCT00568880	Hydroxychloroquine and Bortezomib in Treating Patients with Relapsed or Refractory Multiple Myeloma
Pancreatic cancer	HCQ +Gemcitabine	I/II Closed	NCT01128296	Study of Pre-surgery Gemcitabine + Hydroxychloroquine (GcHc) in Stage IIb or III Adenocarcinoma of the Pancreas
Colorectal cancer	HCQ +Capecitabine +Oxaliplatin +Bevacizumab	II Closed		Hydroxychloroquine, Capecitabine, Oxaliplatin, and Bevacizumab in Treating Patients with Metastatic Colorectal Cancer

Chapter 4

Targeting the Autophagy Process in Breast Cancer Development and Treatment

Joanna Magdalena Zarzynska

Additional information is available at the end of the chapter

http://dx.doi.org/10.5772/61181

Abstract

Autophagy is a homeostatic process that degrades long-lived or damaged proteins and organelles. By recycling intracellular constituents, it is buffering metabolic stress under starvation conditions. The autophagy role in cancer remains unclear and complicated as it appears to be involved in tumorigenesis, cancer development and treatment outcome in different ways. Autophagy can act as both tumor-promoting and tumor-suppressing agent depending on the stage of cancer progression. During the initiation of cancer, autophagy prevents cells from further DNA damage and genomic instability. It could also be a cell death mechanism in cancer cells with apoptotic defect. Autophagy can also promote tumor growth by facilitating oncogene-induced senescence or protecting tumors against necrosis and inflammation. Once the cancer is formed, autophagy can contribute to tumor progression (by allowing cells to survive in stressful conditions) and metastasis. There is evidence that breast cancer could also be controlled by autophagy. Regulation of this process, correlated proteins and active factors are currently under scientific study in the aspect of breast cancer effective therapeutic strategies.

Keywords: Autophagy, breast cancer, tumorigenesis, Beclin-1, miRNA

1. Introduction

Breast cancer (BC) is a potentially life-threatening malignant tumor that still causes high mortality among women. One of the mechanisms through which cancer development could

be controlled is autophagy. This process exerts different effects during the stages of cancer initiation and progression due to the occurring superimposition of signaling pathways of autophagy and carcinogenesis. Chronic inhibition of autophagy or autophagy deficiency promotes cancer due to instability of the genome and defective cell growth as a result of cell stress. However, increased induction of autophagy can become a mechanism which allows tumor cells to survive the conditions of hypoxia, acidosis or chemotherapy. Therefore, in the development of cancer, autophagy is regarded as a double-edged sword. There are different ways of autophagy process control under scientific research for potential therapeutic treatment in anti-cancer strategies.

2. The process of autophagy

Autophagy is a process regulated genetically and/or controlled by a group of evolutionarily conserved genes (ATGs; autophagy-related genes). Initially, autophagy was identified as a cell survival mechanism protecting from nutrient deprivation. It ensures homeostasis by maintaining proteins and organelles turnover. Removing excess or damaged intracellular components in response to stress as well as microorganisms allows cells to restrain damage (including genome instability which limits initiation and progression of cancer) and subsequent inflammation. Cellular stress can be caused by a variety of chemical and physical agents like nutrient starvation, pro-inflammatory state, hypoxia, oxidants, infectious agents and xenobiotics [1, 2]. Under the influence of autophagic pathway, biological and morphological changes have been observed [3]. In certain developmental conditions like in cell's response to metabolic stress or under cytotoxic stimuli, autophagy results in a form of cell death described as programmed cell death type II [4].

Currently, over 35 proteins are believed to be essential for autophagy occurrence and progression [5]. The complete macroautophagy (referred to hereafter as autophagy) is generally divided into the following stages: induction, vesicle nucleation, vesicle elongation and completion, docking and fusion, degradation and then recycling [1, 6]. Vesicle nucleation is the initial step in which proteins and lipids are recruited for construction of the autophagosomal membrane. Nucleation consists of the formation of the phagophore or isolation membrane. In mammalian cells, this process is initiated by activation of the class III PI3K/ Beclin-1 complex including the core members hVps34/PIK3C3, Beclin-1 (BECN1) and p150. Numerous additional binding partners of this complex function as either positive or negative regulators and include BAX-interacting factor-1 (BIF-1), Atg14L, UVRAG (UV irradiation resistance-associated gene), Ambra1 (activating molecule in Beclin-1-regulated autophagy protein 1) and Rubikon [1, 5, 7-9]. Rubicon (RUN domain Beclin 1-interacting cysteine-rich-containing protein) has also been shown to negatively regulate autophagy. Subsequently, the phagophore is elongated by several ATG proteins. During this elongation step, microtubule-associated protein 1 light chain 3 (LC3)-I is lapidated to LC3-II. Then, the phagophore is maturated primarily upon the action of LC3-II and BECN1 proteins. Maturation leads to the formation of autophagosome (enclosed vesicle). The regulation of the maturation process of the autophagosome is multi-factorial and involves Rab GTPase, SNAREs (soluble N-ethylmaleimide-sensitive fusion attachment protein receptors) and ESCRT (endosomal sorting complexes required for transport) proteins, molecules of the acidic lysosomal compartment (e.g. v-ATPase, LAMP proteins- lysosome-associated membrane glycoprteins; lysosomal carriers and hydrolases) and Beclin-1. Finally, the autophagosome fuses with the lysosome to form an autolysosome. The internal material of the autophagic vacuole is degraded by the lysosomal hydrolases.

Basal levels of macroautophagy are kept in check by mTORC1 (mammalian target of rapamycin complex 1) which phosphorylates Atg13 and ULK1 (uncoordinated 51-like kinase 1/ Atg1) or ULK2. This activity in consequence is giving the inhibition of FIP200 (focal adhesion kinase interacting protein of 200 kD/Atg17) phosphorylation by ULK1 [10]. The mTORC1 complex is an important component of a network that accordingly maintains homeostasis by controlling the levels of anabolism and catabolism. For example, high levels of amino acids maintain mTORC1 in an active state by enhancing its binding to regulatory GTPases, Rag (Ras-related GTPase) and Rheb (Ras homolog enriched in brain) [11]. mTORC1 activity could be indirectly induced by insulin and IGF1 (insulin like growth factor 1) [6]. Low glucose levels or high AMP levels (adenosine 5′-monophosphate), indicators of low cellular energy status or stress, could activate AMPK (AMP-activated protein kinase) which in turn inhibits mTORC1 and stimulates autophagy [2,7].

The role of the PI3K/Akt pathway is to suppress autophagy. This pathway activation was shown to decrease autophagy through mTOR activation. It has been considered for cancer treatment. The MAPK pathway also plays a significant role in autophagy. Ras may play a dual role in autophagy [12]. When Ras activates PI3KCA, autophagy is inhibited; however, when it selectively activates the MAPK pathway, autophagy is stimulated.

3. Breast cancer

Breast cancer (BC) is the most common and fatal cancer in women worldwide. Decreasing mortality rates can be observed that result mostly from efficient screening strategies [13] but still BC is ranked on the second place in mortality among cancer types [14]. It has been estimated that approximately 1.3 million females develop BC each year with around 465.000 expected to succumb to the disease [15,16]. It is causing death of about 350.000 women in both developed and developing countries every year (with slightly more cases in less developed than in more developed regions) [17]. According to another data presented by DeSantis et al., there are still 500.000 breast cancer deaths per year worldwide [18]. More than 90% of lethality in patients is caused by metastasis and the occurrence of distant metastases (distinct metastatic pattern involving the regional lymph nodes, bone marrow, lung and liver) severely limits the prognosis [19,20]. The 5-year survival rate for patients with BC drops sharply from 98% for individuals with localized disease to 23% for those with metastatic disease (cancer statistics from 2012) [21]. A significant subpopulation of patients with metastasis risk has a median survival time of 18–30 months [22].

In the pathogenesis and progression of BC are involved many factors including genetic, biological and environmental factors as well as a lifestyle [17]. For example, Bcl-2 protooncogene is overexpressed in half of all human malignancies and more than 60% of BC and exerts its oncogenic role by preventing cells from undergoing apoptosis [23]. Still, the disease background is not fully clear because it has been estimated that 75% of women with sporadic invasive BC have no known epidemiological risk factors [24].

4. Autophagy regulation in breast tumors development

In tumor genesis and treatment responsiveness, autophagy role is complicated and context-dependent. It presumably differs in different stages of cancer development. At the initial stages, autophagy may represent a protective role thanks to its catabolic functions by degrading and/or recycling cell components (e.g. damaged organelles and misfolded proteins) [25-27]. It also protects against the deleterious effects of ROS (reactive oxygen species) in the cells. Proliferation of cells with cancer-linked mutations may be retarded by autophagy. It can also limit propagation of this type of mutations and consequently suppress tumorigenesis by facilitating the cellular senescence phenomenon (biological aging). However, once a tumor develops, the cancer cells can utilize autophagy for their own cytoprotection and use enhanced autophagy to survive under metabolic and therapeutic stress [27]. Autophagy might increase oxidative stress, hence promoting genome instability and malignant transformation [25-27]. As an example, autophagy has been shown to be required for the transformation of mouse embryonic fibroblasts by the Ras oncogene and this effect is linked to its role in nutrients recycling, such as glucose uptake and increased glycolytic flux [28]. What's more, it has been suggested that metastatic cancer cells can escape from anoikis (process of apoptosis induced by lack of correct cell-ECM attachment) through the autophagy induction [29, 30]. The AMP-activated protein kinase (AMPK) stress response pathway is involved in mediating anoikis resistance by inhibiting mTOR and suppression of protein synthesis. The Ras/MAPK and PI3K/Akt pathways are common mechanisms utilized by cancer cells to evade anoikis. Autophagy is necessary for cancer cells survival in hypoxic conditions during the later stages of *in vivo* tumor formation before the vascularization of tumor takes place [31, 32]. However, the induction of autophagy is associated with cell death in normal cells and in some cancer cells [23]. Autophagic cell death has been described e.g. in anti-estrogen-treated cultured human mammary carcinoma MCF-7 cells [33].Some studies have shown that cancer cells express lower levels of the autophagy-related proteins LC3-II and Beclin-1 than normal epithelial cells. There is evidence that heterozygous disruption of BECN1 promotes tumorigenesis and the overexpression inhibits tumorigenesis, which support the assertion, that defective autophagy or inhibition of autophagy playing a role in malignant transformation [23].

The ability of cancer cells to invade and metastasize is closely correlated with the process of epithelial-mesenchymal transition (EMT). It was recently demonstrated that ectopic expression of the DEDD gene in the MDA-MB-231 metastatic BC cell line led to the degradation of the EMT inducers Snail and Twist through autophagy activation [34]. Reversely, knock-down of DEDD in the MCF7 non-metastatic BC cell line led to autophagy reduction and EMT promotion [34].

Regulation of autophagy in tumors is governed by principles similar to normal cells only in a much more complicated manner. Abnormal PI3K activation in cancer cells is frequently observed. The multitude of interactions between the PI3K/Akt/mTOR pathway and other cell signaling cascades could often be deregulated [7]. For example, Ras /Raf /ERK pathway, indicated as one of the most commonly deregulated pathways identified in tumors, frequently are observed activating mutations in Ras or B-Raf oncogenes [4]. ERK activity has been associated with autophagy and autophagic cell death in many cellular models in response to different stresses [4]. It also happens in TNFalpha treatment in MCF-7 cells. A deregulated PI3K/Akt/mTOR axis not only suppresses autophagy process but also induces protein translation, cell growth and proliferation thereby can force tumorigenesis. Tumors with constitutively active PI3K mutations, PTEN loss or Akt activation would be expected to be dependent on autophagy for energy homeostasis and survival. Suppression of autophagy by the PI3K cascade is disadvantageous for rapidly proliferating tumor cells and there are theses that compensatory mechanisms (like deregulated apoptosis and/or metabolism) might be concurrently activated to prevent the negative implications of defective autophagy on tumor cell survival.

Many proteins and active factors correlated with autophagy are reported to be associated with human cancers [35]. Various tumor suppressors (e.g. PTEN, TSC1/2, p53, and DAPK) are autophagy inducers whereas some inhibitors of autophagy (e.g. Akt and Ras) possess onco-genic activity [36]. Some studies [37, 38] showed that the more advanced stages of breast cancer over-express several other oncogenic and signaling proteins such as IGF-1R, Cyclin D1, c myc, pERK, Stat3 and Pak4. Some are known activators of Akt/mTOR pathway. Several other autophagy regulators like mitogen-activated kinases (BNIP3) [39] and HSpin 1 (human homologue of the Drosophila spin gene product) [40] play a critical role in cancer cells. Deletion of the essential autophagy gene *FIP200* impairs oncogene-induced tumorigenesis in a mouse model of breast cancer [41].

PTEN, a critical regulator of the PI3K pathway, has a stimulatory effect on autophagy by downregulating PI3K/Akt signaling through inhibition of the Akt/PKB activation. Akt inhibition leads to mTOR signaling suppression and the induction of autophagy [42]. AMPK (AMP-activated protein kinase) pathway has a negative effect on mTOR signaling and promotes autophagy (e.g. upon starvation conditions by activation of Tuberin -TSC2 and/or mTOR signaling inhibitor).

EI24/PIG8 (etopside induced gene) can be mentioned as another critical factor of autophagic degradation. It remains under control of p53 [43, 44] - well known critical tumor suppressor. p53 effects protectively by cell cycle arrest initiation, removal of cells with incurred DNA damage, senescence and apoptosis. The human EI24 genomic locus is on chromosome 11 in region frequently altered in cancers and was reported to be mutated in aggressive breast cancers. Furthermore, since EI24/ PIG8 (induced by p53) is also known as important apoptotic effector [43, 45], its role may contribute to tumor suppression. EI24/PIG8 loss was associated with tumor invasiveness but not with the development of the primary tumor [45].

mTORC1, class I PI3K, Akt, class III PI3K, Beclin-1 and p53 are critical components of the autophagic pathway that have become major targets of autophagy-related drug design. As an example, rapamycin and its derivatives (e.g. rottlerin, PP242 and AZD8055) target the PI3K/Akt/mTOR signaling pathway to induce autophagy. Spautin-1 and tamoxifen regulate Beclin-1 activity to respectively inhibit and promote autophagy. Oridonin and metformin trigger p53-mediated autophagy and cell death [46].

4.1. Beclin-1 role

The most important evidence linking dysfunctional autophagy and cancer come from studies on mice demonstrating that the inhibition of autophagy by disruption of *BECN1* increases cellular proliferation, the frequency of spontaneous malignancies (i.e. lung cancer, liver cancer and lymphomas) as well as mammary hyperplasia. It also accelerates the development of carcinogen-induced premalignant lesions [23]. Additionally, low levels of Beclin1 in human MCF-7 BC cell line can inhibit tumor cell growth [47].

Human breast cancer cell lines FISH analysis with the Beclin-1-containing PAC 452O8 as a probe revealed that 9 out of 22 cell lines had allelic BECN1 deletions [7]. Monoallelic deletion of BECN1 has been also detected in 40-75% cases of human breast, ovarian and prostate tumors [2, 26]. Thereby BECN1 is considered as a tumor suppressor gene [48]. Deletions of Beclin-1 have recently been found mostly associated with BRCA1 in breast and ovarian human tumors (suggesting that BRCA1 loss is the mutation driver and that Beclin-1 is lost because of its proximity to it) [49]. Many breast carcinoma cell lines, although polyploidal for chromosome 17 (*BECN1* gene is placed in 17q21 loci), exhibit deletions of one or more BECN1 alleles. The aberrant expression of Beclin-1 in many kinds of tumors correlates with poor prognosis [26]. Also, heterozygous deletion of BECN1 in mice (BECN1+/−) resulted in increased incidence of spontaneous tumors [48]. Those mice do not have increased incidence of mammary tumors but rather are susceptible to lymphomas and lung carcinomas. BECN1+/− mice tumors express wild-type BECN1 mRNA and protein indicating that Beclin-1 is a haploinsufficient tumor suppressor [7, 8]. The Beclin-1 loss occurring in BC could have important effects independent of autophagy through its interaction with Bcl-2. Bcl-2 is overexpressed in 50%-70% of cancers including BC. An inverse correlation of Beclin-1 and Bcl-2 expression has been described in breast cancer tissue. Bcl-2 expression was correlated with histological grade, tubule formation, nuclear pleomorphism, mitotic count, ER and distant metastasis [49].

Beclin-1 also alters the expression of several autophagy proteins such as Atg5 and UVRAG [26].

4.2. Adipokines role

Adipokines, auto-/ endocrine and paracrine-acting bioactive molecules secreted by adipose tissue are one of the recently discovered factors correlating with autophagy and BC [50]. Adiponectin (AdipoQ) is the cytokine secreted in greatest abundance. The prevalence correlate low levels of AdipoQ in the blood circulation with higher BC risk and poorer prognosis. In breast tissue, AdipoQ has a direct anti-carcinogenic effect at the site of tumor growth. This cytokine is potentially capable of regulation of autophagy through AMP kinase (5'AMP-activated protein kinase) and its activation has been observed in breast cancer cells [50]. Liu and colleagues observed that AdipoQ caused upregulatation of autophagy in MDA-MB-231 cells *in vitro* and *in vivo* in cholesterol induced mammary tumorigenesis [51].

4.3. microRNA role

MicroRNAs (miRNAs) are endogenous ~22 nucleotide RNAs that suppress gene expression via messenger RNA (mRNA) cleavage and/or translational repression. Unregulated miRNAs of lymphoma, prostate, lung and breast cancers have been also detected in blood plasma and serum. Circulating miRNAs are currently assessed as proxy biomarkers for BC [52]. There is evidence that miRNAs can influence autophagy process in BC cells at many points. MiR-20a, miR-101, miR-106a/b and miR-885-3p have been reported to have direct possibility of targeting ULK1/2 [53]. Also, miR-155 might target multiple players in mTOR signaling including Rheb, RICTOR (RPTOR independent companion of mTOR) and RPS6KB2 (ribosomal protein S6 kinase). MiR-30a and miR-519a can directly target Beclin-1 causing negative regulation in the autophagic flow thereby resulting in decreased autophagic activity. Action of miR-30a was shown in the *in vitro* study on human BC cell lines MDA-MB-468 and MCF-7 [54] by Zhu *et al*. Tumor cells treatment with the mimic of miR-30a decreased the expression of Beclin-1 mRNA and protein whereas administration of the miR-30a antagomir had opposite effects. Furthermore, high expression of miR-30a blunted the rapamycin-induced autophagy activation [54]. Another miRNA, miR-376b also regulates Beclin-1 and it is also targeting directly Atg4C [9, 55] in MCF-7 cells. The antagomir-mediated inactivation of the endogenous miR-376b results in an increased level of Atg4C and Beclin-1 [56]. MiR-374a and miR-630 can modulate the direct regulation of UVRAG. The tumor suppressive miR-101 could act as a potent inhibitor of basal, etoposide-induced and rapamycin-induced autophagy in MCF-7 cells. Also, the miR-101-mediated inhibition of autophagy sensitized BC cells to 4-hydroxytamoxifen-induced apoptotic cell death and thus miR-101 was suggested to modulate the chemosensitivity of cancer cells [9]. Elevated levels of autophagy due to the progressive loss of miR-101, at least in breast cancer cells, have the potential to trigger cancer cell survival [9, 57]. Three components including STMN1 (stathmin1), RAB5A (ras related protein 5A) and Atg4D have been identified as targets of miR-101 among which the over-expression of STMN1 could partially rescue cells from miR-101-mediated inhibition of autophagy. Previously described RAB5A and STMN1 had uncertain roles in autophagy. RAB5A have been shown to regulate ATG5–ATG12 conjugation in the autophagosome completion while STMN1 plays an important role in cell-cycle regulation [58]. Another miRNA, miR-221/222, might inhibit the cell cycle inhibitor, p27Kip1, a downstream modulator of PI3K/Akt thereby leading to autophagic cell death in HER2/neu-positive primary human breast carcinoma MCF-7 cells. The ectopic expression of miR-221/222 renders the parental MCF-7 cells resistant to tamoxifen [59].

4.4. Cancer Stem Cells (CSC) and autophagy

Autophagy is thought to be a critical process for cancer stem cells (CSC) or tumor initiating cell maintenance but the mechanisms through which autophagy supports survival of CSCs remain poorly understood [60]. The CSC theory proposes that heterogeneity within a tumor is driven by a small population of cells which have ability to differentiate and/or self-renewal, increased membrane transporter activity, anchorage independence and ability to migrate, tumorigenic capacities and pluripotency [61, 62]. Breast cancer follows this model since it has been shown that the CD44+/ CD24 low/-phenotype of cell surface markers (which can be found also in normal stem cells in the breast), have an increased ability to form tumors in immuno-suppressed mice than the bulk of the tumor cells. It has been predicted that a quality control mechanism like autophagy is important for maintaining normal and cancer stem cell homeo-stasis [63]. Maycotte *et al.* have previously reported that a subset of BC cell lines enriched in the triple negative (TN) type is particularly dependent on autophagy for survival even in nutrient rich conditions [49]. This process is regulated by autophagy through modulation of STAT3 activity (often activated in TNBC). STAT3 activity is known to be regulated by IL-6 (interleukin 6) paracrine signaling in breast cancer cell lines [64]. The IL-6/ STAT3 pathway has been also shown to be important for TNBC xenograft growth and breast CSC maintenance [65]. Maycotte *et al.* have found that the pathways most affected by autophagy inhibition were related to stem cells, secretion and epithelial to mesenchymal transition [49]. We also show that autophagy regulates the CD44+/CD24 low/-phenotype and mammosphere formation in both the MCF7 and MDA-MB-468 breast cancer cell lines. Although autophagy regulates IL-6 secretion in both the autophagy dependent (MDA-MB-468) and independent (MCF7) cell lines, autophagy inhibition increased IL6 secretion in MCF7 cells while it decreased it in MDA-MB-468 cells. Decreased mammosphere formation in MDA-MB-468 cells induced by autoph-agy inhibition was reversed with conditioned media from autophagy proficient MDA-MB-468 cells or with IL-6 treatment. This identifies a mechanism by which autophagy selectively regulates CSC maintenance in autophagy-dependent breast cancer cells. Maycotte *et al.* had used a flow cytometry based assay to analyze CD24 and CD44 staining in cells with different levels of autophagic flux ("autophagic flux" represents the synthesis of autophagosomes, transportation of different substrates and degradation of autophagy inside the lysosome) [49]. In both MCF7 and MDA-MB-468 cell lines, the cells with low autophagic flux had decreased CD24 staining. Cells expressing a shRNA for ATG7 or BECN1 had lower levels of CD24 staining in both cell lines and no changes in CD44 were observed indicating that cells with lower levels of autophagy also have lower levels of CD24 expression.

CSCs are characteristically resistant to conventional anticancer therapy which may contribute to treatment failure and tumor relapse. CSCs exhibit the potential for regeneration which may promote tumor metastasis [66]. Recently, autophagy has been shown to be a critical factor for CSC survival and drug resistance [67].

5. Authophagy in anti-breast cancer therapies

There are different ways of autophagy process usage and/or influence recognized according to potential therapeutic treatment in anti-cancer strategies.

Inducing protective autophagy and prosurvival mechanism in human cancer cell lines have been shown in a number of currently used antineoplastic therapies including radiation therapy, chemotherapy (e.g. doxorubicin, temozolomide and/or etoposide), histone deaceltylase inhibitors, arsenic trioxide, TNFα, IFNγ, imatinib, rapamycin and anti-estrogen hormonal therapy (e.g. tamoxifen) [12, 23]. In fact, the therapeutic efficacy of these agents can be increased if autophagy is inhibited [23].

The scientific evidence suggests that autophagy leads to cell death in response to several compounds including etoposide, rottlerin, cytosine arabinoside and staurosporine as well as deprivation of growth-factors. A link has been demonstrated between autophagy and related autophagic cell death with usage of pharmacological inhibitors (e.g. 3-MA (3-methyl adenine), CQ (chloroquine), bafilomycin A1 or ammonium chloride) and genetic silencing or knockdown (silencing of *BECN1* and *ATG5*, *ATG7* and/or *ATG12*) approaches for autophagy suppression. This is connected with **cytoprotective form of autophagy** [68].

Autophagy has also been shown to protect against cellular stress induced by the anti-cancer chemotherapeutic drugs (**nonprotective autophagy**) [68]. The cell is apparently carrying out autophagy-mediated degradative functions but where autophagy inhibition does not lead to perceptible alterations in drug or radiation sensitivity. Furthermore, because autophagy is frequently upregulated in tumors in response to therapy, it may protect the tumors against therapy-induced apoptosis [69]. Gewirtz reported that ionizing radiation could promote autophagy in BC cells in cell culture but autophagy inhibition did not alter sensitivity to radiation [68]. Furthermore, the group showed that chloroquine did not sensitize murine breast tumor cells (4T1) to radiation in an immunocompetent animal model. Based on the results obtained, it was impossible to determine whether radiation had promoted autophagy process or the chloroquine actually effectively inhibited autophagy in the tumor-bearing animals. Supposedly, that the lack of sensitization could be related to findings [5] that autophagy inhibition interferes with the immune system's capability for recognition of the tumor undergoing a stress response.

Such disclosures have led to several clinical trials involving the use of the autophagy flux inhibitors as a combination therapy [70] to radiotherapy efficacy improvement in BC patients. For example, to such inhibitors could be included hydroxychloroquine (HCQ). HCQ is a less toxic version of CQ and the best autophagy inhibitor currently commercially available for clinical trials [71]. Irradiated cancer cells can induce damage in neighboring un-irradiated cells by intracellular gap-junction communication or signals released outside of the cells [72]. Huang *et al.* indicated that radiation-induced senescent MDA-MB- 231-2A cells are secreting multiple cytokines and chemokines including CSF2 (colony stimulating factor; expressed in the highest level), CXCL1(C-X-C motif ligand 1), IL-6 and IL-8 (interleukin 8) [73]. These factors are involved in multiple functions during cancer progression. Autophagy inhibition in MDAMB- 231-2A cells significantly decreased the release of CSF2 suggesting that autophagy plays an important role in promoting the secretion of SASPs (senescence-associated secretory phenotypes). In support of this notion, it has been reported that inhibition of autophagy delays the secretion of several senescence-associated cytokines such as IL-6 and IL-8.

Cytotoxic autophagy is the next form of autophagy which should be taken under consideration in the field of cancer treatment. Functionally, this form is associated with a reduction in the number of viable cells and/or reduced clonogenic survival upon treatment [74]. For example,

Bristol *et al.* reported that vitamin D (or the vitamin D analog, EB 1089) can be combined with radiation to promote a cytotoxic form of autophagy in breast tumor cell lines (MCF-7 and ZR-75) [75]. Other research groups also showed that the generation of cytotoxic autophagy may either independently lead to cells death or act as a precursor to apoptosis [76]. Gewirtz identified an additional form of autophagy, termed **cytostatic autophagy**, in non–small cell lung cancer cells (A549 and H460) which was induced in similar conditions to the ones previously described with regards to breast tumor cells [74]. What distinguishes cytostatic autophagy from the cytoprotective form is the failure to detect evidence of cell killing reported in the breast tumor cells [74]. Both Gewirtz and Kroemer's group demonstrated cytoprotective autophagy by radiation alone but the addition of vitamin D or EB 1089 converted cytoprotective autophagy to cytostatic autophagy [74,77]. Kroemer's group observed that the depletion of essential autophagy-relevant gene products such as ATG5 and Beclin-1 increased the sensitivity of human or mouse cancer cell lines (H460, A549 and CT26 cells) to irradiation both *in vitro* (where autophagy inhibition increased radiation-induced cell death and decreased clonogenic survival) *and in vivo* after transplantation of the cell lines into immunodeficient BALB/c nude mice (where autophagy inhibition potentiated the tumor growth-inhibitory effect of radiotherapy) [77].

As Ras/Raf/ERK pathway belongs to the most commonly deregulated pathways identified in tumors and is currently the target of new antitumor strategies based on the inhibition of upstream ERK regulators. Inhibiting ERK activity in combination therapy with classical antitumor compounds might affect the efficiency of such compounds. For example, in MCF-7 human breast adenocarcinoma cell line such combined therapies with: doxorubicin [78], tamoxifen [79], taxol [80] or Δ Raf1 [81] and, TNFα [82] were used. For example, tamoxifen, the most commonly used antiestrogen, exerts its pharmacological action by binding to estrogen receptor alpha (ERα) and blocking the growth promoting action of the estrogen in BC cells. However, the development of antiestrogen resistance has become a major impediment in the treatment of ER-positive BC. It was reported that autophagy plays a critical role in the development of antiestrogen resistance and overexpression of Beclin-1 downregulated estrogenic signaling and growth response [83].

5.1. Autophagic genes silencing

Some studies using gene silencing to receive therapeutic effect via cell death induction could represent genetic therapeutic approaches. For example, the Bcl-2 protooncogene (preventing cells from undergoing apoptosis) is overexpressed in half of all human malignancies and more than 60% of BC. Bcl-2 overexpression not only leads to the resistance of cancer cells towards chemotherapy, radiation and hormone therapy but also causes an aggressive tumor phenotype in patients with a variety of cancers [23]. Recent findings suggested that silencing Bcl-2 expression (by siRNA) in MCF-7 cells led to significant autophagic, not apoptotic, cell death [84]. It has been demonstrated that the knockdown of autophagy genes (e.g. *ATG5* and *BECN1*) significantly inhibited both autophagy and cell death induced by Bcl-2 siRNA after a long-term treatment of up to seven days [84]. MCF-7 cells are known to be caspase 3-deficient providing a higher threshold for the induction of apoptosis potentially rendering the autophagic cell death pathway more important. Furthermore, about 45-75% of tumor tissues from BC patients do not have detectable caspase 3 expression [85]. Akar *et al.* reported that doxorubicin predominantly induced autophagy at low doses and apoptosis at high doses [84]. Furthermore, the combination of Bcl-2 siRNA treatment with a doxorubicin low dose enhanced the autophagic response, tumor growth inhibition and cell death. It was the first evidence that targeted silencing of Bcl-2 induction of autophagic cell death in BC cells so a new path for further research on this type of alternative therapeutic strategies.

There are studies connecting autophagy genes profile with BC prognosis. For example, Gu *et al.* separated BC patients into two groups according to the TP53 mutation status in their study [86]. Then, detected the differential gene expression patterns of autophagy-related genes and investigated the association of autophagy with BC prognosis. Using microarray analysis, they identified a set of eight autophagy genes (*BCL2, BIRC5, EIF4EBP1, ERO1L, FOS, GAPDH, ITPR1 and VEGFA*), which were significantly associated with overall survival in breast cancer. This classifier could accurately predict the clinical outcome of BC independently of other classical clinical factors such as age, tumor size, grade, status of lymph nodes, ER status, PR status and ERBB2 status.

5.2. Pharmacological approach to the autophagic therapies

In pharmacological approach to the anti-tumor autophagic therapies, the aim is to activate or inhibit autophagy. Many drugs and compounds that modulate autophagy are currently receiving considerable attention [26,35]. For example, autophagy inducers such as rapamycin (mTORC1 inhibitor) [26] and its analogs called rapalogs (such as Everolimus; RAD001) are also often used as tools to study autophagy process [6]. Everolimus was shown to enhance the sensitivity of tumors to radiation by induction of autophagy [6].

Also, natural products are considered as potential anti-cancer candidates being direct or indirect sources of new chemotherapy adjuvants to enhance the efficacy of chemotherapy and/ or to ameliorate its side effects [87, 88].

The more challenging issue is the monitoring of autophagic activity in humans, in tissue and blood samples. It seems to be more important to measure autophagic flux than autophagosome number. However, measurements of autophagic flux in paraffin-embedded tissue samples have been unsuccessful till now and even the detection of endogenous LC3-II (commonly used marker for autophagosomes) is problematic in tissue sections [26]

6. Concluding remarks

There has been a tremendous amount of progress in our understanding of the role of autophagy in cancer. But still the molecular mechanisms underlying the regulation of autophagy and the role of autophagy in cancer cells are not fully understood but are progressively revealed. Overall, the data support a dynamic role of autophagy in cancer - both as a tumor suppressor early in progression and later as a pro-tumorigenic process critical for tumor maintenance and therapeutic resistance. The specification of the autophagic cargo in tumors with increased autophagy is important for understanding the changes in metabolism between normal and malignant cells. Undoubtedly, progress in genomics, proteomics and metabolomics will be helpful in this scope. Induction of autophagic cell death may be an ideal approach in resistant cancers therapies. But most experiments regarding BC are carried out on cell lines *in vitro*. Further, functional investigations of autophagy genes using BC cell lines and animal models will increase our understanding of their roles in determining breast cancer prognosis and could thereby provide clinical strategies for the treatment of breast cancer. The first clinical trials where deliberate autophagy inhibition has been attempted in cancer patients are starting to be reported [89-93]. For instance, autophagy inhibition by HCQ in combination with chemotherapy is currently being evaluated in multiple ongoing clinical trials in patients with solid tumors but we should take into account that autophagic effect is context dependent. While tumor cell susceptibility to autophagy may depend on tumor genotype and the therapeutic agents utilized, data are very limited and it remains unclear whether such new strategies will be clinically beneficial.

Author details

Joanna Magdalena Zarzynska

Address all correspondence to: joanna_zarzynska@sggw.pl

Department of Food Hygiene and Public Health, Faculty of Veterinary Medicine, Warsaw University of Life Sciences - SGGW, Warsaw, Poland

References

[1] Frankel LB and Lund AH: MicroRNA regulation of autophagy. Carcinogenesis. 2012; 33, no. 11: 2018–2025.

[2] Awan MU and Deng Y: Role of autophagy and its significance in cellular homeostasis. Applied Microbiology and Biotechnology. 2014; 98, no. 12: 5319–5328.

[3] Andon FT and Fadeel B: Programmed cell death: molecular ´ mechanisms and implications for safety assessment of nanomaterials. Accounts of Chemical Research. 2013; 46, no. 3: 733– 742.

[4] Cagnol S and. Chambard J C: ERK and cell death: Mechanisms of ERK-induced cell death—apoptosis, autophagy and senescence. FEBS Journal. 2010; 277, no. 1: 2–21.

[5] Kroemer G, Marino G, and Levine B:Autophagy and the integrated stress response. Molecular Cell. 2010; 40, no. 2: 280–293.

[6] Benbrook D M and Long A: Integration of autophagy, proteasomal degradation, unfolded protein response and apoptosis. Experimental Oncology. 2012; 34, no. 3: 286–297.

[7] Chen N and Karanza-Wadsworth V: Role and regulation of autophagy in cancer. Biochimica et Biophysica Acta. 2009; 1793, no. 9: 1516–1523.

[8] Sahni S, Merlot AM, Krishan S, Jansson PJ, and Richardson DR: Gene of the month: BECN1. Journal of Clinical Pathology. 2014; 67: 656–660.

[9] Chen Y, Fu LL, Wen X et al.: Oncogenic and tumor suppressive roles of microRNAs in apoptosis and autophagy. Apoptosis. 2014; 19, no. 8: 1177–1189.

[10] Jung CH, Jun CB, Ro S-H et al.: ULK-Atg13-FIP200 complexes mediate mTOR signaling to the autophagy machinery. Molecular Biology of the Cell. 2009; 20, no. 7: 1992–2003.

[11] Sancak Y, Peterson T R, Shaul YD et al.:The rag GTPases bind raptor and mediate amino acid signaling to mTORC1. Science. 2008; 320, no. 5882:1496–1501.

[12] Kondo Y, Kanzawa T, Sawaya R, and Kondo S: The role of autophagy in cancer development and response to therapy. Nature Reviews Cancer. 2005; 5, no. 9: 726–734.

[13] Howard J H and Bland K I: Current management and treatment strategies for breast cancer. Current Opinion in Obstetrics and Gynecology. 2012; 24, no. 1: 44–48.

[14] Gomes LR, Terra LF, Wailemann AM, Labriola L and Sogayar: TGF-β1 modulates the homeostasis between MMPs and MMP inhibitors through p38 MAPK and ERK1/2 in highly invasive breast cancer cells. BMC Cancer. 2012; 12, article 26.

[15] Herranz M and Ruibal A: Optical imaging in breast cancer diagnosis: the next evolution. Journal of Oncology. 2012; Article ID 863747.

[16] Shah NR and Chen H: MicroRNAs in pathogenesis of breast cancer: implications in diagnosis and treatment. World Journal of Clinical Oncology. 2014; 5, no. 2: 48–60.

[17] Porter PL : Global trends in breast cancer incidence and mortality. Salud Publica de Mexico. 2009; 51, no. 2: S141–S146.

[18] DeSantis C, Ma J, Bryan L, Jemal A: Breast cancer statistics, 2013. CA Cancer J Clin. 2014; 61:52-62.

[19] Fang Y, Chen Y, Yu L et al.: Inhibition of breast cancer metastases by a novel inhibitor of TGFβ receptor 1. Journal of the National Cancer Institute. 2013; 105, no. 1: 47–58.

[20] Langlands FE, Horgan K, Dodwell DD, and Smith L: Breast cancer subtypes: response to radiotherapy and potential radiosensitisation. The British Journal of Radiology. 2013; 86, no. 1023, Article ID 20120601.

Chapter 5

The Role of Autophagy and Apoptosis During Embryo Development

M. Agnello, L. Bosco, R. Chiarelli, C. Martino and M. C. Roccheri

Additional information is available at the end of the chapter

http://dx.doi.org/10.5772/61765

Abstract

Programmed cell death (PCD) and cell survival are two sides of the same coin. Autophagy and apoptosis are crucial processes during embryo development of Invertebrates and Vertebrates organisms, as they are necessary for the formation of a new organism, starting from a fertilized egg. Fertilization triggers cell remodeling from each gamete to a totipotent zygote. During embryogenesis, the cells undergo various processes, thus allowing the transformation of the embryo into an adult organism. In particular, cells require the appropriate tools to suddenly modify their morphology and protein content in order to respond to intrinsic and external stimuli. Autophagy and apoptosis are involved in cell proliferation, differentiation and morphogenesis. Programmed cell death is a key physiological mechanism that ensures the correct development and the maintenance of tissues and organs homeostasis in multicellular organisms. PCD has been classified into three types, according to the morphology that the dying cells acquire and the molecular machinery involved: PCD type I or apoptosis; PCD type II or autophagy and PCD type III or necrosis (not involved in physiological development). These different types of cell death have specific features that can be used to be identified and characterized. Apoptosis is a highly conserved, genetically-controlled process through which certain cells destroy themselves. Autophagy is an evolutionarily conserved pathway used by eukaryotes for degrading and recycling various cellular constituents, such as long-lived proteins and entire organelles, that was mainly detected in those tissues where abundant cell death is required. Both autophagy and apoptosis are induced under stress conditions as an adaptive response against stress. Usually, environmental stress cause severe effects on embryonic development. Embryos of different species, exposed to different types of physical or chemical stress, temporarily suspend their development and activate several protective strategies, including PCD II and PCD III. Research has yet to elucidate the interplay between these key processes. Not all types of PCD are always detected in association with a developmental process. Unlike the degeneration of tissues of some invertebrates, the tissues of vertebrates undergo PCD preferentially through an apoptotic mechanisms. In this chapter, we will briefly describe some specific features of apoptotic and autophagic processes. We will focus our attention in some useful model systems of invertebrates and vertebrates organisms, where autophagy and apoptosis occur both in physiological and stress conditions; specifically, we will analyze embryos of: the nematode *Caenorhabditis el-*

egans, the insect *Drosophila melanogaster,* the sea urchin *Paracentrotus lividus,* the fish *Danio Rerio,* the mouse mammalian model, and finally we will consider the differentiation of the male and female embryonic germlines in humans.

Keywords: Cell death, apoptosis-autophagy crosstalk, stress, differentiation, embryo model systems

1. Introduction

Embryonic development is a dynamic and well-coordinated event that includes cell proliferation, differentiation and death influenced by internal and external signals coming from the microenvironment. Research has yet to elucidate the interplay between autophagy and apoptosis, two processes of programmed cell death, and cell proliferation and morphogenesis in embryos of invertebrates and vertebrates.

Programmed cell death is a key physiological mechanism that ensures the correct development and the maintenance of tissues and organs in multicellular organisms [1]. Similar to apoptosis, autophagy is essential for the development, growth and maintenance of homeostasis. It occurs constitutively at basal levels and appears to be increased as an adaptive response to several intracellular and extracellular stimuli. In both lower and higher eukaryotes, autophagy is a crucial event during embryogenesis. It was proposed that autophagy has a key role in insect metamorphosis, representing a dramatic developmental change associated with widespread cell death and complete disappearance of whole tissues.

In this chapter, we will discuss an emerging research field: programmed cell death and cell survival through apoptosis and/or autophagy under physiological and stressful conditions during the development of invertebrates and vertebrates.

Recently, cell death (CD) has been classified into three types according to the morphology and molecular machinery involved: PCD type I or apoptosis; PCD type II or autophagy; and PCD type III or necrosis, not involved in embryo development [2]. These different types of cell death have specific features that can be used for their identification and characterization. Not all types of PCD are always detected in association with a developmental process. Autophagy is mainly detected in those tissues where abundant cell death is required. Vertebrates' tissues undergo PCD preferentially through apoptotic mechanisms in contrast with the degeneration of tissues in some invertebrates [3].

Although many features are specific among different types of death, some overlapping exists between these different mechanisms. It is noteworthy that this crosstalk often allows the conversion of autophagy into apoptosis or vice versa. Thus, if one pathway is blocked, a cell may still die through a second biological pathway.

1.1. Apoptosis

Apoptosis is a cellular phenomenon that orchestrates cell suicide following two main pathways: cytochrome *c* liberation from the mitochondria or activation of death receptors. This genetically controlled process is highly conserved during the evolution from nematodes to mammals, playing critical roles in both homeostasis and development during the morphogenesis and metamorphosis of invertebrates and vertebrates. Cells undergoing apoptosis show a series of physical and biochemical changes such as plasma membrane blebbing, loss of mitochondrial membrane potential, caspase-activation, DNA fragmentation in distinct ladders and, finally, cell disintegration into apoptotic bodies subsequently engulfed by specialized cells. Phosphatidylserine (PS), a phospholipid normally asymmetrically expressed in the inner leaflet of the plasma membranes in living cells during the final stages of apoptosis, is actively extruded from the internal face of the cell membrane of the dying cell; its exteriorization represents one of the markers that identify the cell as a target for phagocytosis [4] (Figure 1).

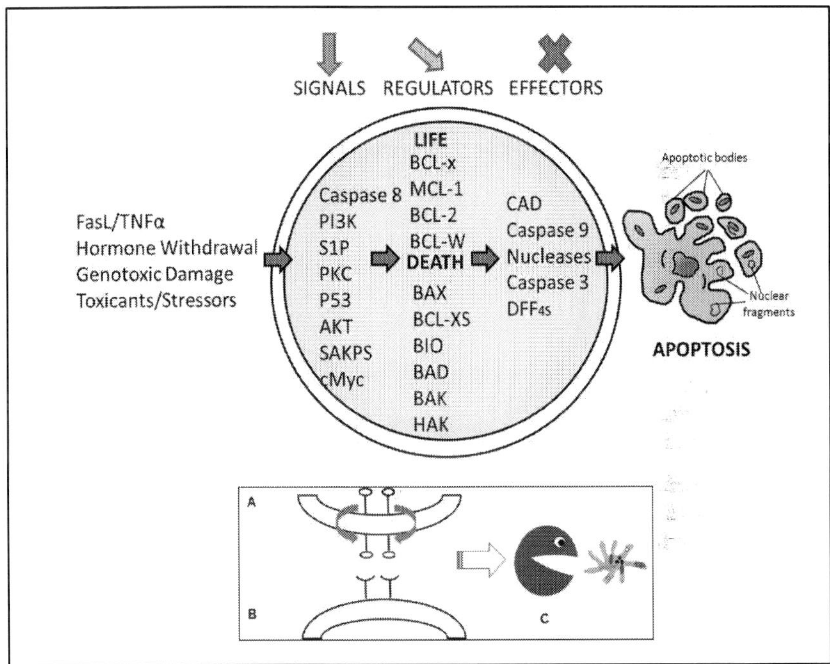

Figure 1. *Top*: signals, regulators and effectors involved in the balance between life and death. *Left*, the main inductors of apoptosis. *Bottom*, exemplification of the exposure mechanism of PS from the inner to the outer membrane of an apoptotic cell. (**A**) PS exposure in the apoptotic cell and (**B**) PS receptors of the phagocyte. (**C**) This specific recruitment consents the engulfment of the apoptotic cell or apoptotic bodies and their clearance.

This type of cell death can be greatly affected by ATP levels: if the energy level is not sufficient, cells can undergo partial apoptosis [5]. It is well known that PCD-I is required to remove transitory structures, to sculpt tissues and to eliminate damaged cells that can be harmful to the organism. On the other hand, apoptosis is also employed in response to environmental stimuli to remove cells damaged by chemical, physical and mechanical stress.

1.2. Autophagy

Autophagy is an evolutionarily conserved pathway used by eukaryotes for degrading and recycling various cellular constituents such as long-lived proteins and entire organelles [6]. Autophagy, contrary to apoptosis, can induce cell survival or cell death: it is a process of cell survival if the cellular damage is not too extensive; alternatively, it is a process of cell death if the damage/stress is irreversible. In addition, autophagy can act in association with apoptosis or as an independent pathway.

In higher eukaryotes, the lysosome compartmentalizes a range of hydrolytic enzymes and maintains a highly acidic pH in order to decompose into small molecules and then recycle the organelles and the components of cytosol targeted to them [7]. This recycling mechanism allows the cells to conserve their limited resources by minimizing the costs associated with biosynthesis or acquiring resources from the environment.

Depending on how the substrates are delivered to the lysosomal compartment, autophagy is classified into macroautophagy, microautophagy and chaperone-mediated autophagy [8,9]. Macroautophagy involves the formation of autophagosomes that will subsequently fuse with the lysosome. The molecular mechanism governing macroautophagy is highly conserved among eukaryotes. At first, a cup-shaped membranous organelle emerges and encircles a portion of the cytosol which sometimes includes various organelles. This activity results in the formation of spherical bodies, the autophagosomes, in which a double membrane sequesters the compartmentalized material. The outer membrane of autophagosomes fuses with the limiting membrane of lysosomes to release the sequestered materials into the lumen of an autophagolysosome.

In microautophagy, lysosomal compartments engulf a portion of the cytosol along with organelles, forming membrane-bound spherical bodies within lysosomes. In chaperone-mediated autophagy, the substrates, such as proteins, are translocated across the lysosome membrane and delivered directly into the lumen.

In lower eukaryotes, autophagy functions as a cell death mechanism or as a stress response during development. Autophagy's significance and the role (if any) of vertebrate-specific factors in its regulation remain unclear. In particular, in mammals, autophagy may be involved in specific cytosolic rearrangements needed for proliferation and differentiation during embryogenesis and postnatal development. Thus, autophagy is a process of cytosolic "renovation", crucial for cell fate decisions [10]. However, in both invertebrate and vertebrate organisms, it is generally thought that autophagy plays an essential dual role both in the adaptation to stress and in the starvation occurring during morphogenesis, as well as in cell elimination in concert with the apoptotic machinery.

Genetic studies have revealed the importance of autophagy during the early stages of embryogenesis; most of the genes involved, the so-called autophagy-related (ATGs) genes, have been discovered in *Saccharomyces cerevisiae* [11], and their orthologues have been isolated and functionally characterized in higher eukaryotes, indicating that autophagy is an evolutionarily conserved process [12]. Although many apoptosis and autophagy regulatory genes have been discovered and characterized and some of them are conserved during evolution, the relationship between autophagy and apoptosis still remains rather obscure.

1.3. Cell death in stress conditions

Environmental stress can cause severe effects on embryonic development, affecting the phenotype as a result of some emergency responses and adaptive modifications. Embryos of different species, exposed to various toxicants or to physical or chemical stresses, temporarily slow down or suspend their development, eliminating the affected cells throughout apoptosis and thus altering the normal developmental program. In the long run, embryos with several accumulated damages could also die if the stressful conditions persist.

On the other hand, embryos have the ability to activate a general protective strategy against many stress-inducing agents. The accumulation of damaged proteins acts as an inductor signal that activates the stress response and the apoptotic program.

The autophagic process, similar to apoptosis, is triggered as an adaptive response to several intracellular and extracellular stimuli such as toxic stimuli, radiation, nutrient deprivation (starvation), accumulation of misfolded proteins and damaged organelles, hormonal treatments and bacterial/viral infections. Generally, autophagy seems to be crucial for cell survival in stress conditions because it promotes the recycling of damaged proteins and organelles.

A few years ago, we studied both the apoptotic and the autophagic processes in *Paracentrotus lividus* sea urchin embryos, investigating whether these events are activated as a defense strategy after cadmium exposure, a heavy metal recognized as an environmental contaminant [13,14]. Our model suggests that the temporal choice of the apoptotic or autophagic mechanism depends on the persistence of the stressful event: initially, the embryo tries to face the stress conditions using a defense strategy that is less deleterious, namely, autophagy, in an attempt to preserve the developmental program; if this strategy is not sufficient to offset the stress-induced damage, the embryo activates the mechanism of apoptosis. In addition, it can be assumed that autophagy could provide the ATP necessary for apoptosis during development, by recycling damaged cellular components. In light of these and other evidences, it can be hypothesized that there is a close interplay between autophagy and apoptosis [15].

In conclusion, apoptotic and autophagic processes may be used as alternative and/or combined defense strategies by cells exposed to many kinds of stresses. Nowadays, there is a growing interest in cell death via autophagy, which could substitute or act synergistically to the apoptotic pathway.

In this review, we will describe and compare various eukaryotic model systems that use apoptosis and autophagy during development both under physiological and stress conditions. We will also focus on the research methods employed to study the cascade of events involved in these two processes. The purpose of discussing the data in this chapter is not to review all the work in the field but rather to focus on a few arguments with the intent of re-examining some ideas and concepts.

2. Apoptosis and autophagy during embryogenesis of eukaryotes

Recent studies have shown that cell death mechanisms are used for specific purposes: morphogenesis during embryogenesis, histogenesis in the progression of metamorphosis and phylogenesis for the elimination of vestigial or larval organs. Like proliferation and differentiation, programmed cell death, PCD-I and PCD-II, play a conspicuous role during normal development as well as during disease conditions. It is essential for the removal of undesirable cells and it is critical both for restricting cell number and for tissue patterning during development.

In both lower and higher eukaryotes, autophagy seems to be crucial during embryo development by acting in tissue remodeling, in parallel with apoptosis. An increase of autophagy is observed in the embryonic stages characterized by massive cell elimination. Moreover, autophagy protects cells during metabolic stress and nutrient paucity occurring during tissue remodeling.

The study of autophagy-defective model systems has highlighted the contribution of PCD-II in the development of invertebrates, for example, during the complex events occurring in the metamorphosis of flies and worms [16]. Furthermore, it has been well documented in the early stages of the development of invertebrates that the activation of apoptotic processes contributes to the formation of different body parts and multiple organs of an organism. Using *Caenorhabditis elegans*, as well as *Drosophila* and mice, it has been demonstrated that developmental cell death is under genetic regulation as shown by mutagenesis experiments.

In vertebrates, on the other hand, there are many examples in which autophagy and apoptosis are involved in embryogenesis. For example, autophagy defects can be lethal for the animal if the mutated gene is involved in the early stages of development or it can lead to severe phenotypes if the mutation affects later stages [17].

Cell death starts at a very early stage in mammalian development. Inhibition of caspase activity leads to the arrest of embryonic development. During gastrulation, apoptosis allows the generation of a pro-amniotic cavity by the removal of the inner ectodermal cells.

Autophagy also has a crucial role during cavitation in the early stages of mammalian development [18]. Furthermore, evidence outlines the importance of autophagy during tissue differentiation in mammals [19].

Both PCD-I and PCD-II are well-controlled biological processes that play fundamental functions during development, differentiation, morphogenesis, tissue homeostasis as well as disease. The different modes of execution of cell death were investigated as separate events from each other. However, in recent times, several findings suggest that these two types of death are often regulated by similar pathways and, depending on the cellular context, can cooperate in a complementary fashion to facilitate cellular destruction. Interactions among components of the two pathways show that there is a complex crosstalk that may be induced by similar stimuli: PCD-I and PCD-II can cooperate, antagonize or assist each other affecting cell fate.

3.2. Insects: *Drosoph*

D. melanogaster prov
roles of PCD during
nematodes and man
and autophagic resp
continuing to invigo

The development of
cell proliferation and
circumstances. The ic
the tools for the exp
allowed the study of
D. melanogaster, all t
genes: reaper, grim a
opmental apoptosis t
of apoptosis and the
been outlined [40].

The initiator genes
including caspases, a
characterized from n
seven caspases have
new insect genome se
across an evolutiona
and regulation. This
where a single caspa
death in some cell ty
suggesting that, simi
of insects [41].

The methods used to
well as in cultured ce
orange staining, frag
activity assays and
analysis of apoptosis

In *Drosophila,* many
formation. Two bcl-2
their requirement du
in fruit flies are shap
ment. By irradiating 1
does not rely upon tl
apoptotic response to

Cells damaged by er
homeostasis in meta

3. Apoptosis and autophagy in the development of the invertebrate model system

3.1. Nematodes: *Caenorhabditis elegans*

C. elegans has been widely recognized as a suitable model system in developmental research because of some important features: it is a simple and transparent animal with a highly reproducible development and an invariant cell lineage, it is self-fertilizing, it is easy to culture and it has a short reproductive cycle of about 3 days [20]. Recently, it was also recognized as a valuable model organism to study apoptosis and autophagy, two processes redundantly required during its embryogenesis.

Apoptosis can be observed during two stages of *C. elegans* life in two different types of tissues: the "developmental cell death" occurring during embryonic and post-embryonic development of the soma and the "germ cell death" occurring in the gonad of adult hermaphrodites. Developmental cell death can be divided into two moments: between 250 and 450 min after fertilization, the first apoptotic event removes almost a fifth (113/628) of the cells that are generated during embryonic development [21]; the second event occurs during the larval stage L2 and removes some of the newly generated neurons.

During the development of the *C. elegans* hermaphrodite soma, the embryo generates 1090 cells and exactly 131 of these undergo programmed cell death in a highly reproducible manner [22]. It was demonstrated that these dying cells are essentially invariant among individuals since they can be easily identified because of some changes in their morphology and because of their high refractivity under differential interference contrast optics [23]. To unravel the genetic pathway involved in these cell death events, a biochemical characterization of the genes was carried out. The identification of the apoptotic machinery suggested that cell suicide is performed by an evolutionarily conserved molecular program innate in all metazoan cells. Indeed, the key cell death genes of *C. elegans* have one or more mammalian homologs and all the interactions among proteins have also been described for their mammalian counterparts [24].

The first event needed to induce the apoptotic process is the transcription of the egl-1 (egg-laying defective-1) gene coding for a BH3-only protein and directly regulated in a cell-specific manner by transcription factors of the Hox family. This protein will then bind to the Bcl-2 (B cell lymphoma-2)-like anti-apoptotic protein CED-9 (cell death defective-9), which normally protects cells from undergoing apoptosis. This will activate CED-4, the nematode ortholog of the mammalian Apaf-1 [25], which mediates the activation of the caspase CED-3 (CEll Death abnormal) from the inactive zymogen (proCED-3) into the mature protease [26,27]. The activity of CED-4 and CED-3 is essential for the execution of the apoptotic cell process. It is worthy to note that autophagy contributes to the removal of embryonic apoptotic cell corpses by promoting phagosome maturation.

Besides, almost half of the female germ cells undergo apoptosis just before exiting the pachytene stage of meiotic prophase I [28], but egl-1 has no role in this case. During physiological germ cell apoptosis, the nuclei of the apoptotic cells are rapidly cellularized away from

the syncitiu
indicate tha
trigger diffe

In 2012, C. e
for monitor
a useful mul
the high cor
single orthc
and Atg8.

Autophagy
ooplasm tri
prone prote
spermatid-s
the lysosom
autophagy
occurring r
developmer
regulating r
of the miRN

Autophagy
tions such a
these stress
the Dauer d
allow long-
neosynthes
was also su
hermaphro
execute gen

It is worthy
essential fo
embryos de
with super
atg genes d
that, strikin
are unable
tions sugge
share essen
mutants em
developme

creased germ cell apoptosis. Journal of Clinical Endocrinology and Metabolism 2007;92(8) 3292–3304.

[121] Meyer-Ficca ML, Lonchar J, Credidio C, Ihara M, Li Y, Wang ZQ, Meyer RG. Disruption of poly(ADP-ribose) homeostasis affects spermiogenesis and sperm chromatin integrity in mice. Biology of Reproduction 2009;81(1) 46–55.

[122] Aitken RJ and De Iuliis GN. On the possible origins of DNA damage in human spermatozoa. Molecular Human Reproduction 2010;16(1) 3–13.

[123] Adler ID. Spermatogenesis and mutagenicity of environmental hazards: extrapolation of genetic risk from mouse to man. Andrologia 2000;32(4–5) 233–237.

[124] Ruvolo G, Roccheri MC, Brucculeri AM, Longobardi S, Cittadini E, Bosco L. Lower sperm DNA fragmentation after r-FSH administration in functional hypogonadotropic hypogonadism. Journal of Assisted Reproduction and Genetics 2013;30(4) 497–503.

Chapter 6

Mitophagy Regulated by the PINK1-Parkin Pathway

Taku Arano and Yuzuru Imai

Additional information is available at the end of the chapter

http://dx.doi.org/10.5772/61284

Abstract

Mitochondria play key roles in the cellular metabolism of lipids and iron as well as in cell death signaling. Mitochondrial dysregulation produces reactive oxygen species (ROS), which results in oxidative stress. Moreover, the accumulation of damaged mitochondria leads to cell death and tissue dysfunction. Mitochondrial maintenance involves mitophagy, a selective autophagy process that removes abnormal mitochondria. Parkinson's disease (PD) is a movement disorder caused by the specific loss of dopaminergic neurons in the substantia nigra of the midbrain. Two genes implicated in PD, *PINK1* and *Parkin*, regulate mitophagy in cultured cells. Reduction of the $\Delta\Psi$m leads to activation of PINK1, which stimulates the recruitment of Parkin to the mitochondrial outer membrane of damaged mitochondria and activates Parkin's ubiquitin-ligase activity. Activated mitochondrial Parkin leads to the ubiquitination of mitochondrial proteins and subsequent mitophagy. This elaborate molecular mechanism was recently uncovered and the findings demonstrate the physiological and pathological roles of the PINK1-Parkin pathway. Here, we review these key findings on the molecular mechanism and ideas relevant to neurodegeneration caused by dysregulation of the PINK1-Parkin pathway.

Keywords: Dopaminergic neurons, mitochondria, Parkinson's disease, protein kinase, ubiquitin ligase

1. Introduction

In eukaryotic cells, mitochondria are highly efficient power-generating systems that perform aerobic respiration. Injured mitochondria leak ROS, resulting in oxidative stress and reduction of energy supply; this dysfunction eventually leads to cell death. Therefore, appropriate regulation of mitochondria is critical for vital activity and anti-aging. Mitochondrial dysregulation has indeed been implicated in various human diseases, including cancer, diabetes, myopathy, and a variety of neurodegenerative disorders such as amyotrophic lateral sclerosis

(ALS), Huntington's disease, neuropathy, and Parkinson's disease (PD). PD is a progressive neurodegenerative disorder characterized by the degeneration of dopaminergic neurons in the midbrain. The motor symptoms of PD include tremor, rigidity, slowness of movement, and difficulty with ambulation. Although familial forms of PD are relatively rare cases, the identification of genes responsible for PD enables a better understanding of the molecular mechanisms underlying neurodegeneration. *Parkin* and *PINK1* mutations are associated with autosomal recessive forms of early-onset PD [1, 2]. A series of studies on *PINK1* and *Parkin* indicates that these two genes work in a coordinated manner in functions related to mitochondrial maintenance including mitochondria motility, proteasomal degradation of mitochondrial proteins, and selective mitochondrial autophagy (also known as mitophagy). These results strongly imply that dysregulation of mitochondria is one of the major factors in the etiology of PD. In this chapter, we focus on the latest studies that have made significant progress in elucidating the molecular mechanisms of mitochondrial quality control via the PINK1-Parkin pathway.

The selective degradation of mitochondria via an autophagic process was originally reported as mitophagy by J.J. Lamasters et al. [3]. In yeast, loss of the *MDM38* gene product, a component of the mitochondrial protein export machinery, reduces the content of respiratory chain complexes, elicits morphological mitochondrial changes, and disturbs mitochondrial K$^+$ homeostasis, resulting in mitophagy [4]. When mammalian reticulocytes mature into erythrocytes, mitochondria are removed by mitophagy [5]. In fertilized *C. elegans* oocytes, sperm-contributed mitochondria are selectively degraded by mitophagy [6, 7]. These observations indicate that mitophagy plays important roles in mitochondrial maintenance, differentiation, and developmental processes in eukaryotes.

2. Mitochondrial segregation and mitophagy

Damaged mitochondria are selectively segregated and degraded by mitophagy [3]. Mitochondrial morphology is maintained by mitochondrial fusion and fission. Mitofusin (Mfn) regulates or mediates mitochondrial fusion whereas Drp1 and Fis1 promote mitochondrial fission in mammals as well as in yeast [8]. Inhibition of Drp1 or Fis1 activity results in the suppression of mitophagy and the accumulation of oxidized mitochondrial proteins, leading to reduced respiration and impaired insulin secretion [9]. Orthologs of *Parkin* and *PINK1* have been identified in *Drosophila* [10–12]. Loss of *Drosophila PINK1* causes mitochondrial degeneration, resulting in male sterility, apoptotic muscle degeneration, and increased sensitivity to multiple stresses, including oxidative stress. Loss of *Drosophila Parkin* produces phenotypes similar to those elicited by the loss of *PINK1*, and Parkin overexpression rescues the mitochondrial defects observed in PINK1 mutant flies [11–13]. The *Drosophila* Parkin and PINK1 phenotypes are suppressed by increased Drp1 activity and are exacerbated by Opa1 or Mfn [14–16]. PINK1 and Parkin collaboratively ubiquitinate Mfn and the steady-state abundance of Mfn is inversely correlated with the activity of PINK1 and Parkin in *Drosophila* [17, 18]. Parkin also ubiquitinates Mfn1 and Mfn2 in mammalian cells, leading to proteasome- and p97/VCPdependent degradation [19–21]. These reports indicate that PINK1 and Parkin positively regulate mitochondrial fission, which may facilitate mitochondrial clearance via mitophagy.

3. Parkin E3 ligase and ubiquitination

Parkin contains a ubiquitin-like (Ubl) domain at the N-terminus, RING-between-RING (RBR) domains at the C-terminus, and an atypical RING domain, RING0, in its linker region (Figure 1) [22–24].

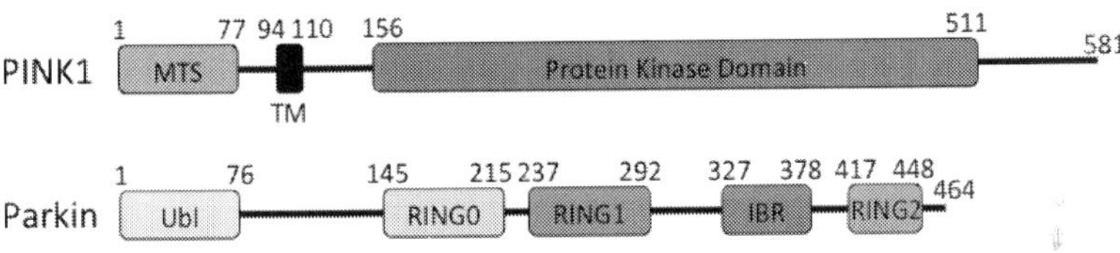

Figure 1. Schematic of PINK1 and Parkin proteins MTS: mitochondrial targeting sequence, TM: transmembrane region, Ubl: ubiquitin-like domain, RING: really interesting new gene domain, IBR: in between RING domain. Numbers indicate the positions in the amino acid sequence.

E3 ubiquitin ligases are roughly divided into two groups: RING finger-type E3 ligases and homologous to the E6AP carboxyl terminus (HECT)-type E3 ligases. HECT-type E3 ligases form a thioester intermediate between ubiquitin and a catalytic cysteine residue before transferring ubiquitin from E2 to a substrate. By contrast, RING finger-type E3 ligases mediate the direct transfer of ubiquitin from E2 to the substrate. Parkin, which was formerly classified as a RING finger-type E3 ligase, is now categorized as a HECT-RING hybrid E3-ligase [25, 26]. To activate Parkin, a ubiquitin-charged E2 associates with Parkin RING1 and ubiquitin is transferred from E2 to Cys431 in the RING2 domain of Parkin to form the HECT-like thioester intermediate [25–27]. Similar molecular behaviors were observed in other RBR proteins such as HHARI, and proteins containing a RBR domain are thought to be HECT-RING hybrid E3-ligases [25, 26].

PINK1 encodes a serine–threonine protein kinase with a mitochondrial targeting signal at the N-terminus (Figure 1) [2]. PINK1 is constitutively processed by mitochondrial proteases at the mitochondrial membrane of healthy mitochondria, resulting in proteasomal degradation [28–30]. The reduction in mitochondrial membrane potential ($\Delta\Psi$m) in damaged mitochondria leads to the accumulation and activation of PINK1 on the outer mitochondrial membrane [29]. Activated PINK1 recruits Parkin from the cytosol to mitochondria in response to decreased $\Delta\Psi$m. This action stimulates Parkin E3 activity, thereby promoting mitochondrial degradation via mitophagy [29, 31–35]. The Ubl domain in the N-terminal region of Parkin inhibits the E3 activity of Parkin by interacting with the RBR region [36]. PINK1 phosphorylates Ser65 in the Ubl domain of Parkin to activate Parkin E3 activity (Figure 2) [37–42]. Activated PINK1 also phosphorylates monomeric ubiquitin at Ser65 in the cytosol. Transient interaction with phosphorylated ubiquitin leads to a conformational change in Parkin and subsequent activation of Parkin E3 activity [41, 43–45].

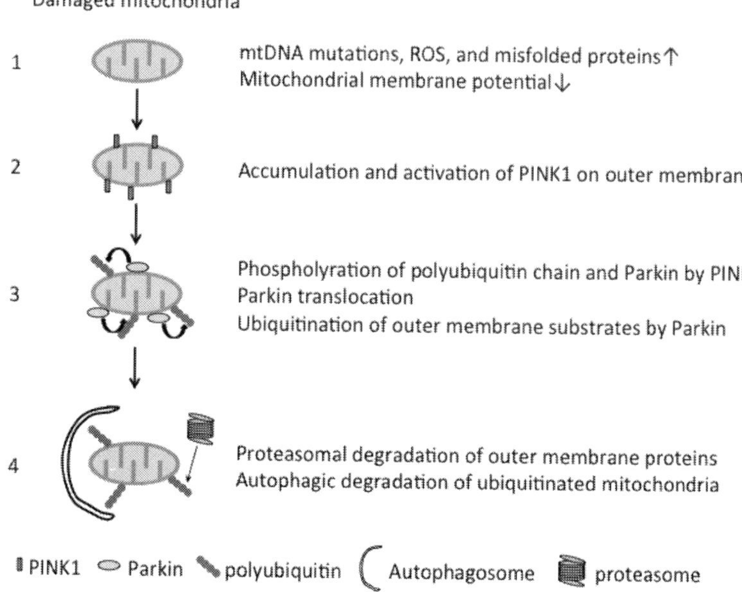

Figure 2. PINK1-Parkin-mediated mitophagy (1) Mitochondrial DNA (mtDNA) mutations, ROS overproduction and misfolded protein accumulation cause a reduction in $\Delta\Psi$m. (2) PINK1, which is constitutively degraded under steady-state conditions, accumulates on damaged mitochondria. (3) Accumulated PINK1 activates itself and elicits mitochondrial translocation and Parkin activation through phosphorylation. Activated Parkin ubiquitinates substrates on mitochondria. (4) Polyubiquitinated proteins on the mitochondrial outer membrane are degraded by the proteasome, and damaged mitochondria are eliminated concurrently by mitophagy.

It has been reported that Parkin binds to four tandem-repeated mitochondrial ubiquitin chains, which mimic Lys63-linked polyubiquitin chains only when PINK1 is activated [46]. Subsequent reports have revealed that PINK1 phosphorylates mitochondrial polyubiquitin, resulting in Parkin activation and mitochondrial relocation (Figure 3) [41, 47]. While PINK1 phosphorylates both monoubiquitin and polyubiquitin, including Lys48- and Lys63-linked polyubiquitin chains, activated Parkin preferentially associates with Lys63-linked phosphorylated polyubiquitin chains on mitochondria [41]. Lys48-linked polyubiquitin chains are generally utilized as signals for proteasomal degradation [48], whereas Lys63-linked ubiquitin chains were first identified in yeast as atypical ubiquitin chains that respond to stress [49]. A variety of functions of Lys63-linked polyubiquitin chains were subsequently characterized, including the regulation of kinase activity, DNA damage response, signal transduction scaffolding, vesicular trafficking, and endocytosis [50]. Thus, the formation of Lys63-linked ubiquitin chains during mitophagy might have a critical role beyond Parkin recruitment [51].

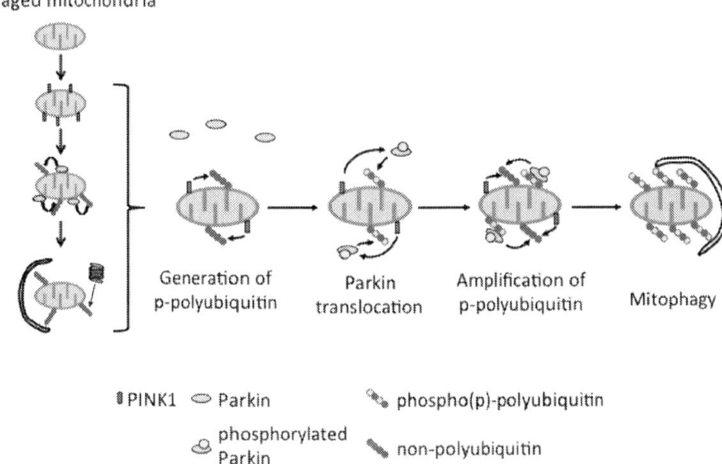

Figure 3. Amplification of phospho-polyubiquitin chain production achieves rapid Parkin translocation and activation Phospho-polyubiquitin chains on mitochondria are produced in collaboration with PINK1 and Parkin. The mechanism responsible for the formation of initial ubiquitin chains on mitochondria remains unresolved. The ubiquitin chains might be attached to outer membrane proteins via mitochondrial ubiquitin ligases other than Parkin. Alternatively, Parkin, which is activated by phospho-monoubiquitin in the cytosol, could attach ubiquitin chains to mitochondrial proteins. Phosphorylation of polyubiquitin chains by PINK1 promotes further Parkin activation and relocation to the mitochondrial outer membrane, amplifying the generation of phospho-polyubiquitin chains and subsequently recruiting autophagy machinery to ubiquitinated mitochondria.

Several reports have demonstrated that Parkin-interacting E2 enzymes mediate the ubiquitination reaction of Parkin. UBE2N is related to Parkin-mediated Lys63-linked ubiquitination [51, 52], whereas UBE2N, UBE2L3, and UBE2D2/3 synergistically contribute to Parkin-mediated mitophagy [53]. Knockdown of UBE2N, UBE2L3, or UBE2D2/3 but not UBE2A, UBE2S, or UBE2T significantly reduces autophagic clearance of depolarized mitochondria or Parkin E3 activity [53, 54]. However, recent reports indicate that the linkage property of polyubiquitination depends on Parkin itself rather than involved E2s [47], and that atypical Lys6- and Lys11-mediated polyubiquitination chains are also generated by Parkin and contribute to mitophagy [47, 55].

Deubiquitinating enzymes (DUBs) are also involved in PINK1-Parkin-mediated mitophagy. Because USP30 preferentially removes Lys6- and Lys11-linked ubiquitin chains generated by Parkin on damaged mitochondria, USP30 knockdown rescues the defective mitophagy caused by pathogenic mutations in Parkin [55, 56]. Moreover, knockdown of USP30 improves mitochondrial morphology in *Parkin-* or *PINK1-*deficient flies and protects them from the paraquat-induced reduction in dopamine, motor dysfunction, and shortened lifespan [56]. Conversely, USP8 removes Lys6-linked polyubiquitin on Parkin, which activates Parkin-mediated mitophagy [57]. As USP15 does not affect Parkin autoubiquitination and translocation to mitochondria, knockdown of CG8334, the closest homolog of USP15 in *Drosophila*,

Chapter 7

The Role of Cell Autophagy in Cancer and Its Application in Drug Discovery

Ming Hong, Ning Wang and Yibin Feng

Additional information is available at the end of the chapter

http://dx.doi.org/10.5772/61780

Abstract

Autophagy is a vital basic phenomenon that widely exists in eukaryotic cells. As one type of programmed cell death, autophagy has gained much more attention in the past few years. Recent studies suggest that the alterations in autophagy are associated with the genesis and development of cancers. It can affect cell apoptosis, angiogenesis, and treatment of tumor. Others' and our studies have found that some herbal medicines can induce autophagic cell death in cancer cell models. As herbal medicines are very important recourses for drug discovery and lead compounds of anticancer drugs, we have summarized the role of autophagy in inhibitive effect of natural products in cancer cell growth and metastasis. Finally, we present summary and critical comments on problems in current autophagy study and its future prospect.

Keywords: Autophagy, cancer, autophagic cell death, drug discovery, natural products

1. Introduction

Autophagy is a highly conserved cellular process that widely exists in eukaryotic cells. As one type of programmed cell death autophagy has gained great attention in the past few years. Recent studies suggest that autophagy is associated with the tumoriogenesis and progression of cancers. It can affect tumour cell apoptosis, angiogenesis as well as the treatment outcome of chemotherapeutics. In this chapter, we have summarized the up-to-date research about cell autophagy, including but not limited to the molecular mechanisms underlying autophagy initiation and the potential targets that autophagy regulates. As many studies have demonstrated that autophagy plays an important role in cancer initiation and progression, we also critically reviewed the double-edged sword effect of autophagy in cancer, which is reflected as the cytoprotective autophagy and cytotoxic autophagy. The role of

autophagy in treatment and resistance of several present conventional anticancer agents is also discussed in our chapter. As Chinese medicines have a long history in treating cancer in Asian countries, we also introduced some anticancer natural products from Chinese medicines that target on autophagy in this chapter.

1.1. Cell autophagy overview through the up-to-date research information

Autophagy means "self-eating" in cells. It is a highly conserved cellular process among eukaryotes by which long-lived proteins and damaged organelles are packaged in the double-membrane autophagosomes and transported to lysosomes for degradation. Autophagy is one type of programmed cell death (PCD) for maintaining cell homeostasis by removing damaged or abnormal cells. It is a vital basic phenomenon that widely exists in eukaryotic cells, which has gained much more attention in the recent decades. In the process of autophagy, the intracellular damaged proteins or organelles are wrapped by the double-membrane structure autophagy vesicles and sent into the lysosome (for animal) or vacuoles (for yeast and plant) for further degradation and recycle. As autophagosome belongs to subcellular structures, normal optical microscope cannot observe it, so transmission electron microscope is needed for observing the process of autophagy. Generally, the features of phagophore are described as cup-shaped or crescent-shaped double-membrane or multilayer-membrane structure with the tendency of wrapping the cytoplasmic components. The characteristics of autophagosomes are described as double-layer or multilayer vacuole-like structure with cytoplasmic components such as mitochondria, endoplasmic reticulum, and ribosome. Autophagy lysosome is a monolayer-membrane structure in which the cytoplasmic components undergo degradation. In the study of the relationship between autophagy and cell death, the presence of a large number of autophagosomes or autophagic lysosomes were found in the cytoplasm before cell death, but these cells lack typical apoptosis features such as nuclear pyknosis and karyorhexis or cell shrinkage and the formation of apoptotic bodies. So this is a new type of programmed cell death which is different from apoptosis. In order to distinguish between these two types of cell death, autophagy is also called Type II cell death. In contrast, apoptosis is called Type I cell death. However, whether autophagy is the direct cause of cell death is still controversial. Some researchers believe that cell death is caused by the process of autophagy, while some think that autophagy is not the direct reason for cell death.

As a main intracellular degradation and recycling process, autophagy is important for maintaining energy homeostasis and cellular remodeling during physiological process. Recent studies have suggested that alterations in autophagy can affect the genesis and development in several diseases such as inflammatory disease, heart disease, neurodegenerative disease, and cancer. A common characteristic in these diseases is a dysfunction of autophagy, which influenced the susceptibility of programmed cell death. Our knowledge of the functions and regulation of this programmed cell death pathway has increased substantially in recent years from studies conducted in Drosophila and yeast, as well as in mammalian cells and tissues.

Generally, there are three types of autophagy: chaperone-mediated autophagy, microautophagy, and macroautophagy. For chaperone-mediated autophagy, single intracellular proteins are recognized by a chaperone complex that improves target-motif-governed transport, lysosomal membrane binding, migration, and subsequent internalization. For microautophagy, it utilizes lysosomal limiting membrane invagination, septation, or protrusion to transport cytoplasmic materials. Macroautophagy is a highly conserved cellular process by which cells sequester a part of their cytoplasm or organelles into the autophagosomes that will fuse with lysosomes for degradation and recycling of the enclosed materials. Of the above three types of autophagy, most of our current studies are focused on macroautophagy (hereafter referred to as autophagy), which is also the focus of this chapter. Although all these three types of autophagy differ in the way by which cargo material is transported to the lysosome, they are similar in the last step of lysosomal protein degradation by hydrolase exposure.

Autophagy is very important for cellular homeostasis. It can promote the catabolism of denatured proteins or damaged organelles and the elimination of long-lived proteins. In addition, autophagy is an important adaptive mechanism. It is augmented obviously when the cells need to cope with certain cellular stresses such as starvation. Recent studies have demonstrated that autophagy also plays pathological roles sometimes. But the molecular mechanism of these pathological effects remains unknown.

2. The molecular mechanisms of autophagy

2.1. The initial stage of autophagy

In the initial stage, serine–threonine protein kinase ULK1 promotes the autophagy process. Autophagy inducers such as the defective nutrition or exposure to some chemotherapy agents cause dephosphorylation of ULK1; the ULK1 complex consists of ULK1, ATG13, FIP200 and receives stress signals from mTOR complex 1. When mTORC1 kinase activity is inhibited, autophagosome formation occurs. ULK1 can activate the phosphorylation of ATG13 and FIP200 to start the initial stage of autophagy.

2.2. Double-membrane vesicle formation stage of autophagy

In this stage, the Beclin-1–Vps34 complex, which consists of Vps34, Beclin-1, ATG14L, and Vps15, takes part in the process of vesicle nucleation. Under certain circumstances that induce autophagy, the dephosphorylated ULK1 complex musters the Beclin-1–Vps34 complex to form the autophagosome by phosphorylation of Ambra1. In this stage, TRAF6 interacts with Ambra1, which acts as an E3 ubiquitin ligase for ULK1, which results in stabilization and self-association of ULK1 by K63 ubiquitination. The dephosphorylated ULK1 complex also facilitates the activity of the ATG14L-containing Vps34 complex by phosphorylation of Beclin-1. Furthermore, EGFR and AKT regulate autophagy by phosphorylation of Beclin-1 without the impact of mTORC1. Next, the phosphatidylinositol-3-phosphate created by Vps34 musters an effector protein to facilitate the formation of double-membrane vesicle. The WIPI protein, which is the transcription product of a member of the ectopic P-granule subset of the metazoan-specific autophagy gene family, also participates in this stage.

2.3. Double-membrane vesicle elongation stage of autophagy

ATG proteins participate in the accomplishing of the autophagosome and fit out into two ubiquitin-like conjugation systems, ATG8 (LC3)–phosphatidylethanolamine and ATG12–ATG5–ATG16L. For the ATG12–ATG5–ATG16L system, ATG12 is connected to ATG5 by the synergistic action of ATG107 and ATG. The ATG12–ATG5–ATG16L complex consists of ATG12–ATG5 and ATG16L by conjugate binding, which takes part in the LC3–phosphatidylethanolamine conjugation. LC3, created by the ATG4 protease from Pro-LC3, is connected to phosphatidylethanolamine by ATG3, ATG7, and the ATG12–ATG5–ATG16L complex. After the above procedure, the lipid-conjugated LC3, which is positioned to the double-membrane of the autophagosomes, takes part in the formation and elongation of autophagosomes.

2.4. Fusion and degradation stage of autophagy

The captured proteins or organelles waiting for degradation in an engulfing or developing autophagosome are promoted by autophagy adaptor proteins, or receptors act as a connection. The complete autophagosomes will fuse with lysosomes to become autolysosomes. In these structures, the captured organelles and materials will be digested by lysosomal enzymes.

2.5. The final stage of autophagy

In the final stage, an essential component of the mTORC1 complex (mTOR) is reactivated by nutrients created by the autolysosomes. This process is very important for depressing the redundant activation of autophagy when the cells are in the periods of starvation. The mTOR generated from this stage can produce raw materials to form new lysosomes so that the lysosomes do not reduce during the autophagy procedure.

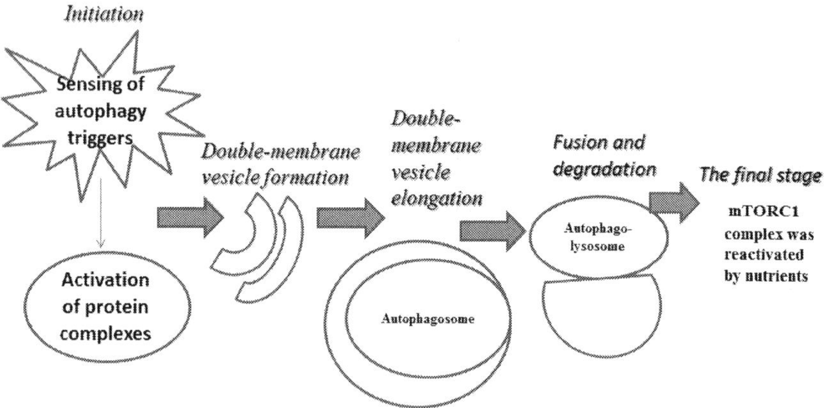

Figure 1. The process of autophagy

3. Major autophagy regulation target

cAMP-dependent protein kinases A (PKA), mammalian target of rapamycin complex 1 (MTORC1), and AMP-activated protein kinase (AMPK) are three major kinases for regulating cell autophagy. These kinases, along with some other molecules such as CAMKK2/CaMKKβ and TSC1/2, participate in a wide range of intracellular or extracellular signal pathways to regulate autophagy.

3.1. cAMP-dependent Protein Kinase A (PKA)

When the cell faces some stresses such as absence of growth factors, nutrient starvation, hypoxia, or endoplasmic reticulum stress, the survival of these cells can be dependent on autophagy. Recent research has shown that Ras/PKA (cAMP-dependent protein kinase) signaling pathway was associated with the early stage of autophagy in yeast. The studies showed that Ras/PKA signaling pathway can suppress the early stage of autophagy in *Saccharomyces cerevisiae* cells when faced with nutrient-rich conditions [20]. In mammals, this suppression is conducted at least partially by the phosphorylation of microtubule-associate protein 1 light chain 3 by PKA.

3.2. Mammalian Target of Rapamycin Complex 1 (MTORC1)

Studies have shown that MTORC1 is activated by the presence of amino acids. Amino acids can activate RAG (RAS-GTPases) proteins that regulate MTORC1. Recent research has demonstrated that in *Drosophila melanogaster* S2 cells, knockdown of Rag gene inhibited the activation effect of amino acids on TORC1. The constitutively active (GTP-bound) Rag gene expression can activate TORC1 without the presence of amino acids signals while expression of dominant-negative Rag can inhibit the activation effects of amino acids signals on TORC1. Researchers believe that there is some cross talk between these pathways. For example, PKA can directly activate MTORC1 and also indirectly activate MTORC1 by suppression of the AMPK.

3.3. AMP-activated protein Kinase (AMPK)

AMPK can respond to the change of ATP/AMP levels in cells and act as an intracellular ATP/AMP-sensing kinase. It is also a substrate of PKA and participates in regulating autophagy. AMP can activate the activity of AMPK, while ATP binding suppresses this kinase activity. AMPK can phosphorylate the TSC1/2 complex when activated by low ATP/AMP levels. Then, the phosphorylated TSC1/2 complex can indirectly or directly suppress the activity of MTORC1. Several studies have also demonstrated that AMPK can induce autophagy by activating ULK1.

3.4. Other regulating mechanisms

Some studies have proved that ER stress can cause autophagy by leading to Ca^{2+} concentrations' upregulation and inducing calcium/calmodulin-dependent protein kinase 2, beta

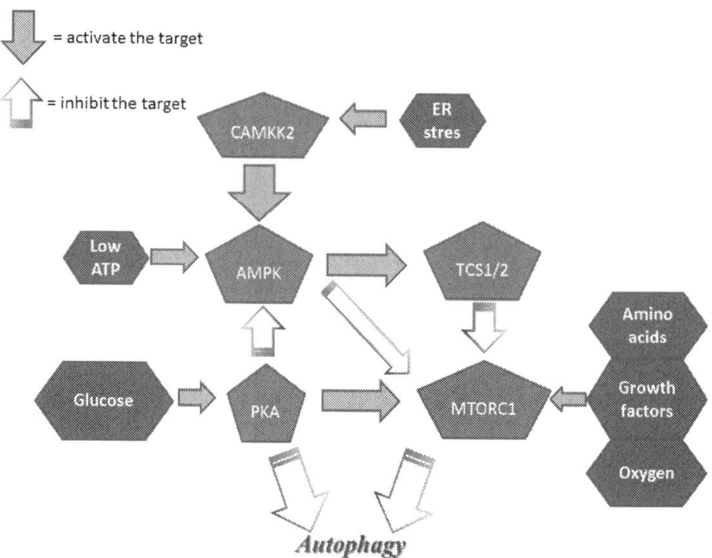

(Modified from: Antioxid Redox Signal. 2014 Jan 20; 20(3): 460–473)

Figure 2. The regulation mechanisms of autophagy

(CAMKK2/CaMKKβ), to activate AMPK [29]. (UPR) signaling also plays an important role in ER stress-related autophagy. This signal pathway has been proved in both mammals and yeast models. However, the final results of autophagy in response to ER stress are still not clear. Some researchers insist that it will improve cell survival while others believe that it may lead to cell death.

4. The relationship of cell autophagy and cancer

4.1. Autophagy: A double-edged sword in cancer

Recent studies have demonstrated that autophagy plays an important role in cancer initiation and progression stages. In some studies, autophagy has shown tumor inhibition effect by inducing autophagic cell death. But some studies have also proved that autophagy can facilitate cell survival during cancer treatment or other stresses. Many cancer researchers have recognized that autophagy is a double-edged sword in cancer. Usually, the role of autophagy in cancer depends on the tumor stage, the tumor genotype and microenvironment, as well as the type of therapy. For example, some recent studies have found that different kinds of autophagy-inducing drugs can play different roles in cancer treatment. In one study, the researchers used TRAIL and Fas ligand to treat cell lines with different autophagic levels respectively. These two compounds can both activate the cell death receptors and induce programmed cell death.

4.2. The cytoprotective autophagy in cancer

Previous studies have confirmed that autophagy can lead to resistance against metabolic stress and protect tumor cells through recycling organelles or proteins. In a study of Kirsten rat model, the sarcoma-activating viral oncogene homolog (K-Ras) variations can upregulate autophagy and facilitate tumor cell survival by autophagy. Some studies also demonstrated that both radiation therapy and chemotherapeutic agents can lead to autophagy and facilitate tumor cell survival. Although these studies have proved that the conventional cancer therapy may induce autophagy and reduce tumor cell apoptosis, no direct evidence has proved that inhibition of cell autophagy can increase the sensitivity of chemotherapy and radiation for cancer treatment. So the cytoprotective role by autophagy in cancer still needs more supportive evidence. If we can prove this cytoprotective form of autophagy in cancer, we can reduce the therapy resistance by inhibiting tumor cell autophagy during conventional cancer treatment [37].

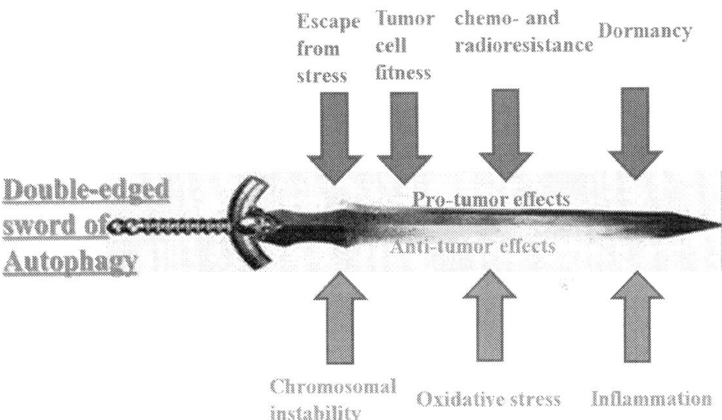

(Modified from: Semin Cancer Biol. 2013 Aug;23(4):252-61.)

Figure 3. Autophagy: a double-edged sword in cancer

4.3. The cytotoxic and cancer inhibitory autophagy

As we have discussed above, many studies have shown the cytotoxic form of autophagy in cancer. One study has combined vitamin D with radiation therapy for breast tumor model. The results showed that autophagy induced by the therapy can facilitate cell death in breast tumor. According to recent studies on autophagy, the cytotoxic form of autophagy was generated on directly killing cells by itself or further inducing apoptosis-related cell death. Some recent studies have also discovered that the autophagy-related genes such as Beclin 1

[64] Ko, H., et al., *Induction of autophagy by dimethyl cardamonin is associated with proliferative arrest in human colorectal carcinoma HCT116 and LOVO cells.* J Cell Biochem, 2011. 112(9): p. 2471-9.

[65] Law, B.Y., et al., *Alisol B, a novel inhibitor of the sarcoplasmic/endoplasmic reticulum Ca(2+) ATPase pump, induces autophagy, endoplasmic reticulum stress, and apoptosis.* Mol Cancer Ther, 2010. 9(3): p. 718-30.

[66] Guo, J., et al., *Cucurbitacin B induced ATM-mediated DNA damage causes G2/M cell cycle arrest in a ROS-dependent manner.* PLoS One, 2014. 9(2): p. e88140.

[67] Wang, N., et al., *Fangchinoline induces autophagic cell death via p53/sestrin2/AMPK signalling in human hepatocellular carcinoma cells.* Br J Pharmacol, 2011. 164(2b): p. 731-42.

[68] Turei, D., et al., *Autophagy Regulatory Network – A systems-level bioinformatics resource for studying the mechanism and regulation of autophagy.* Autophagy, 2015. 11(1): p. 155-65.

[69] Zalckvar, E., et al., *A systems level strategy for analyzing the cell death network: implication in exploring the apoptosis/autophagy connection.* Cell Death Differ, 2010. 17(8): p. 1244-53.

Chapter 8

Modulation of Autophagy by Free Fatty Acids

Jiuhui Wang and Daotai Nie

Additional information is available at the end of the chapter

http://dx.doi.org/10.5772/61484

Abstract

Fatty acids are important molecules with multiple biological properties. Emerging evidence suggests that fatty acids can modulate autophagy. Saturated fatty acids contribute to pancreatic β-cell dysfunction in type 2 diabetes. Palmitic acid, one of the long-chain saturated fatty acids (LCFA), induces autophagy of β-cells which protects them from dysfunctions and apoptotic cell death. Short-chain fatty acids (SCFA) possess antitumor activity in colon cancer cells by promoting autophagy. SCFAs can induce autophagy by suppressing the activity of mTOR signaling. As the most common monosaturated fatty acid (MUFA) in daily nutrition, oleic acid could induce autophagy, which is responsible for the regulation of lipids metabolism in hepatocytes. The ω-3 and ω-6 polyunsaturated fatty acids (PUFA) are essential in normal physiology and metabolism and play a contributory role in the incidence and progress of a series of disease including cancer. Autophagy triggered by ω-3 PUFAs contributes to the cytotoxicity in cancer cells by enhancing apoptosis, while autophagy mediated by ω-6 PUFAs led to the increase in *Caenorhabditis elegans* lifespan. The recent findings illustrate the potential involvement of autophagy regulation by fatty acids in a number of biological and pathological processes.

Keywords: Autophagy, fatty acids, lipids, apoptosis, cancer

1. Introduction

Fatty acids are aliphatic carboxylic acids consisting of a hydrocarbon chain and a terminal carboxyl group. Fatty acids are classified as several groups with respect to their structure and biological functions. Saturated fatty acids, which have no double bond between individual carbon atoms of the hydrocarbon chain, are divided into short-chain fatty acids (SCFA), medium-chain fatty acids (MCFA), long-chain fatty acids (LCFA), and very long-chain fatty

Chapter 9

Autophagy as a Therapeutic Target in Gastrointestinal Cancer

Michiko Shintani

Additional information is available at the end of the chapter

http://dx.doi.org/10.5772/61523

Abstract

Autophagy is a bulk protein and organelle degradation system and is an important homeostatic cellular recycling mechanism. The following kinds are the three types of autophagy: macroautophagy, microautophagy, and chaperone-mediated autophagy. In general, the term "autophagy" indicates macroautophagy. Autophagy is mediated by double-membrane-bound structures called autophagosomes. During the autophagic process, cytoplasmic components are sequestered and engulfed by autophagosomes. Autophagosomes then fuse with lysosomes to form autolysosomes where the sequestered components are digested by lysosomal hydrolases. Microtubule-associated protein 1 light chain 3 (LC3) is an autophagosomal ortholog of the yeast protein ATG8. Autophagy stimulates the upregulation of LC3 expression, and a cytosolic form of LC3 (LC3-I) is conjugated to phosphatidylethanolamine to form LC3-II which is recruited to autophagosomal membranes. Subsequently, LC3-II is degraded by lysosomal hydrolases after the fusion of autophagosomes with lysosomes. Therefore, LC3 is a specific marker of autophagosome formation. Additionally, beclin 1, the mammalian ortholog of the yeast protein ATG6, has been known to play a crucial role in autophagy. Beclin 1 acts in conjunction with the phosphoinositide-3 kinase pathway to enhance the formation of the autophagic vacuole.

Recently, autophagy has been reported to play roles in both cell death and survival. Autophagy is a multifaceted process, and alterations in autophagic signaling pathways are frequently observed in cancer. Cancer is a disease caused by mutation, selection, and genome instability in tumor tissues, and the role of autophagy in cancer is unclear.

One anticancer treatment strategy is to trigger tumor-selective cell death. Apoptosis is regarded as the central mediator of programmed cell death in response to radiation and chemotherapy. Our previous report suggested that different cell-death pathways are activated in gastric and colorectal carcinomas and the extrinsic and

intrinsic apoptotic pathways could be mutually regulated in gastric adenocarcinomas. In contrast, in colorectal carcinomas, autophagy may function as a cellular guardian to prevent caspase-9-dependent apoptosis (intrinsic apoptotic pathway). LC3 positivity was less frequent in gastric adenocarcinomas than in colorectal adenocarcinomas. Therefore, we suggested that LC3 expression in colorectal carcinomas is likely to aid cancer therapy, owing to its involvement in apoptosis and/or autophagy.

In this chapter, we discuss the following: (1) the detection of autophagy using immunohistochemistry, (2) autophagy and tumor suppression and/or progression, and (3) autophagy as a therapeutic target in gastrointestinal carcinomas.

Keywords: Gastric carcinoma, colorectal carcinoma, immunohistochemistry, cancer therapy

---

1. Introduction

Autophagy is a bulk protein and organelle degradation system and is an important homeostatic cellular recycling mechanism. The following are the three types of autophagy: macroautophagy, microautophagy, and chaperone-mediated autophagy. In general, macroautophagy is believed to be the major type of autophagy. Autophagy is mediated by double-membrane-bound structures called autophagosomes [1–3]. During the autophagic process, cytoplasmic components are sequestered and engulfed by autophagosomes. Autophagosomes then fuse with lysosomes to form autolysosomes where the sequestered components are digested by lysosomal hydrolases. Microtubule-associated protein 1 light chain 3 (LC3) is an autophagosomal ortholog of the yeast protein ATG8. LC3 exists in two forms, LC3-I and LC3-II. LC3-I is localized in the cytoplasm. Autophagy stimulates the upregulation of LC3 expression, and LC3-I conjugates with phosphatidylethanolamine to form LC3-II. LC3-II binds to autophagosomes and it is degraded by lysosomal hydrolases after the fusion of autophagosomes with lysosomes [4–6]. Therefore, LC3 is a specific marker of autophagosome formation. The autophagic pathway includes several phases: initiation, vesicle elongation, maturation, fusion, and degradation (Figure 1). Recent studies have suggested that additional membranes are derived from the Golgi complex, mitochondria, and plasma membrane; however, this phenomenon has not been confirmed [1–2, 7–10].

Current studies are examining the molecular regulation and function of autophagy. Additionally, autophagy is believed to play a role in various diseases such as cancer, infectious diseases, cardiovascular diseases, metabolic diseases, pulmonary diseases, and neurodegenerative disorders [11–20]. Recently, in clinical trials, several autophagic inhibitors, including hydroxychloroquine and chloroquine, have been examined as targets in diseases. In cancer, these autophagic components are being studied to enhance chemotherapeutic efficacy. Thus, autophagy is now an important and widely studied topic in human health and disease [11–15].

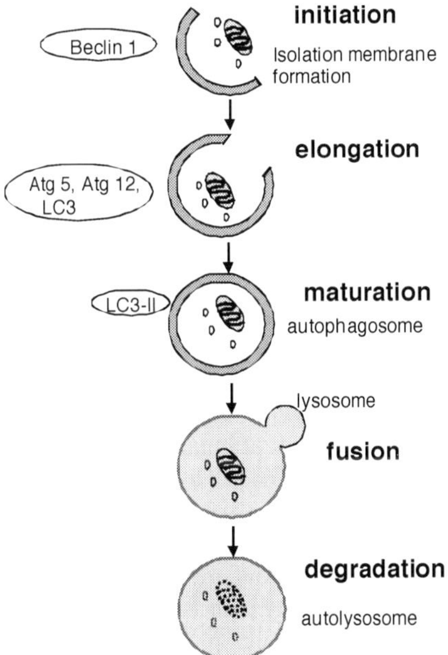

Figure 1. Cellular mechanism of autophagy. The autophagic pathway includes several phases: initiation, vesicle elongation, maturation, fusion, and degradation.

2. Detection of autophagy using immunohistochemistry

The role of autophagy in cancer development and progression has been studied by investigating autophagy-related proteins, LC3, beclin 1, and p62 using immunohistochemistry [21–23].

2.1. LC3

LC3 is an autophagosomal ortholog of the yeast protein ATG8 and is a specific marker of autophagosome formation. LC3-I is localized in the cytoplasm whereas LC3-II binds to autophagosomes. LC3 is presently used as an autophagy marker [21–26].

2.2. Beclin 1

Beclin 1 is the mammalian homolog of the yeast protein ATG6 and it has a central role in autophagy. The expression of beclin 1 has been reported in tumors such as breast, ovarian, prostate, lung, brain, stomach, and colorectal tumors. Beclin 1 may play a role in the tumorigenesis and/or progression of human cancers. However, beclin 1 has several physiological functions other than autophagy [21–22, 27–31].

[33] Yang SY, Winslet MC. Dual role of autophagy in colon cancer cell survival. Ann Surg Oncol 2011;18(3):S239. doi: 10.1245/s10434-011-1789-x. Epub 2011 May 21.

[34] Li J, Hou N, Faried A, Tsutsumi S, Takeuchi T, Kuwano H. Inhibition of autophagy by 3-MA enhances the effect of 5-FU-induced apoptosis in colon cancer cells. Ann Surg Oncol 2009;16(3):761–71. doi: 10.1245/s10434-008-0260-0.

[35] Levine B, Klionsky DJ. Development by self-digestion: molecular mechanisms and biological functions of autophagy. Cell 2004;6(4):463–77. doi:10.1016/S1534-5807(04)00099-1.

[36] Baehrecke EH. Autophagy: dual roles in life and death? Nat Rev Mol Cell Biol 2005;6(6):505–10.

[37] Amaravadi RK, Thompson CB. The roles of therapy-induced autophagy and necrosis in cancer treatment. Clin Cancer Res 2007;13(24):7271–9.

[38] Turcotte S, Giaccia AJ. Targeting cancer cells through autophagy for anticancer therapy. Curr Opin Cell Biol 2010;22(2):246–51.

[39] Edinger AL, Thompson CB. Death by design: apoptosis, necrosis and autophagy. Curr Opin Cell Biol 2004;16(6):663–9. doi:10.1016/j.ceb.2004.09.011.

[40] Degenhardt K, Mathew R, Beaudoin B, Bray K, Anderson D, Chen G, Mukherjee C, Shi Y, Gélinas C, Fan Y, Nelson DA, Jin S, White E. Autophagy promotes tumor cell survival and restricts necrosis, inflammation, and tumorigenesis. Cancer Cell 2006;10(1):51–64. doi: 10.1016/j.ccr.2006.06.001.

[41] Maiuri MC, Zalckvar E, Kimchi A, Kroemer G. Self-eating and self-killing: crosstalk between autophagy and apoptosis. Nat Rev Mol Cell Biol 2007;8(9):741–52. doi: 10.1038/nrm2239.

[42] Park JM, Huang S, Wu TT, Foster NR, Sinicrope FA. Prognostic impact of Beclin 1, p62/sequestosome 1 and LC3 protein expression in colon carcinomas from patients receiving 5-fluorouracil as adjuvant chemotherapy. Cancer Biol Ther 2013;14(2):100–7. doi: 10.4161/cbt.22954.

[43] Li J, Hou N, Faried A, Tsutsumi S, Kuwano H. Inhibition of autophagy augments 5-fluorouracil chemotherapy in human colon cancer in vitro and in vivo model. Eur J Cancer 2010;46(10):1900–9. doi: 10.1016/j.ejca.2010.02.021.

[44] Bijnsdorp IV, Peters GJ, Temmink OH, Fukushima M, Kruyt FA. Differential activation of cell death and autophagy results in an increased cytotoxic potential for trifluorothymidine compared to 5-fluorouracil in colon cancer cells. Int J Cancer 2010;126(10):2457–68. doi: 10.1002/ijc.24943.

[45] Huang S, Sinicrope FA. Celecoxib-induced apoptosis is enhanced by ABT-737 and by inhibition of autophagy in human colorectal cancer cells. Autophagy 2010;6(2):256–69. Epub 2010 Feb 6.

[46] Yuan CX, Zhou ZW, Yang YX, He ZX, Zhang X, Wang D, Yang T, Wang NJ, Zhao RJ, Zhou SF. Inhibition of mitotic Aurora kinase A by alisertib induces apoptosis and autophagy of human gastric cancer AGS and NCI-N78 cells. Drug Des Develop Ther 2015;9:487–508. doi: 10.2147/DDDT.S74127.

[47] Chun J, Kang M, Kim YS. A triterpenoid saponin from Adenophora triphylla var. japonica suppresses the growth of human gastric cancer cells via regulation of apoptosis and autophagy. Tumour Biol 2014;35(12):12021–30. doi: 10.1007/s13277-014-2501-0.

[48] Xie B, Zhou J, Shu G, Liu DC, Zhou J, Chen J, Yuan L. Restoration of klotho gene expression induces apoptosis and autophagy in gastric cancer cells: tumor suppressive role of klotho in gastric cancer. Cancer Cell Int 2013;13(1):18. doi: 10.1186/1475-2867-13-18.

[49] Rasul A, Yu B, Zhong L, Khan M, Yang H, Ma T. Cytotoxic effect of evodiamine in SGC-7901 human gastric adenocarcinoma cells via simultaneous induction of apoptosis and autophagy. Oncol Rep 2012;27(5):1481–7. doi: 10.3892/or.2012.1694.

[50] Song X, Kim SY, Zhang L, Tang D, Bartlett DL, Kwon YT, Lee YJ. Role of AMP-activated protein kinase in cross-talk between apoptosis and autophagy in human colon cancer. Cell Death Dis 2014;5:e1504. doi: 10.1038/cddis.2014.463.

[51] Çoker-Gürkan A, Arisan ED, Obakan P, Palavan-Unsal N. Lack of functional p53 renders DENSpm-induced autophagy and apoptosis in time dependent manner in colon cancer cells. Amino Acids 2015;47(1):87–100. doi: 10.1007/s00726-014-1851-7.

[52] Lee Y, Sung B, Kang YJ, Kim DH, Jang JY, Hwang SY, Kim M, Lim HS, Yoon JH, Chung HY, Kim ND. Apigenin-induced apoptosis is enhanced by inhibition of autophagy formation in HCT116 human colon cancer cells. Int J Oncol 2014;44(5):1599–606. doi: 10.3892/ijo.2014.2339.

[53] Kim AD, Kang KA, Kim HS, Kim DH, Choi YH, Lee SJ, Kim HS, Hyun JW. A ginseng metabolite, compound K, induces autophagy and apoptosis via generation of reactive oxygen species and activation of JNK in human colon cancer cells. Cell Death Dis 2013;4:e750. doi: 10.1038/cddis.2013.273.

[54] Gandesiri M, Chakilam S, Ivanovska J, Benderska N, Ocker M, Di Fazio P, Feoktistova M, Gali-Muhtasib H, Rave-Fränk M, Prante O, Christiansen H, Leverkus M, Hartmann A, Schneider-Stock R. DAPK plays an important role in panobinostat-induced autophagy and commits cells to apoptosis under autophagy deficient conditions. Apoptosis 2012;17(12):1300–15. doi: 10.1007/s10495-012-0757-7.

[55] Zhan Y, Gong K, Chen C, Wang H, Li W. P38 MAP kinase functions as a switch in MS-275-induced reactive oxygen species-dependent autophagy and apoptosis in human colon cancer cells. Free Radic Biol Med 2012;53(3):532–43. doi: 10.1016/j.freeradbiomed.2012.05.018.

[56] Zhang N, Chen Y, Jiang R, Li E, Chen X, Xi Z, Guo Y, Liu X, Zhou Y, Che Y, Jiang X. PARP and RIP 1 are required for autophagy induced by 11'-deoxyverticillin A, which precedes caspase-dependent apoptosis. Autophagy 2011;7(6):598–612.

[57] Nishikawa T, Tsuno NH, Okaji Y, Sunami E, Shuno Y, Sasaki K, Hongo K, Kaneko M, Hiyoshi M, Kawai K, Kitayama J, Takahashi K, Nagawa H. The inhibition of autophagy potentiates anti-angiogenic effects of sulforaphane by inducing apoptosis. Angiogenesis 2010;13(3):227–38. doi: 10.1007/s10456-010-9180-2.

Apoptosis

Chapter 10

Apoptotic Molecular Advances in Breast Cancer Management

Pontsho Moela and Lesetja R. Motadi

Additional information is available at the end of the chapter

http://dx.doi.org/10.5772/61654

Abstract

Breast cancer is the most common cancer type amongst women, accounting for most female cancer deaths second to cervical cancer worldwide. It is, therefore, highly crucial to understand the molecular biology and explore other pathways involved in carcinogenesis in order to select appropriate treatment not only for breast cancer but for other cancers as well. Cancer progression is favoured by DNA damage and in most cases a consequent disruption of the apoptotic pathway, thus leading to uncontrolled cell proliferation. Therefore, current therapeutic strategies aim at targeting the apoptotic pathways in order to combat cancer. In this manuscript, we discuss the ways in which evasion of apoptosis during carcinogenesis occurs and the types of current therapeutic strategies as well as promising future approaches against breast cancer.

Keywords: Breast cancer, apoptosis, small molecules, p53, RBBP6

1. Introduction

The human body is composed of trillions of cells that behave and function to provide structure of the body, convert nutrients into energy and carry out specialised functions [1, 3]. Growing, dividing, differentiating and dying are the cells' behavioural mechanisms to maintain tissue homeostasis [3]. However, molecular disturbances that disrupt this balance may potentially lead to disease. Such molecular disturbances include mutations, among others, during which any change to the DNA sequence might result in abnormality in the cell or tissue [4]. With a population of more than a trillion cells, the human body is prone to mutations that may give one cell a selective advantage of growing and dividing more vigorously to become a growing mutant clone [4, 5]. Such mutations, in which a mutant clone of cells grows and divides out of control at an expense of neighbouring wild-type cell populations, serve as a prerequisite for the development of cancer [3].

Cancer is defined as uncontrolled cell proliferation that leads to the formation of abnormal cells and invasion of other adjacent tissues [1, 3-5]. The migration of cells from the origin of tumour to another part of the body is referred to as metastasis. Tumours can either be malignant or benign. While malignant tumours have the ability to invade surrounding tissue, benign tumours cannot invade other tissues and are therefore not as life-threatening [3]. Efficient treatment against malignant tumours is therefore necessary in cancer management. In this chapter, we discuss current anticancer strategies that are targeted on the apoptosis pathway in breast cancer management.

In order to understand breast cancer, it is necessary to understand the normal anatomy of the female breast [3, 20]. The female breast is made up of milk-producing glands called lobules which are connected to ducts that transport milk from the glands to the nipples. The ducts and lobules are surrounded by connective tissue, fatty tissue, blood vessels and lymphatic vessels. In most cases, breast cancer starts in cells surrounding the ducts or the lobules [23]. Metastatic breast cancer is as a result of migration of cancerous cells from ducts and/or lobules via lymphatic vessels to the lymph nodes of the lymphatic system [3, 20, 23].

Breast cancer is the most common cancer type amongst women accounting for many cancer deaths, second to cervical cancer. Risk factors of breast cancer are divided into non-modifiable and modifiable factors [39]. Advanced age, female gender, menarche before the age of 12, menopause after the age of 45, genetic mutations and family history are the major non-modifiable risk factors associated with breast cancer [6, 11, 26, 46, 56, 57]. Breast cancer risk factors that can be controlled include hormone replacement therapy, oral contraceptives, pregnancy, breast feeding and high breast density [31]. Behavioural and life-style risk factors associated with the development of breast cancer include poor diet, i.e. high fat, low vegetable/ fruit, low fibre and high in simple carbohydrates; overweight and obesity; and decreasing physical activity [29, 39].

Nearly 80% of human breast cancers are hormone-positive (estrogen and progesterone), followed by human epidermal growth factor receptor 2 (HER2)-positive, then vascular endothelial growth factor (VEGF)-positive breast tumours [8, 9]. Targeting estrogen receptor (ER) pathway, VEGF and HER2 are the long-established breast cancer therapeutic approaches responsible for the improvements of breast cancer prevention and treatment. However, resistance to these endocrine and cell-growth-inhibiting treatments is the main drawback that reduces the benefits of these novel treatment approaches [8, 9, 20, 23]. It is therefore highly crucial to understand the molecular biology and explore other pathways involved in carcino-genesis in order to select appropriate treatment not only for breast cancer but for other cancers as well. In this chapter we discuss different ways of targeting apoptosis in breast cancer management.

2. Targeting apoptosis in breast cancer treatment

During the process of breast cancer progression, normal cells transform into malignant types as a result of genetic alterations [12]. This leads to dysregulation of cellular processes such as angiogenesis, cell cycle and apoptosis [17]. Therefore, current therapeutic strategies aim at targeting these pathways, more especially apoptosis, in order to combat cancer [18]. Apoptosis is a form of programmed cell death in which cells are programmed to die if found to be cellular damaged [21, 25]. Apoptosis is made up of two major pathways called the death receptor pathway and the mitochondrial pathway, which are both propagated by a caspase cascade that ultimately leads to apoptosis induction [27, 34]. Evasion of apoptosis during carcinogenesis occurs by three distinct mechanisms: disrupted signalling of death receptors, loss of caspase activity as well as impaired balance between anti-apoptotic and pro-apoptotic proteins [14, 42, 50, 59]. Targeting the caspase cascade, Bcl-2 family proteins as well as other factors associated with apoptosis signalling have thus become the major strategy in anticancer therapeutics (table 1).

Reagent	Target	Technology	Function	Status
Apoptin	Caspases in the extrinsic pathway	Vector-based (adenoviral and virus vectors)	Caspase 3 and 8 activation	Preclinical
Flavipirodol, gossypol, depsipeptide, ABT-737, ABT-264, fenretinide, HA 14-1, GX15-070	Anti-Bcl-2 family proteins	Small molecule	Inhibit BCl-2 family proteins by reducing their expression	Phase I/II
ABT 737	Anti-apoptotic proteins	Small molecule	Inhibit expression of anti-apoptotic proteins such as Bcl-xL, Bcl-2 and Bcl-W	Phase I
Oblimersen Sodium	Anti-Bcl-2 targeted drug	Antisense	Bcl-2 antisense increases survival rates in chronic myeloid leukaemia patients when combined with chemotherapy	Phase II
ONYX-015 drug	p53-based gene therapy	Adenoviral	Genetically engineered adenovirus that has been modified to infect and lyse p53-deficient cells	Phase III
CD8+ cytotoxic T-lymphocytes (CTLs)	Tumour associated antigens (mutant p53)	Vaccine	Recognize TAA-derived peptides that are processed and presented on the tumours cell surface in association with MHC class I molecules, leading to killing of tumour cells	Phase I

Chapter 11

Between Armour and Weapons — Cell Death Mechanisms in Trypanosomatid Parasites

Rubem Figueiredo Sadok Menna-Barreto and Solange Lisboa de Castro

Additional information is available at the end of the chapter

http://dx.doi.org/10.5772/61196

Abstract

Among the pathogenic protozoa, trypanosomatids stand out due to their medical and economic impact, especially for low-income populations in tropical countries. Together, sleeping sickness, Chagas disease and leishmaniasis affect millions of humans and animals worldwide, yet are neglected by the pharmaceutical industry. The current drugs for trypanosomatid infections are limited and unsatisfactory, with severe side effects leading to reduced quality of life and, in several instances, to the abandonment of treatment. An intense search for alternative compounds has been performed, aiming at specific parasite targets by cellular, molecular and biochemical approaches. One interesting strategy could be interference with the protozoan cell death pathways. However, these pathways are poorly understood in unicellular eukaryotes, with the controversial existence and uncertain biological relevance of programmed cell death (PCD). This chapter will discuss apoptosis-like and autophagic cell death and necrosis in *Trypanosoma brucei*, *Trypanosoma cruzi* and *Leishmania* sp. and the possible implications of these pathways for the parasite life cycle and infection persistence. It will also revisit the genomic and proteomic metadata of these trypanosomatids in the literature to rebuild the map of cell death proteins expressed under different conditions. The interaction of leading candidates with parasite-specific molecules, especially with enzymes that regulate key steps in the cell death process, is a rational and attractive alternative for drug development for these neglected diseases.

Keywords: Cell death, apoptosis-like, autophagy, necrosis, *Leishmania* sp, *T. cruzi*, *T. brucei*

1. Introduction

Neglected tropical diseases (NTDs) are a group of the seventeen mostly life-threatening infections, which affect more than a billion people worldwide. They affect poor populations,

often in underdeveloped and developing countries (low-income countries) [1]. Among NTDs, infections caused by the so-called "protozoan" parasites, such as African trypanosomiasis, Chagas disease and leishmaniasis, are responsible for a high annual death toll among the poor populations of tropical countries. New safe and affordable medicines are urgently needed. These diseases all present therapeutic difficulties by developing resistance to existing therapies and/or by toxic side effects.

1.1. Neglected tropical diseases and trypanosomatids

1.1.1. Sleeping sickness

Human African trypanosomiasis (HAT), or sleeping sickness, is caused by extracellular protozoa belonging to the genus *Trypanosoma* and the species *T. brucei*. Two subspecies of *T. brucei* cause diseases with different epidemiological and clinical patterns: *T. b. gambiense*, a chronic disease present in western and central Africa accounting for 98% of the cases, and *T. b. rhodesiense*, an acute zoonosis located in eastern and southern Africa that occasionally infects humans. In 2001, WHO launched a major initiative to reinforce disease control and surveillance. After 10 years, the number of new cases of HAT decreased by 73.4%. Presently, the estimated the number of cases is 30, 000, and 70 million people are at risk [2, 3]. HAT clinically evolves in two stages. In the first stage, parasites are found in the lymphatic system and bloodstream. After a variable period of time, which is much shorter for the rhodesiense form, the second stage begins, with the parasites penetrating the blood-brain barrier and invading the central nervous system, leading to progressive neurological damage [4]. HAT is usually fatal if left untreated. Rhodesiense HAT usually progresses to death within six months, while gambiense HAT has a more chronic progressive course with an average duration of almost three years [5].

T. brucei is transmitted by the tsetse fly Glossina spp when it takes a blood meal. Non-dividing metacyclic forms enter the bloodstream of the mammalian host and differentiate into a rapidly dividing slender form able to evade antibody responses through antigenic variation [6]. Most of these forms undergo cell cycle arrest and develop into short-stumpy forms. When the tsetse fly bites an infected host, only the short-stumpy parasites survive in the insect's midgut and develop into a procyclic form, which undergoes multiple developmental phases on its way to the salivary gland, finally culminating in the infective metacyclic form [7, 8].

The drug of choice for treatment depends on the infecting species and the stage of infection. In early stages, *T. b. gambiense* and *T. b. rhodesiense* infections can be treated with pentamidine and suramin, respectively [9]. If the disease has progressed, treatment relies on melarsoprol or eflornithine. Melarsoprol, an arsenical drug, is extremely toxic. Eflornithine is less toxic, but is expensive, and has a difficult administration than melarsoprol and lacks efficacy against *T. b. rhodesiense* [2]. Since 2001, this drug has been combined with nifurtimox (NECT) for first-line treatment for CNS-stage *T. b. gambiense* HAT. It is the most recent breakthrough in anti-trypanosomiasis drug research and was added to the World Health Organisation's list of essential medicines in 2009. A major problem related to the treatment of HAT is the development of resistance to melarsoprol and the other drugs [10].

1.1.2. Chagas disease

Chagas disease is caused by the intracellular obligatory parasite *Trypanosoma cruzi* and affects approximately eight million individuals in Latin America [11]. The transmission of this disease occurs through the faeces of sucking triatominae insects, blood transfusions, organ transplantation, oral contamination, laboratory accidents and congenital routes [12, 13]. Current major concerns are the outbreaks of acute Chagas disease associated with the ingestion of contaminated food and its emergence in non-endemic areas, such as North America and Europe, due to the immigration of infected individuals [14-16]. This disease is characterised by two clinical phases. The acute phase appears shortly after infection and is defined by patent parasitaemia. If left untreated, symptomatic chronic disease develops in about one-third of individuals after a long latent period (10-30 years), which is known as the indeterminate form. The main clinical manifestations of Chagas disease include digestive and/or cardiac alterations. The chronic cardiac form of the disease is the most significant clinical manifestation. Consequences include dilated cardiomyopathy, congestive heart failure, arrhythmias, cardioembolism and stroke [17].

The life cycle of *T. cruzi* involves four major developmental stages during its passage through vertebrate and invertebrate hosts [18]. The infective stage of the parasite, the metacyclic trypomastigote, enters the mammalian host from insect faeces through wound openings or mucous membranes. In the mammalian host, the metacyclic trypomastigote differentiates into the amastigote form. After several rounds of replication in the host cells, the amastigote differentiates into the bloodstream trypomastigote, which can enter new cells and perpetuate the infection. When the insect bites an infected host, the bloodstream trypomastigote differentiates into the replicative epimastigote that lives in the insect's gut. Finally, in the rectum of the insect, the epimastigote differentiates into the infective metacyclic trypomastigote, which is ready to infect its host again.

The available chemotherapy for this illness includes two nitroheterocyclic agents, nifurtimox and benznidazole, which are effective against acute infections, but show poor activity in the late chronic phase, with severe collateral effects and limited efficacy against different parasitic isolates. These drawbacks justify the urgent need to identify better drugs to treat chagasic patients, and several new compounds are currently in preclinical development involving *in vitro* parasite phenotype screens and target-based drug discovery [19-21]. Recently, clinical trials with the azoles posaconazole and E1224 (ravuconazole prodrug) led to higher percentages of treatment failure in chronic patients than benznidazole [22, 23], suggesting their potential use in combination therapy [24].

1.1.3. Leishmaniasis

Leishmaniasis, which is caused by different species of *Leishmania*, is a vector-borne disease, with an estimated 12 million cases worldwide. Infection is caused by the bite of infected female sand flies of the genera *Phlebotomus* (Europe, Asia, Africa) and *Lutzomyia* (America) [25]. *Leishmania* parasites live a digenetic life cycle as either a promastigote flagellar or an amastigote form. The type of clinical manifestation depends on the infecting species and host factors, such as general health and genetic and immune constitution [26]. It is a disease complex with three clinical manifestations, visceral (VL, kala-azar), cutaneous (CL) and muco-cutaneous (MCL), which arise from parasite replication in the mononuclear phagocyte system, dermis and naso-oropharyngeal mucosa, respectively [27]. Some post-treated *L. donovani*-infected patients develop the diffuse cutaneous form named post-kala-azar dermal leishmaniasis (PKDL) [28, 29]. VL, after initial skin lesions, takes 2-8 months to develop gross inflammatory reactions within the viscera (liver and spleen in particular) and is usually fatal unless treated. CL manifests as an open sore at the site of the insect bite and will frequently self-heal, leaving a scar. The diffuse form of CL is more problematic, causing lepromatous type lesions disseminated across the skin that can be difficult to heal. The MCL form, endemic in parts of Latin America, starts with skin sores that spread to the mucosal membranes of the face. Profound inflammatory damage can lead to the erosion of the nostrils and mouth in particular [29].

In the *Leishmania* life cycle, there are two principal parasite forms: amastigotes and motile promastigotes. In the alimentary tract of the insect vector, the parasite exists as multiplicative, non-infective procyclic promastigotes and non-multiplicative, infective metacyclic promastigotes [30]. Upon injection into the mammalian host, promastigotes are taken up by macrophages where the metacyclic forms differentiate into small multiplicative, non-motile amastigotes that live in a lysosomal compartment known as the parasitophorous vacuole [31]. These developmental forms are distinguished by their nutritional requirements, their growth rate and ability to divide, the regulated expression of their surface molecules, and their morphology. Metacyclic promastigotes are pre-adapted for survival in the mammalian host, as they are complement-resistant. Amastigotes are intracellular, non-motile forms that have adapted to the low pH of this compartment and have an adapted energy metabolism.

The current drugs are highly toxic, resistance is common and compliance of patients to treatment is low, as the treatment is long and the drug price is high. Although recent initiatives have improved the antileishmanial drug arsenal by combining current medicines or using new formulations of old ones, none are ideal for treatment due to their high toxicity, resistance issues, prohibitive prices, long treatment length and need of intravenous administration [32-34]. Pentavalent antimonials (glucantime and pentostan) are first-line drugs for both VL and CL. However, they present several limitations, including variable efficacy, need for daily injectable administration for approximately one month, and severe side effects. Many patients are unable to complete the treatment, increasing the risk of drug resistance development. Amphotericin B is a systemic antifungal that is used as a second-line drug for VL. It is highly toxic, requiring careful and slow intravenous administration. Lipid formulations of amphotericin B have been developed to improve its bioavailability and pharmacokinetic properties, reducing toxicity [35]. Miltefosine is the most recent antileishmanial drug on the market and the first effective oral treatment against VL [36]. However, it has common gastrointestinal side effects and is also limited by its relatively high cost [34], potential teratogenicity and growing concerns in relation to increases in clinical isolate susceptibility [37]. Paromomycin is an aminoglycoside antibiotic that is used in topical treatment for CL and as a parenteral drug for VL. Pentamidine was used as a second-line drug in antimony-resistant VL treatment. However, its high toxicity combined with decreased efficacy led to the abandonment of this drug to treat VL in India, but it is valuable for combined therapies [38].

2. Cell death: State of art

As used for whole organisms, the term death is employed to describe a sequence of events culminating in the breakdown of all biological functions. However, more than one century after the first citation [39], cell death still represents a crucial gap in our understanding of cellular physiology. It can be triggered by natural processes or induced by extrinsic factors (exposure to chemicals or physical stresses). The consequent tissue injury usually leads to a state of disease [40]. On the other hand, many studies pointed to cell death playing a fundamental role in the physiology of multicellular organisms, especially in processes such as metamorphosis and embryogenesis [41]. In this context, in 1964, the term programmed cell death (PCD) was created, proposing a sequence of well-controlled steps regulating a non-accidental cell death process in the absence of an inflammatory response [42]. Currently, it is known that distinct death mechanisms and phenotypes participate in PCD, with apoptosis and autophagy being the most prominent [43].

2.1. Apoptosis

The apoptotic pathway was first described in the early 1970s as a fundamental step for proper embryo development [44]. This process is crucial during tissue development, especially in immune response regulation and removal of infected or damaged cells [45, 46]. Apoptosis is involved not only in growth regulation in multicellular organisms [47, 48] but also in their defence against viral, bacterial or parasitic infections [49-53] and even against cancer development [54-57]. The removal of non-functional cells by the apoptotic pathway is efficient and prevents the inflammatory response [58].

During apoptosis in multicellular organisms, the cell activates death machinery that culminates in chromosomal condensation and nuclear DNA fragmentation [59, 60]. Biochemically, apoptosis is orchestrated by the activation of a family of cysteine proteases, named caspases, that are activated by extrinsic and intrinsic factors [45, 46]. The extrinsic pathway is activated by the interaction of death ligands with their respective cell surface receptor (i.e., FasL/Fas, TNF-α/TNFR) [61-63]. Such binding triggers the cleavage of procaspase 8 into active caspase 8, which cleaves procaspase 3. Executioner caspase 3 activates endonuclease G (EndoG), starting the characteristic DNA fragmentation, a distinctive marker of apoptosis [63-65]. On the other hand, the intrinsic pathway can be triggered by two distinct mechanisms with mitochondrion or endoplasmic reticulum (ER) dependency. In the mitochondrial pathway, activation occurs by membrane permeabilization, releasing cytochrome c, apoptosis induction factor (AIF), EndoG and regulators of the B-cell lymphoma 2 (Bcl2) protein family into the cytosol. In the cytosol, the apoptosome is formed by the interaction of released cytochrome c with apoptotic protease activating factor 1 (APAF-1) and procaspase 9, activating caspase 9, which subsequently activates the effector caspase 3 [66-70]. The ER pathway is mainly caspase 12-dependent and occurs in this organelle during stress conditions. Because this pathway was described in the mouse and humans lack functional caspase 12, the relevance of ER-mediated apoptosis is still debatable [71-73].

Undoubtedly, the caspase cascade represents a central point in the apoptotic process. Its regulation is well-controlled by pro- and anti-apoptotic molecules from the Bcl-2 family [74]. The apoptotic morphological and biochemical phenotypes include cell shrinkage, membrane blebbing (formation of apoptotic bodies), chromatin condensation and typical internucleosomal DNA fragmentation, externalization of phosphatidylserine (PS), loss of mitochondrial membrane potential ($\Delta\Psi$m), and target protein degradation by caspase activation [75-79]. The characterization of apoptosis is experimentally based on the detection of apoptotic markers. The loss of $\Delta\Psi$m (labelling with rhodamine 123 derivatives, such as TMRE), PS exposure (binding to labelled annexin V), chromatin condensation (DAPI labelling) and DNA fragmentation (TUNEL technique) are usually quantified by fluorescence microscopy or flow cytometry. DNA fragmentation can also be assessed by agarose gel electrophoresis, presenting a laddering pattern that represents internucleosomal cleavage. Analysis of caspase activity using labelled specific substrates and/or inhibitors can be performed by immunotechniques such as ELISA [80].

2.2. Autophagy

In the 1950s, acidic organelles involved in the intracellular degradation of macromolecules were described and termed lysosomes by Dr. Christian de Duve. In a subsequent study [81], he proposed the term autophagy for a self-degrading process [82]. Currently, the autophagic pathway is considered to be the main cellular mechanism for the degradation of non-functional organelles and/or macromolecules and is fundamental for homeostasis in eukaryotic cells [83]. In other words, autophagy is a housekeeping self-digestion mechanism that is crucial for cellular turnover and recycling and occurs by the engulfment of cytosolic portions containing material that should be degraded. Degradation starts immediately after the fusion of autophagosomes to lysosomes in an organelle named the autophagolysosome [84, 85].

In multicellular organisms, autophagy is involved in many physiological situations, including development, cell growth and cell differentiation. Autophagy sustains cell survival under 'extracellular stress', such as nutrient starvation, hypoxia, acidic pH and high temperature. It acts as a housekeeping device under 'intracellular stress' by removing damaged or redundant cytoplasmic components, including organelles [86]. Increased autophagic activity is observed in pathological states and in host defences against pathogens [87-92]. Despite the relevant role of autophagy for the maintenance of the regular cell cycle, prolonged starvation periods or other strong autophagic stimuli induce a cellular misbalance and promote autophagic cell death [93, 94].

The autophagic molecular machinery was first assessed in the yeast model *Saccharomyces cerevisiae*, and 30 proteins, called Atgs (AuTophaGy-related), were described and associated with different steps of the pathway [95]. Atg orthologues were identified in all eukaryotes, with Atg8 (LC3 in mammals) being one of the most studied [82]. Autophagy can be a selective or non-selective process, degrading specific or random cellular components. Examples of selective routes are mitophagy, pexophagy or reticulophagy, in which mitochondria (or part of the organelle), peroxisomes and ER are degraded, respectively [82].

Additionally, there are three types of autophagy: macroautophagy, microautophagy and chaperone-mediated autophagy (CMA). The most common is macroautophagy, a process that involves the engulfment of cytosolic portions by a double membrane structure called the phagophore. The double-membrane vesicle formed from phagophore engulfment is named the autophagosome and is directed to lysosomes for degradation by lysosomal hydrolases. These steps are regulated by Atgs [92, 96-98]. The chronological events related to macroautophagy are (a) autophagic induction; (b) cargo selection; (c) phagophore elongation; (d) autophagosome formation; (e) fusion to lysosomes; and (f) cargo degradation [99]. The early steps in this process depend on the serine/threonine protein kinase TOR (target of rapamycin), which is essential for autophagic regulation. TOR complexes 1 and 2 work as sensors of nutritional availability (especially amino acids). The autophagic enzyme Atg6 (Beclin 1 in mammals) is a phosphatidylinositol 3-kinase (PI-3K) and shares its signalling function with other cellular pathways. For autophagy, these kinases present a critical role for autophagosome formation [82].

In contrast, there are no autophagosomes in the microautophagic pathway. Invagination of the lysosomal membrane occurs, resulting in a single-membrane small vesicle inside the lysosomes that will be degraded. Interestingly, both macro- and microautophagy could be selective or non-selective processes. Indeed, CMA appears to be the most selective type of autophagy. The proteins that will be degraded contain pentapeptide motifs (KFERQ, QREFK or VDKFQ), the binding sites of a cytosolic chaperone. Such a chaperone-substrate complex binds to a LAMP-2A receptor in the lysosomal membrane, promoting receptor dimerization. A membrane channel is formed, and the specific protein reaches the lysosomal lumen to be degraded [82, 100].

For many years, electron microscopy was the only tool available for the identification of autophagic morphological features, especially the presence of double-membrane vesicles (autophagosomes). In the last 20 years, advances in the molecular description of autophagy allowed the detection, localization and quantification of Atgs by molecular, biochemical and morphological approaches. Currently, the gold-standard method to monitor autophagy is Atg8/LC3 detection by different techniques: (a) Western blotting (presence of two isoforms); (b) confocal or fluorescence microscopy (identification of LC3 puncta); (c) knock down or knock out (deletion and analysis of the phenotype); and (d) pharmacological induction/ inhibition (rapamycin and/or PI-3K inhibitors). These techniques can also be employed *in vitro* or *in vivo* for other Atgs, indicating autophagic activity [101].

2.3. Necrosis

Necrosis is a term that is extensively employed as synonymous with cell death. In the Greek aetiology, it signifies the "stage of dying". In this death type, strong cellular damage occurs caused by external stimuli (drugs, infection, mechanical trauma), promoting the random degradation of the whole cell, with plasma membrane disruption. Necrosis is defined as an accidental cell death process, differing from PCD (especially apoptosis) [102]. One of the main differences between apoptosis and necrosis is the induction of the inflammatory response in the latter. The release of intracellular material into the extracellular environment during necrotic cell death triggers intense inflammation in the surrounding cells and tissues [103]. Classical necrotic features are the loss of plasma membrane integrity, cytosolic vacuolization, disruption of calcium homeostasis, general degradation by lysosomal hydrolases and induction of the inflammatory response.

Necrosis can also be a regulated process. Necroptosis is a programmed and non-accidental death pathway. Surprisingly, the activation of this pathway can occur by TNF-α or FasL, classical apoptotic ligands. Necroptosis depends on the participation of the receptor-interacting protein kinases 1 and 3 (RIPK1 and RIPK3), which are kinases that regulate this pathway. RIPK1 is pharmacologically inhibited by a small molecule named necrostatin-1 (Nec-1) [104-106].

2.4. Others

In addition to apoptosis, autophagy and necrosis (accidental or not), other non-canonical death styles can take place in eukaryotic cells. In an inflammatory context, pyroptosis and NETosis are prominent. Pyroptosis, primarily observed in macrophages after bacterial infection, is caspase 1-dependent. This caspase promotes an increase in the inflammatory cytokine levels (IL-1β and IL-18) and the formation of plasma membrane pores, leading to the release of cellular material to the extracellular matrix. The main difference between pyroptosis and apoptosis is the participation of caspase 1, which is only involved in the pyroptotic death pathway, a proinflammatory PCD [106-108]. Another type of cell death that plays a crucial role in the innate immune response is the neutrophil extracellular trap (NETosis), where neutrophilic death leads to the release of a neutrophil DNA network coated with histones and elastase to the extracellular environment to capture pathogens. However, the direct antimicrobial effect of the NETs is still controversial [109, 110]. Currently, DNA release has also been described in other immune cells, such as eosinophils, basophils, macrophages and mast cells, but its precise role deserves further analysis [110-114].

Other cell death types not involved in inflammation have been characterized. Ferroptosis is iron-dependent cell death that has been identified in some mammalian cells and involves oxidative stress induced by a small molecule named erastin, which is inhibited by ferrostatin 1. Despite that lack of complete understanding of the erastin mechanism, the X_C–Cys/Glu antiporter system is inhibited in ferroptosis, leading to a misbalance of these amino acids inside the cell [106, 115]. Additionally, there is another non-canonical cell death pathway in cancer cells (*in vitro* and *in vivo* models) called autoschizis, which involves oxidative stress induced by treatment with ascorbate and menadione. Autoschizic cell death presents remarkable morphological evidence, with electron microscopy as the best technique for its identification. Among the autoschizic features are cell shrinkage, extrusion of large portions of the cytosol (without any organelles), random DNA fragmentation and the subsequent deterioration of all cellular structures [116, 117]. Interestingly, annexin V (AV) and propidium iodide (PI) assays (gold standards for apoptosis detection in mammals) of cells treated with ascorbate and menadione demonstrate high percentages of AV-/PI+ cells [117], which are not discussed in almost all apoptotic studies, suggesting that these membrane shedding events could occur in a large variety of cell models. Table 1 summarizes the main types of cell death discussed herein.

Cell death	Features	References
apoptosis	cell shrinkage	[44, 76, 78]
	membrane blebbing	
	DNA fragmentation	
	externalization of PS	
	activity of caspases	
	regulation by Bcl-2 family proteins	
	loss of $\Delta\Psi$	
	release of cytochrome c	
	no inflammatory response	
autophagy	presence of autophagosomes	[82, 101]
	participation of Atgs	
	regulation by PI-3K and TORC	
	degradation by lysosomes	
	presence of KFERQ, QREFK or VDKFQ motifs in the protein to be degraded (only in CMA)	
necrosis	disruption of plasma membrane	[102, 103]
	cytoplasmic vacuolization	
	imbalance of Ca2+ homeostasis	
	release of lysosomal enzymes	
	induction of inflammatory response	
necroptosis	participation of RIP1 and RIP3	[104, 106]
	inhibition by Nec-1	
pyroptosis	participation of caspase 1	[106-108]
	increase in IL-1β and IL-18 levels	
	induction of inflammatory response	
NETosis	formation of NETs	[109, 110]
	participation of elastase and histones	
	occurrence in neutrophils, macrophages, mast cells, eosinophils and basophils	
ferroptosis	participation of iron	[106, 115]
	presence of oxidative stress	
	induction by erastin	
	blockage of X_C–Cys/Glu antiporter system	
	inhibition by ferrostatin 1	
autoschizis	cell shrinkage	[116, 117]
	random DNA fragmentation	
	extrusion of large cytosolic portions (without organelles)	
	degradation of cellular components	
	increase in the AV-/PI+ population	

Table 1. Types of cell death

3. Cell death in trypanosomatids: An overview

The term PCD was employed for decades to exclusively describe cell death in metazoans and its involvement in embryogenesis and maintenance of homeostasis. Indeed, the relevance of PCD for lower eukaryotes is unclear. In an evolutionary scenario, these regulated processes could allow clonal selection in the parasite population, guaranteeing the propagation of identical genetic information even in adverse environmental conditions. However, differences in the cell death mechanisms observed between metazoans and protozoans must be considered [78, 118]. In the following sections, we will discuss the role of different death styles described in pathogenic trypanosomatids.

3.1. Apoptosis-like

In trypanosomatids, the first PCD report was published in 1995 by Ameisen and coworkers describing apoptotic characteristics (DNA fragmentation and cytoplasmic and nuclear morphological alterations) in *T. cruzi* epimastigotes during differentiation to trypomastigotes [119]. In the last two decades, a variety of stimuli were reported to induce the appearance of the apoptotic phenotype in this parasite, including exposure to fresh human serum (FHS), heat shock and drugs [76, 119-128]. Curiously, the apoptosis-like phenotype was also associated with the regulation of the *T. cruzi* life cycle [129]. These cell death phenotypes in pathogenic trypanosomatids have been characterized by the use of classical apoptotic markers (see item 2.1) [76, 79, 118, 123, 130-133]. Among the apoptotic hallmarks identified, we found (a) loss of $\Delta\Psi m$, (b) cytochrome c release, (c) PS externalization, and (d) abnormal DNA condensation and fragmentation [76, 119, 129, 130, 134] (Table 3, Figure 1).

In *Leishmania* sp., apoptotic features (nuclear condensation, DNA fragmentation, cell shrink-age, loss of $\Delta\Psi m$, and release of cytochrome c) were also observed in stress conditions induced by heat, starvation, oxidative agents and drugs [118, 134, 136-139, 135]. *L. donovani*, *L. major* and *L. mexicana* stationary phase promastigotes and axenic amastigotes exhibited DNA fragmentation with a laddering electrophoretic profile, suggesting oligonucleosomal cleavage. These data were corroborated by the description of a non-canonical, Ca^{2+}- and Mg^{2+}-inde-pendent 45-59 kDa endonuclease [76, 136, 140].

As in other pathogenic trypanosomatids, apoptotic features were also identified in *T. brucei* under non-physiological conditions, such as incubation with drugs, cytokines or ROS [129, 133, 141-143]. Interestingly, the gene for prohibitin and the receptor for activated protein kinase C have been correlated with the apoptotic process, suggesting convergence between these pathways in protozoa and mammals (Table 2) [129]. Despite several reports about caspase-like activity in trypanosomatids [75, 134, 136, 144], the exact role of these proteases in protozoa is not clear. Metacaspases are structurally similar to mammalian orthologues, but their catalytic activity on caspase substrates is quite controversial [145-147]. Despite their presence in *T. cruzi*, *T. brucei* and *Leishmania* sp., only *L. major* metacaspase shows *in vitro* self-proteolytic activity (Table 2) [146]. In fact, the participation of metacaspases cleaving vital substrates in the cell death cascade has not yet been described [148, 149]. Surprisingly, experimental

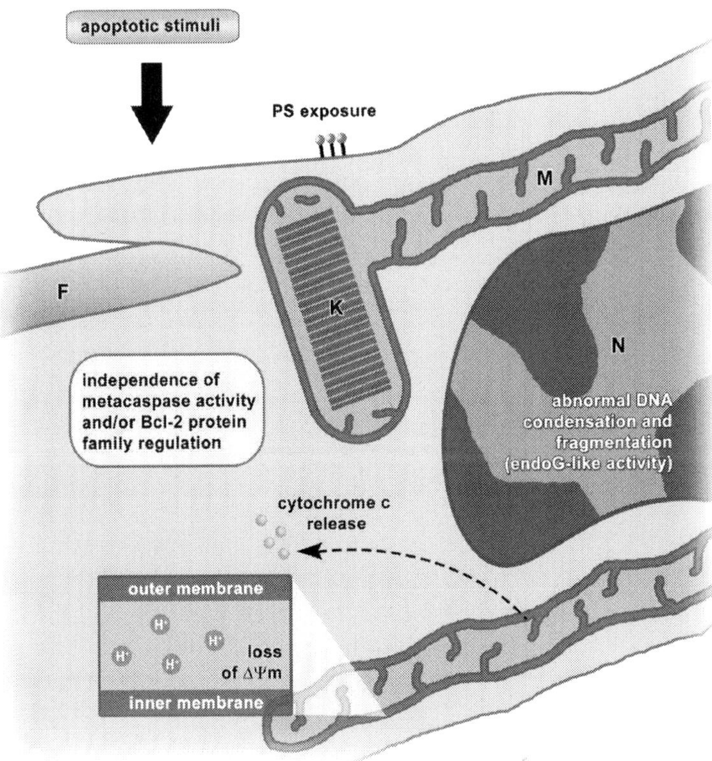

Figure 1. Schematic representation of apoptosis-like PCD in pathogenic trypanosomatids. Apoptotic stimuli induce loss of ΔΨm, release of mitochondrial cytochrome c to the cytosol, PS externalization and DNA fragmentation by EndoG activity. Apoptotic regulators from the Bcl-2 family were not found until now, and the role of metacaspases is controversial, suggesting that apoptosis-like PCD in trypanosomatids is a caspase-like- and Bcl-2-independent pathway. N: nucleus; M: mitochondrion; K: kinetoplast; F: flagellum.

evidence pointed to the involvement of these proteases in cell cycle control and metacyclogenesis, not in death [145, 150-153].

In unicellular organisms, the mitochondrion is a central organelle in cell death pathways, leading to ROS production [125]. In *T. brucei* procyclic forms, mitochondrial Ca^{2+} influx misbalance culminates in ROS generation [154]. Additionally, prostaglandin D2-induced ROS production in both the bloodstream and procyclic forms led to the labelling of different apoptotic markers, with the death phenotype reverted by oxidative scavengers, such as N-acetyl cysteine [130, 155, 156]. In *L. donovani*, hydrogen peroxide induced classical apoptotic features (DNA fragmentation, loss of ΔΨm and caspase-like activity). This phenotype was partially reverted by caspase inhibitors [134, 137]. Oxidative stress plays a crucial role not only in apoptosis-like PCD but also in autophagy and necrosis, as we will discuss later [78, 157].

Molecule	Organism	References
Prohibitin RACK	*T. brucei*	[129]
Elongation factor 1 $\propto$	*T. cruzi*	[161]
Metacaspases 1	*L. donovani* *T. brucei*	[147, 162, 163]
Metacaspases 2	*L. donovani* *T. brucei*	[145, 147, 162]
Metacaspases 3	*T. cruzi* *T. brucei*	[145, 150, 153, 162]
Metacaspases 4	*T. brucei*	[162, 164]
Metacaspases 5	*T. cruzi* *T. brucei* *L. major*	[145, 150, 162, 165]
Metacaspase Z-DEVD-FMK -sensitive	*T. cruzi* *L. donovani*	[124, 134, 136]
Endonuclease G	*L. major* *T. brucei* *L. infantum* *L. donovani*	[132, 158, 166]
LdFEN-1 LdTatD-like nuclease	*L. donovani*	[158]

Table 2. Apoptotic molecules described in pathogenic trypanosomatids

The participation of EndoG-like in mitochondrial-mediated cell death has been reported, but the process is metacaspase-independent (Table 2) [132, 158, 159]. *L. infantum* submitted to heat stress also presents an apoptotic pattern, but without caspase-like activity, which was partially reversed by the expression of the anti-apoptotic mammalian gene Bcl-XL [160]. On the other hand, the overexpression of mammalian anti-apoptotic Bcl-2 in *T. brucei* caused no reversion of the mitochondrial damage induced by ROS [154]. However, members of the Bcl-2 protein family have not been described in trypanosomatids [129]. More studies regarding the regulation steps of apoptosis-like processes in trypanosomatids need to be performed.

3.2. Autophagy

Almost forty years ago, the first morphological autophagic evidence was described in trypanosomatids by electron microscopy of *T. brucei* [170]. In the last four decades, many studies have described recurrent autophagosome formation (initially named autophagic vacuoles), multivesicular bodies as well as myelin-like structures in pathogenic trypanosomatids treated with different classes of drugs (Figure 2) [169, 168, 171-178]. Such autophagosomes showed distinct levels of degradation depending on the degree of cellular structure damage inside the organelle. Myelin-like structures are one of the most frequent ultrastructural alterations detected in drug-treated parasites and are suggestive of the cellular recycling of damaged structures. Currently, it is postulated that myelin-like structures are phagophores (or pre-autophagosomal structures, PAS), an early step in the formation of doubled-membrane autophagosomes (Table 3). In *T. cruzi*, ER profiles were reported as the main origin of phago-phores (Figure 2). These profiles usually surround a pre-lysosomal compartment, named the reservosome, suggesting the participation of this organelle in autophagolysosome formation in epimastigote forms [82, 178].

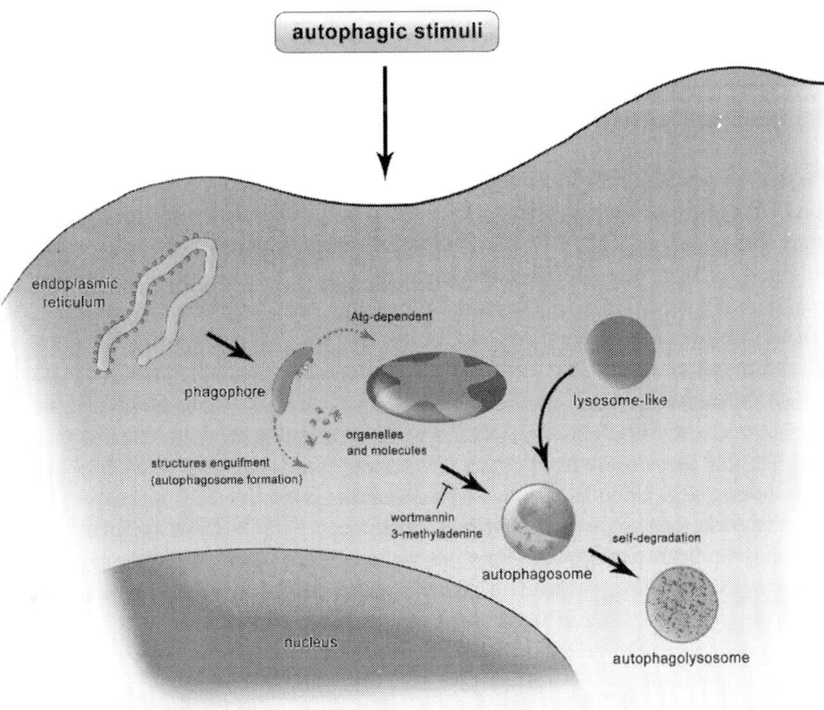

Figure 2. Schematic representation of autophagy in pathogenic trypanosomatids. Autophagic stimuli induce the formation of phagophores from ER profiles. The phagophore engulfs organelles and molecules, generating autophagosomes. Targeting and engulfment are Atg-dependent processes. These autophagosomes fused with lysosomes generate autophagolysosomes. Continuous autophagic stimuli lead to autophagic cell death, which is inhibited by the pre-treatment of the parasite with autophagic inhibitors (wortmannin or 3-methyladenine).

The existence of cross-talk among different cell death pathways, especially autophagy and apoptosis, has been proposed (Figure 3) [93, 234]. In unicellular parasites, different cell death types have been described to be induced by physical and/or chemical stress conditions (drugs, heat shock, and nutritional deprivation, among others), resulting in a non-classical cell death phenotype. The total absence of commercial typical PCD markers, such as antibodies and enzyme activity kits, for protozoa and of key autophagic and apoptotic-like molecules reinforce the hypothesis of an interplay of distinct death mechanisms, suggesting their convergence, leading to necrosis. Likewise, the possibility of the occurrence of other PCD forms cannot be excluded [74, 78, 168, 178]. A better molecular characterization of cell death in pathogenic trypanosomatids is essential for advances in novel alternatives for therapeutic intervention.

Author details

Rubem Figueiredo Sadok Menna-Barreto* and Solange Lisboa de Castro

*Address all correspondence to: rubemsadok@gmail.com

Laboratory of Cell Biology, Oswaldo Cruz Institute, Oswaldo Cruz Foundation, Manguinhos, Rio de Janeiro, Brazil

References

[1] WHO. Neglected tropical diseases, http://www.who.int/neglected_diseases/en/ accessed in 12/01/2015, 2015a.

[2] Simarro PP, Diarra A, Ruiz Postigo JA, Franco JR, Jannin JG. The Human African Trypanosomiasis Control and Surveillance Programme of the World Health Organization 2000–2009: The Way Forward. PLoS Negl Trop Dis 2011;5(2):e1007.

[3] WHO, http://www.who.int/trypanosomiasis_african/en/index.html, accessed in 12/01/2015, 2015b.

[4] Kennedy PG. Clinical features, diagnosis, and treatment of human African trypanosomiasis (sleeping sickness). Lancet Neurol 2013;12(2):186-94.

[5] Franco JR, Simarro PP, Diarra A, Jannin JG. Epidemiology of human African trypanosomiasis. Clin Epidemiol 2014;6:257-75.

[6] Horn D. Antigenic variation in African trypanosomes. Mol Biochem Parasitol 2014;195(2):123-9.

[7] Vickerman K. Developmental cycles and biology of pathogenic trypanosomes. Br Med Bull 1985;41(2):105-14.

[8] Matthews KR. The developmental cell biology of *Trypanosoma brucei*. J Cell Sci 2005;118(Pt 2):283-90.

[9] Steverding D. The development of drugs for treatment of sleeping sickness: a historical review. Parasit Vectors 2010;3(1):15.

[10] Barrett MP, Vincent IM, Burchmore RJ, Kazibwe AJ, Matovu E. Drug resistance in human African trypanosomiasis. Future Microbiol 2011;6(9):1037-47.

[11] WHO. Working to overcome the global impact of neglected tropical diseases: first WHO report on neglected tropical diseases. Geneva: WHO; 2010. Available from: http://www.who.int/neglected_diseases/2010report/en/

[12] Steindel M, Kramer Pacheco L, Scholl D, Soares M, de Moraes MH, et al. Characterization of *Trypanosoma cruzi* isolated from humans, vectors, and animal reservoirs following an outbreak of acute human Chagas disease in Santa Catarina State, Brazil. Diagn Microbiol Infect Dis 2008;60(1):25-32.

[13] Dias JC, Amato Neto V, Luna EJ. Alternative transmission mechanisms of *Trypanosoma cruzi* in Brazil and proposals for their prevention. Rev Soc Bras Med Trop 2011;44:375-9.

[14] Coura JR, Dias JCP. Epidemiology, control and surveillance of Chagas disease - 100 years after its discovery. Mem Inst Oswaldo Cruz 2009;104 (Suppl 1):31-40.

[15] Gascon J, Bern C, Pinazo MJ. Chagas disease in Spain, the United States and other non-endemic countries. Acta Trop 2010;115:22-7.

[16] Schmunis GA, Yadon ZE. Chagas disease: a Latin American health problem becoming a world health problem. Acta Trop 2010;115(1-2):14-21.

[17] Rassi A Jr, Rassi A, Marcondes de Rezende J. American trypanosomiasis (Chagas disease). Infect Dis Clin North Am 2012;26(2):275-91.

[18] De Souza W. From the cell biology to the development of new chemotherapeutic approaches against trypanosomatids: dreams and reality. Kinetoplastid Biol Dis 2002;1(1):3.

[19] McKerrow JH, Doyle PS, Engel JC, Podust LM, Robertson SA, Ferreira R, et al. Two approaches to discovering and developing new drugs for Chagas disease. Mem Inst Oswaldo Cruz 2009;104 Suppl 1:263-9.

[20] Soeiro MN, De Castro SL. Screening of potential anti-*Trypanosoma cruzi* candidates: *in vitro* and *in vivo* studies. Open Med Chem J 2011;5:21-30.

[21] Moraes CB, Giardini MA, Kim H, Franco CH, Araujo-Junior AM, Schenkman S, et al. Nitroheterocyclic compounds are more efficacious than CYP51 inhibitors against *Trypanosoma cruzi*: implications for Chagas disease drug discovery and development. Sci Rep 2014;4:4703.

[22] Urbina JA. Recent clinical trials for the etiological treatment of chronic Chagas disease: advances, challenges and perspectives. J Eukaryot Microbiol 2015;62(1):149-56.

[23] Chatelain E. Chagas disease drug discovery: toward a new era. J Biomol Screen 2015;20(1):22-35.

[24] Molina I, Gómez-Prat J, Salvador F, Treviño B, Sulleiro E, Serre Net al. Randomized trial of posaconazole and benznidazole for chronic Chagas' disease. N Engl J Med 2014;370(20):1899-908.

[25] WHO. Global report for research on infectious diseases of poverty, available at http://whqlibdoc.who.int/publications/2012/9789241564489_eng.pdf, 2012.

[26] Murray HW, Berman JD, Davies CR, Saravia NG. Advances in leishmaniasis. Lancet 2005;366:1561-77.

[27] Desjeux P. Leishmaniasis. Nat Rev Microbiol 2004;2(9):692.

[28] Zijlstra EE, Musa AM, Khalil EA, el-Hassan IM, el-Hassan AM. Post-kala-azar dermal leishmaniasis. Lancet Infect Dis 2003;3(2):87-98.

[29] Barrett MP, Croft SL. Management of trypanosomiasis and leishmaniasis. Br Med Bull 2012;104:175-96.

[30] Bates PA, Rogers ME. New insights into the developmental biology and transmission mechanisms of Leishmania. Curr Mol Med 2004;4(6):601-9.

[31] Chang KP, Dwyer DM. Multiplication of a human parasite (Leishmania donovani) in phagolysosomes of hamster macrophages *in vitro*. Science 1976;193:678-80.

[32] Croft SL, Olliaro P. Leishmaniasis chemotherapy - challenges and opportunities. Clin Microbiol Infect 2011;17(10):1478-83.

[33] Sundar S, Chakravarty J. Leishmaniasis: an update of current pharmacotherapy. Expert Opin Pharmacother 2013;14(1):53-63.

[34] Sundar S, Chakravarty J. Investigational drugs for visceral leishmaniasis. Expert Opin Investig Drugs 2015;24(1):43-59.

[35] Yardley V, Croft SL. Activity of liposomal amphotericin B against experimental cutaneous leishmaniasis. Antimicrob Agents Chemother 1997;41(4):752-6.

[36] Sundar S, Singh A, Rai M, Prajapati VK, Singh AK, Ostyn B, et al. Efficacy of miltefosine in the treatment of visceral leishmaniasis in India after a decade of use. Clin Infect Dis 2012;55(4):543-50.

[37] Prajapati VK, Mehrotra S, Gautam S, Rai M, Sundar S. *In vitro* antileishmanial drug susceptibility of clinical isolates from patients with Indian visceral leishmaniasis - status of newly introduced drugs. Am J Trop Med Hyg 2012;87(4):655-7.

[38] Olliaro PL. Drug combinations for visceral leishmaniasis. Curr Opin Infect Dis 2010;23(6):595-602.

[39] Crile G, Dolley DH. On the effect of complete anemia of the central nervous system in dogs resuscitated after relative death. J Exp Med 1908;10(6):782-810.

[40] Smirlis D, Duszenko M, Ruiz AJ, Scoulica E, Bastien P, Fasel N, Soteriadou K. Targeting essential pathways in trypanosomatids gives insights into protozoan mechanisms of cell death. Parasit Vectors 2010;3:107.

[41] Lockshin RA, Zakeri Z. Programmed cell death and apoptosis: origins of the theory. Nat Rev Mol Cell Biol 2001;2(7):545-50.

[42] Lockshin RA, Williams CM. Programmed cell death. I. Cytology of degeneration in the intersegmental muscles of the pernyi silkmoth. J Insect Physiol 1965;11:123-33.

[43] Kroemer G, Galluzzi L, Vandenabeele P, Abrams J, Alnemri ES, Baehrecke EH, et al., Nomenclature Committee on Cell Death 2009: Classification of cell death: recommendations of the Nomenclature Committee on Cell Death 2009. Cell Death Differ 2009;16:3-11.

[44] Kerr JF. History of the events leading to the formulation of the apoptosis concept. Toxicology 2002;181/182:471-4.

[45] Green DR. Introduction: apoptosis in the development and function of the immune system. Semin Immunol 2003;15(3):121-3.

[46] Danial NN, Korsmeyer SJ. Cell death: critical control points. Cell 2004;116(2):205-19.

[47] Evan IG. Better dead than red. Ther Immunol 1994;1(6):343-8.

[48] Vaux DL, Haecker G, Strasser A. An evolutionary perspective on apoptosis. Cell 1994;76(5):777-9.

[49] Williams GT. Programmed cell death: a fundamental protective response to pathogens. Trends Microbiol 1994;2(12):463-4.

[50] Shen Y, Shenk TE. Viruses and apoptosis. Curr Opin Genet Dev 1995;5:105-11.

[51] Roulston A, Marcellus RC, Branton PE. Viruses and apoptosis. Annu Rev Microbiol 1999;53:577-628.

[52] Weinrauch Y, Zychlinsky A. The induction of apoptosis by bacterial pathogens. Annu Rev Microbiol 1999;53:155-87.

[53] DosReis GA, Barcinski MA. Apoptosis and parasitism: from the parasite to the host immune response. Adv Parasitol 2001;49:133-61.

[54] Williams GT. Programmed cell death: apoptosis and oncogenesis. Cell 1991;65:1097-98.

[55] Raff MC. Social controls on cell survival and cell death. Nature 1992;356(6368): 397-400.

[56] Reed JC. Bcl-2 and the regulation of programmed cell death. J Cell Biol 1994;124(1-2): 1-6.

[57] Rich T, Allen RL, Wyllie AH. Defying death after DNA damage. Nature 2000;407(6805):777-83.

[58] Voll RE, Herrmann M, Roth EA, Stach C, Kalden JR, Girkontaite I. Immunosuppressive effects of apoptotic cells. Nature 1997;390:350-1.

[59] Wyllie AH. Glucocorticoid-induced thymocyte apoptosis is associated with endoGenous endonuclease activation. Nature 1980;284:555-556.

[60] Steller H. Mechanisms and genes of cellular suicide. Science 1995;267:1445-1449.

[61] Kischkel FC, Hellbardt S, Behrmann I, Germer M, Pawlita M, Krammer PH, Peter ME. Cytotoxicity-dependent APO-1 (Fas/CD95)-associated proteins form a death-inducing signaling complex (DISC) with the receptor. EMBO J 1995;14(22):5579-88.

[62] Muzio M, Chinnaiyan AM, Kischkel FC, O'Rourke K, Shevchenko A, Ni J, et al. FLICE, a novel FADD-homologous ICE/CED-3-like protease, is recruited to the CD95 (Fas/APO-1) death-inducing signaling complex. Cell 1996;85(6):817-27.

[63] Peter ME, Krammer PH. The CD95(APO-1/Fas) DISC and beyond. Cell Death Differ 2003;10(1):26-35.

[64] Jänicke RU, Sprengart ML, Wati MR, Porter AG. Caspase-3 is required for DNA fragmentation and morphological changes associated with apoptosis. J Biol Chem 1998;273(16):9357-60.

[65] Opferman JT, Korsmeyer SJ. Apoptosis in the development and maintenance of the immune system. Nat Immunol 2003;4(5):410-5.

[66] Newmeyer DD, Farschon DM, Reed JC: Cell-free apoptosis in Xenopus egg extracts: inhibition by Bcl-2 and requirement for an organelle fraction enriched in mitochondria. Cell 1994;79:353-64.

[67] Kluck RM, Bossy-Wetzel E, Green DR, Newmeyer DD. The release of cytochrome c from mitochondria: a primary site for Bcl-2 regulation of apoptosis. Science 1997;275:1132-6.

[68] Susin SA, Lorenzo HK, Zamzami N, Marzo I, Snow BE, Brothers GM, et al. Molecular characterization of mitochondrial apoptosis-inducing factor. Nature 1999;397:441-6.

[69] Kroemer G, Reed JC: Mitochondrial control of cell death. Nat Med 2000;6:513-9.

[70] Parrish J, Li L, Klotz K, Ledwich D, Wang X, Xue D. Mitochondrial endonuclease G is important for apoptosis in *Caenorhabditis elegans*. Nature 2001;412:90-94.

[71] Nakagawa T, Zhu H, Morishima N, Li E, Xu J, Yankner BA, Yuan J. Caspase-12 mediates endoplasmic-reticulum-specific apoptosis and cytotoxicity by amyloid-β. Nature 2000;403(6765):98-103.

[72] Fischer H, Koenig U, Eckhart L, Tschachler E. Human caspase 12 has acquired deleterious mutations. Biochem Biophys Res Commun 2002;293:722-6.

[73] Momoi T. Caspases involved in ER stress-mediated cell death. J Chem Neuroanat 2004;28(1-2):101-5

[74] Guimaraes CA, Linden R. Programmed cell deaths. Apoptosis and alternative death-styles. Eur J Biochem 2004;271:1638-50.

[75] Arnoult D, Tatischeff I, Estaquier J, Girard M, Sureau F, Tissier JP, et al. On the evolutionary conservation of the cell death pathway: mitochondrial release of an apoptosis-inducing factor during *Dictyostelium discoideum* cell death. Mol Biol Cell 2001;12(10):3016-30.

[76] Debrabant A, Lee N, Bertholet S, Duncan R and Nakhasi HL. Programmed cell death in trypanosomatids and other unicellular organisms International J Parasitol 2003, 33:257-267.

[77] Carmen JC, Sinai AP. Suicide prevention: disruption of apoptotic pathways by protozoan parasites. Mol Microbiol 2007;64(4):904-16

[78] Bruchhaus I, Roeder T, Rennenberg A, Heussler VT: Protozoan parasites: programmed cell death as a mechanism of parasitism. Trends Parasitol 2007, 23:376-383.

[79] Kaczanowski S, Sajid M, Reece SE. Evolution of apoptosis-like programmed cell death in unicellular protozoan parasites. Parasit Vectors 2011;4:44.

[80] Deponte M. Programmed cell death in protists. Biochim Biophys Acta 2008;1783(7): 1396-405.

[81] De Duve C. The lysosome. Sci Am 1963;208:64-72.

[82] Duszenko M, Ginger ML, Brennand A, Gualdrón-López M, Colombo MI, Coombs GH, et al. Autophagy in protists. Autophagy 2011;7(2):127-58.

[83] Reggiori F, Klionsky DJ. Autophagosomes: biogenesis from scratch? Curr Opin Cell Biol 2005;17(4):415-22.

[84] Yoshimori A, Takasawa R, Tanuma S. A novel method for evaluation and screening of caspase inhibitory peptides by the amino acid positional fitness score. BMC Pharmacol 2004;4:7.

[85] Yorimitsu T, Klionsky DJ. Autophagy: molecular machinery for self-eating. Cell Death Differ 2005;12 Suppl 2:1542-52.

[86] He C, Klionsky DJ. Regulation mechanisms and signaling pathways of autophagy. Annu Rev Genet 2009;43:67-93.

[87] Kirkegaard K, Taylor MP, Jackson WT. Cellular autophagy: surrender, avoidance and subversion by microorganisms. Nat Rev Microbiol 2004;2(4):301-14.

[88] Mizushima N. The pleiotropic role of autophagy: from protein metabolism to bactericide. Cell Death Differ 2005;12 Suppl 2:1535-41.

[89] Ogawa M, Sasakawa C. Shigella and autophagy. Autophagy 2006;2(3):171-4.

[90] Swanson MS. Autophagy: eating for good health. J Immunol 2006;177(8):4945-51.

[91] Colombo MI. Autophagy: a pathogen driven process. IUBMB Life 2007;59(4-5): 238-42.

[92] Alvarez VE, Kosec G, Sant Anna C, Turk V, Cazzulo JJ, Turk B. Blocking autophagy to prevent parasite differentiation: a possible new strategy for fighting parasitic infections? Autophagy 2008;4(3):361-3.

[93] Levine B, Yuan J. Autophagy in cell death: an innocent convict? J Clin Invest 2005;115(10):2679-88.

[94] Maiuri MC, Zalckvar E, Kimchi A, Kroemer G: Self-eating and self-killing: crosstalk between autophagy and apoptosis. Nature Reviews 2007;8:741-752.

[95] Klionsky DJ, Cregg JM, Dunn WA Jr, Emr SD, Sakai Y, Sandoval IV, et al. A unified nomenclature for yeast autophagy-related genes. Dev Cell 2003;5:539-45.

[96] Klionsky DJ, Ohsumi Y. Vacuolar import of proteins and organelles from the cytoplasm. Annu Rev Cell Dev Biol 1999;15:1-32.

[97] Shintani T, Klionsky DJ. Autophagy in health and disease: a double-edged sword. Science 2004;306(5698):990-5.

[98] Klionsky DJ. Autophagy. Curr Biol 2005;15(8):R282-3.

[99] Brennand A, Gualdrón-López M, Coppens I, Rigden DJ, Ginger ML, Michels PA. Autophagy in parasitic protists: unique features and drug targets. Mol Biochem Parasitol 2011;177(2):83-99.

[100] Bejarano E, Cuervo AM. Chaperone-mediated autophagy. Proc Am Thorac Soc 2010;7(1):29-39.

[101] Klionsky DJ, Abdalla FC, Abeliovich H, Abraham RT, Acevedo-Arozena A, Adeli K, et al., Guidelines for the use and interpretation of assays for monitoring autophagy. Autophagy 2012;8(4):445-544.

[102] Proskuryakov SY, Konoplyannikov AG, Gabai VL. Necrosis: a specific form of programmed cell death?. Exp Cell Res 2003;283(1):1-16.

[103] Zong WX, Thompson CB. Necrotic death as a cell fate. Genes Dev 2006;20(1):1-15.

[104] Holler N, Zaru R, Micheau O, Thome M, Attinger A, Valitutti S, et al. Fas triggers an alternative, caspase-8-independent cell death pathway using the kinase RIP as effector molecule. Nat Immunol 2000;1(6):489-95.

[105] Christofferson DE, Yuan J. Necroptosis as an alternative form of programmed cell death. Curr Opin Cell Biol 2010;22(2):263-8.

[106] Vanden Berghe T, Linkermann A, Jouan-Lanhouet S, Walczak H, Vandenabeele P. Regulated necrosis: the expanding network of non-apoptotic cell death pathways. Nat Rev Mol Cell Biol 2014;15(2):135-47.

[107] Li P, Allen H, Banerjee S, Franklin S, Herzog L, Johnston C et al. Mice deficient in IL-1 β-converting enzyme are defective in production of mature IL-1 beta and resistant to endotoxic shock. Cell 1995;80:401-11.

[108] Fink SL, Cookson BT. Apoptosis, pyroptosis, and necrosis: mechanistic description of dead and dying eukaryotic cells. Infect Immun 2005;73(4):1907-16.

[109] Brinkmann V, Reichard U, Goosmann C, Fauler B, Uhlemann Y, Weiss DS, Weinrauch Y, Zychlinsky A. Neutrophil extracellular traps kill bacteria. Science 2004;303(5663):1532-5.

[110] Yipp BG, Kubes P. NETosis: how vital is it? Blood 2013;122(16):2784-94.

[111] von Köckritz-Blickwede M, Goldmann O, Thulin P, Heinemann K, Norrby-Teglund A, Rohde M, Medina E. Phagocytosis-independent antimicrobial activity of mast cells by means of extracellular trap formation. Blood 2008;111(6):3070-80.

[112] Aulik NA, Hellenbrand KM, Czuprynski CJ. Mannheimia haemolytica and its leukotoxin cause macrophage extracellular trap formation by bovine macrophages. Infect Immun 2012; 80(5):1923-33.

[113] Schorn C, Janko C, Latzko M, Chaurio R, Schett G, Herrmann M. Monosodium urate crystals induce extracellular DNA traps in neutrophils, eosinophils, and basophils but not in mononuclear cells. Front Immunol 2012;3:277.

[114] Ueki S, Melo RC, Ghiran I, Spencer LA, Dvorak AM, Weller PF. Eosinophil extracellular DNA trap cell death mediates lytic release of free secretion-competent eosinophil granules in humans. Blood 2013;121(11):2074-83.

[115] Dixon SJ, Lemberg KM, Lamprecht MR, Skouta R, Zaitsev EM, Gleason CE, et al. Ferroptosis: an iron-dependent form of nonapoptotic cell death. Cell 2012;149(5):1060-72.

[116] Jamison JM, Gilloteaux J, Taper HS, Calderon PB, Summers JL. Autoschizis: a novel cell death. Biochem Pharmacol 2002;63(10):1773-83.

[117] Gilloteaux J, Jamison JM, Arnold D, Neal DR, Summers JL. Morphology and DNA degeneration during autoschizic cell death in bladder carcinoma T24 cells induced by ascorbate and menadione treatment. Anat Rec A Discov Mol Cell Evol Biol 2006;288(1):58-83.

[118] Nguewa PA, Fuertes MA, Valladares B, Alonso C, Pérez JM. Programmed cell death in trypanosomatids: a way to maximize their biological fitness? Trends Parasitol 2004;20(8):375-80.

[119] Ameisen JC, Idziorek T, Billaut-Mulot O, Loyens M, Tissier JP, Potentier A, Ouaissi A. Apoptosis in a unicellular eukaryote (*Trypanosoma cruzi*): implications for the evolutionary origin and role of programmed cell death in the control of cell proliferation, differentiation and survival. Cell Death Differ 1995;2(4):285-300.

[120] Barcinski MA, DosReis GA. Apoptosis in parasites and parasite-induced apoptosis in the host immune system: a new approach to parasitic diseases. Braz J Med Biol Res 1999;32(4):395-401.

[121] De Souza EM, Araújo-Jorge TC, Bailly C, Lansiaux A, Batista MM, Oliveira GM, Soeiro MN. Host and parasite apoptosis following *Trypanosoma cruzi* infection in *in vitro* and *in vivo* models. Cell Tissue Res. 2003;314(2):223-35.

[122] De Souza EM, Menna-Barreto R, Araújo-Jorge TC, Kumar A, Hu Q, Boykin DW, Soeiro MN. Antiparasitic activity of aromatic diamidines is related to apoptosis-like death in *Trypanosoma cruzi*. Parasitology 2006;133(Pt 1):75-9.

[123] Ouaissi A. Apoptosis-like death in trypanosomatids: search for putative pathways and genes involved. Kinetoplastid Biol Dis 2003;2(1):5.

[124] Piacenza L, Peluffo G, Radi R. L-arginine-dependent suppression of apoptosis in *Trypanosoma cruzi*: contribution of the nitric oxide and polyamine pathways. Proc Natl Acad Sci USA 2001;98(13):7301-6

[125] Piacenza L, Irigoín F, Alvarez MN, Peluffo G, Taylor MC, Kelly JM, Wilkinson SR, Radi R. Mitochondrial superoxide radicals mediate programmed cell death in *Trypanosoma cruzi*: cytoprotective action of mitochondrial iron superoxide dismutase overexpression. Biochem J 2007;403(2):323-34.

[126] Jimenez V, Paredes R, Sosa MA, Galanti N. Natural programmed cell death in *Trypanosoma cruzi* epimastigotes maintained in axenic cultures. J Cell Biochem 2008;105(3): 688-98.

[127] Benitez D, Pezaroglo H, Martínez V, Casanova G, Cabrera G, Galanti N, González M, Cerecetto H. Study of *Trypanosoma cruzi* epimastigote cell death by NMR-visible mobile lipid analysis. Parasitology 2012;139(4):506-15.

[128] Veiga-Santos P, Reignault LC, Huber K, Bracher F, De Souza W, De Carvalho TM. Inhibition of NAD+-dependent histone deacetylases (sirtuins) causes growth arrest and activates both apoptosis and autophagy in the pathogenic protozoan *Trypanosoma cruzi*. Parasitology 2014;141(6):814-25.

[129] Welburn SC, Murphy NB. Prohibitin and RACK homologues are up-regulated in trypanosomes induced to undergo apoptosis and in naturally occurring terminally differentiated forms. Cell Death Differ 1998;5(7):615-22.

[130] Duszenko M, Figarella K, Macleod ET, Welburn SC. Death of a trypanosome: a self-ish altruism. Trends Parasitol 2006;22(11):536-42.

[131] Lüder CG, Campos-Salinas J, Gonzalez-Rey E, van Zandbergen G. Impact of protozo-an cell death on parasite-host interactions and pathogenesis. Parasit Vectors 2010;3:116.

[132] Gannavaram S, Vedvyas C, Debrabant A. Conservation of the pro-apoptotic nuclease activity of endonuclease G in unicellular trypanosomatid parasites. J Cell Sci 2008;121(Pt 1):99-109.

[133] Gannavaram S, Debrabant A. Programmed cell death in *Leishmania*: biochemical evi-dence and role in parasite infectivity. Front Cell Infect Microbiol 2012;2:95

[134] Das M, Mukherjee SB, Shaha C. Hydrogen peroxide induces apoptosis-like death in *Leishmania donovani* promastigotes. J Cell Sci 2001;114(Pt 13):2461-9.

[135] Moreira ME, Del Portillo HA, Milder RV, Balanco JM, Barcinski MA. Heat shock in-duction of apoptosis in promastigotes of the unicellular organism *Leishmania (Leish-mania) amazonensis*. J Cell Physiol 1996;167(2):305-13.

[136] Lee N, Bertholet S, Debrabant A, Muller J, Duncan R, Nakhasi HL. Programmed cell death in the unicellular protozoan parasite *Leishmania*. Cell Death Differ 2002;9(1): 53-64.

[137] Mukherjee SB, Das M, Sudhandiran G, Shaha C. Increase in cytosolic Ca2+ levels through the activation of non-selective cation channels induced by oxidative stress causes mitochondrial depolarization leading to apoptosis-like death in *Leishmania do-novani* promastigotes. J Biol Chem 2002;277(27):24717-27.

[138] Marquis JF, Drolet M, Olivier M. Consequence of Hoechst 33342-mediated Leishma-nia DNA topoisomerase-I inhibition on parasite replication. Parasitology 2003;126(Pt 1):21-30

[139] Sen N, Das BB, Ganguly A, Mukherjee T, Tripathi G, Bandyopadhyay S, et al. Camp-tothecin induced mitochondrial dysfunction leading to programmed cell death in unicellular hemoflagellate *Leishmania donovani*. Cell Death Differ 2004;11(8):924-36.

[140] Zangger H, Mottram JC, Fasel N. Cell death in *Leishmania* induced by stress and dif-ferentiation: programmed cell death or necrosis? Cell Death Differ 2002;9(10):1126-39.

[141] Welburn SC, Macleod E, Figarella K, Duzensko M. Programmed cell death in African trypanosomes. Parasitology 2006;132Suppl:S7-S18.

[142] Murphy NB, Welburn SC. Programmed cell death in procyclic *Trypanosoma brucei rhodesiense* is associated with differential expression of mRNAs. Cell Death Differ 1997;4(5):365-70.

[143] Rosenkranz V, Wink M. Alkaloids induce programmed cell death in bloodstream forms of trypanosomes (*Trypanosoma brucei brucei*). Molecules 2008;13(10):2462-73.

[144] Sen N, Banerjee B, Das BB, Ganguly A, Sen T, Pramanik S, et al. Apoptosis is induced in leishmanial cells by a novel protein kinase inhibitor withaferin A and is facilitated by apoptotic topoisomerase I-DNA complex. Cell Death Differ 2007;14(2):358-67.

[145] Helms MJ, Ambit A, Appleton P, Tetley L, Coombs GH, Mottram JC. Bloodstream form *Trypanosoma brucei* depend upon multiple metacaspases associated with RAB11-positive endosomes. J Cell Sci 2006;119(Pt 6):1105-17.

[146] González IJ, Desponds C, Schaff C, Mottram JC, Fasel N. *Leishmania major* metacaspase can replace yeast metacaspase in programmed cell death and has arginine-specific cysteine peptidase activity. Int J Parasitol 2007;37(2):161-72.

[147] Lee N, Gannavaram S, Selvapandiyan A, Debrabant A. Characterization of metacaspases with trypsin-like activity and their putative role in programmed cell death in the protozoan parasite *Leishmania*. Eukaryot Cell 2007;6(10):1745-57.

[148] Meslin B, Zalila H, Fasel N, Picot S, Bienvenu AL. Are protozoan metacaspases potential parasite killers? Parasit Vectors 2011;4:26.

[149] Proto WR, Coombs GH, Mottram JC. Cell death in parasitic protozoa: regulated or incidental? Nat Rev Microbiol 2013;11(1):58-66.

[150] Kosec G, Alvarez VE, Agüero F, Sánchez D, Dolinar M, Turk B, et al. Metacaspases of *Trypanosoma cruzi*: possible candidates for programmed cell death mediators. Mol Biochem Parasitol 2006;145(1):18-28.

[151] Moss CX, Westrop GD, Juliano L, Coombs GH, Mottram JC. Metacaspase 2 of *Trypanosoma brucei* is a calcium-dependent cysteine peptidase active without processing. FEBS Lett 2007;581(29):5635-9.

[152] González IJ. Metacaspases and their role in the life cycle of human protozoan parasites. Biomedica 2009;29(3):485-93.

[153] Laverrière M, Cazzulo JJ, Alvarez VE. Antagonic activities of *Trypanosoma cruzi* metacaspases affect the balance between cell proliferation, death and differentiation. Cell Death Differ 2012;19(8):1358-69.

[154] Ridgley EL, Xiong ZH, Ruben L. Reactive oxygen species activate a Ca2+-dependent cell death pathway in the unicellular organism *Trypanosoma brucei brucei*. Biochem J 1999;340(Pt 1):33-40.

[155] Figarella K, Rawer M, Uzcategui NL, Kubata BK, Lauber K, Madeo F, et al. Prostaglandin D2 induces programmed cell death in *Trypanosoma brucei* bloodstream form. Cell Death Differ 2005;12(4):335-46.

[156] Figarella K, Uzcategui NL, Beck A, Schoenfeld C, Kubata BK, Lang F, Duszenko M. Prostaglandin-induced programmed cell death in *Trypanosoma brucei* involves oxidative stress. Cell Death Differ 2006;13(10):1802-14.

[157] Le Bras M, Clément MV, Pervaiz S, Brenner C. Reactive oxygen species and the mitochondrial signaling pathway of cell death. Histol Histopathol 2005;20(1):205-19.

[158] BoseDasgupta S, Das BB, Sengupta S, Ganguly A, Roy A, Dey S, et al. The caspase-independent algorithm of programmed cell death in *Leishmania* induced by baicalein: the role of LdEndoG, LdFEN-1 and LdTatD as a DNA 'degradesome'. Cell Death Differ 2008;15(10):1629-40.

[159] Alvarez VE, Niemirowicz GT, Cazzulo JJ. The peptidases of *Trypanosoma cruzi*: digestive enzymes, virulence factors, and mediators of autophagy and programmed cell death. Biochim Biophys Acta 2012;1824(1):195-206

[160] Alzate JF, Arias AA, Moreno-Mateos D, Alvarez-Barrientos A, Jiménez-Ruiz A. Mitochondrial superoxide mediates heat-induced apoptotic-like death in *Leishmania infantum*. Mol Biochem Parasitol 2007;152(2):192-202.

[161] Billaut-Mulot O, Fernandez-Gomez R, Loyens M, Ouaissi A. *Trypanosoma cruzi* elongation factor 1-alpha: nuclear localization in parasites undergoing apoptosis. Gene 1996;174(1):19-26.

[162] McLuskey K, Moss CX, Mottram JC. Purification, characterization, and crystallization of Trypanosoma metacaspases. Methods Mol Biol 2014;1133:203-21.

[163] Mottram JC, Helms MJ, Coombs GH, Sajid M. Clan CD cysteine peptidases of parasitic protozoa. Trends Parasitol 2003;19(4):182-7.

[164] Szallies A, Kubata BK, Duszenko M. A metacaspase of *Trypanosoma brucei* causes loss of respiration competence and clonal death in the yeast *Saccharomyces cerevisiae*. FEBS Lett 2002;517(1-3):144-50.

[165] Ambit A, Fasel N, Coombs GH, Mottram JC. An essential role for the Leishmania major metacaspase in cell cycle progression. Cell Death Differ 2008;15(1):113-22.

[166] Rico E, Alzate JF, Arias AA, Moreno D, Clos J, Gago F, Moreno I, Domínguez M, Jiménez-Ruiz A. *Leishmania infantum* expresses a mitochondrial nuclease homologous to EndoG that migrates to the nucleus in response to an apoptotic stimulus. Mol Biochem Parasitol 2009;163(1):28-38.

[167] Herman M, Gillies S, Michels PA, Rigden DJ. Autophagy and related processes in trypanosomatids: insights from genomic and bioinformatic analyses. Autophagy 2006;2(2):107-18.

[168] Menna-Barreto RF, Salomão K, Dantas AP, Santa-Rita RM, Soares MJ, Barbosa HS, De Castro SL. Different cell death pathways induced by drugs in *Trypanosoma cruzi*: an ultrastructural study. Micron. 2009a Feb;40(2):157-68.

[169] McGwire BS, Kulkarni MM. Interactions of antimicrobial peptides with *Leishmania* and trypanosomes and their functional role in host parasitism. Exp Parasitol 2010;126(3):397-405.

[170] Vickerman K, Tetley L. Recent ultrastructural studies on trypanosomes. Ann Soc Belg Med Trop 1977;57(4-5):441-57.

[171] Merkel P, Beck A, Muhammad K, Ali SA, Schönfeld C, Voelter W, Duszenko M. Spermine isolated and identified as the major trypanocidal compound from the snake venom of *Eristocophis macmahoni* causes autophagy in *Trypanosoma brucei*. Toxicon 2007;50(4):457-69.

[172] Rodrigues JC, Seabra SH, De Souza W. 2006. Apoptosis-like death in parasitic protozoa. Braz J Morphol Sci 20056;23(2):87-98.

[173] Santa-Rita RM, Lira R, Barbosa HS, Urbina JA, de Castro SL. Anti-proliferative synergy of lysophospholipid analogues and ketoconazole against *Trypanosoma cruzi* (Kinetoplastida: Trypanosomatidae): cellular and ultrastructural analysis. J Antimicrob Chemother 2005;55:780-4.

[174] Bera A, Singh S, Nagaraj R, Vaidya T. Induction of autophagic cell death in *Leishmania donovani* by antimicrobial peptides. Mol Biochem Parasitol 2003;127(1):23-35.

[175] Delgado M, Anderson P, Garcia-Salcedo JA, Caro M, Gonzalez-Rey E. Neuropeptides kill African trypanosomes by targeting intracellular compartments and inducing autophagic-like cell death. Cell Death Differ 2009;16(3):406-16.

[176] Uzcategui NL, Carmona-Gutierrez D, Denninger V, Schoenfeld C, Lang F, Figarella K, Duszenko M. Antiproliferative effect of dihydroxyacetone on *Trypanosoma brucei* bloodstream forms: cell cycle progression, subcellular alterations, and cell death. Antimicrob Agents Chemother 2007;51:3960-8.

[177] Menna-Barreto RF, Corrêa JR, Pinto AV, Soares MJ, de Castro SL. Mitochondrial disruption and DNA fragmentation in Trypanosoma cruzi induced by naphthoimidazoles synthesized from β-lapachone. Parasitol Res 2007;101(4):895-905.

[178] Menna-Barreto RF, Corrêa JR, Cascabulho CM, Fernandes MC, Pinto AV, Soares MJ, De Castro SL. Naphthoimidazoles promote different death phenotypes in *Trypanosoma cruzi*. Parasitology 2009b;136(5):499-510.

[179] Klionsky DJ. What can we learn from trypanosomes? Autophagy 2006;2(2):63-4.

[180] Kiel JA. Autophagy in unicellular eukaryotes. Philos Trans R Soc Lond B Biol Sci 2010;365:819-30.

[181] Rigden DJ, Herman M, Gillies S, Michels PA. Implications of a genomic search for autophagy-related genes in trypanosomatids. Biochem Soc Trans 2005;33(Pt 5):972-4.

[182] Besteiro S, Williams RA, Morrison LS, Coombs GH, Mottram JC. Endosome sorting and autophagy are essential for differentiation and virulence of *Leishmania major*. J Biol Chem 2006;281(16):11384-96.

[183] Williams RA, Woods KL, Juliano L, Mottram JC, Coombs GH. Characterization of unusual families of ATG8-like proteins and ATG12 in the protozoan parasite *Leishmania major*. Autophagy 2009;5(2):159-72.

[184] Koopmann R, Muhammad K, Perbandt M, Betzel C, Duszenko M. *Trypanosoma brucei* ATG8: structural insights into autophagic-like mechanisms in protozoa. Autophagy 2009;5(8):1085-91.

[185] Li FJ, Shen Q, Wang C, Sun Y, Yuan AY, He CY. A role of autophagy in *Trypanosoma brucei* cell death. Cell Microbiol 2012;14(8):1242-56.

[186] Barquilla A, Crespo JL, Navarro M. Rapamycin inhibits trypanosome cell growth by preventing TOR complex 2 formation. Proc Natl Acad Sci USA 2008;105(38):14579-84.

[187] Saldivia M, Barquilla A, Bart JM, Diaz-González R, Hall MN, Navarro M. Target of rapamycin (TOR) kinase in *Trypanosoma brucei*: an extended family. Biochem Soc Trans 2013;41(4):934-8.

[188] Denninger V, Koopmann R, Muhammad K, Barth T, Bassarak B, Schönfeld C, Kilunga BK, Duszenko M. Kinetoplastida: model organisms for simple autophagic pathways? Methods Enzymol 2008;451:373-408.

[189] Besteiro S, Williams RA, Coombs GH, Mottram JC. Protein turnover and differentiation in *Leishmania*. Int J Parasitol 2007;37(10):1063-75.

[190] Williams RA, Smith TK, Cull B, Mottram JC, Coombs GH. ATG5 is essential for ATG8-dependent autophagy and mitochondrial homeostasis in *Leishmania major*. PLoS Pathog. 2012;8(5):e1002695.

[191] Cazzulo JJ, Stoka V, Turk V. Cruzipain, the major cysteine proteinase from the protozoan parasite *Trypanosoma cruzi*. Biol Chem 1997;378(1):1-10.

[192] Brennand A, Rico E, Michels PA. Autophagy in trypanosomatids. Cells 2012;1(3): 346-71.

[193] Williams RA, Tetley L, Mottram JC, Coombs GH. Cysteine peptidases CPA and CPB are vital for autophagy and differentiation in *Leishmania mexicana*. Mol Microbiol 2006;61(3):655-74.

[194] Besteiro S, Coombs GH, Mottram JC. The SNARE protein family of *Leishmania major*. BMC Genomics 2006b;7:250.

[195] Herman M, Pérez-Morga D, Schtickzelle N, Michels PA. Turnover of glycosomes during life-cycle differentiation of Trypanosoma brucei. Autophagy 2008;4(3): 294-308.

[196] Li FJ, He CY. Acidocalcisome is required for autophagy in *Trypanosoma brucei*. Autophagy 2014;10(11):1978-88.

[197] Williams RA, Mottram JC, Coombs GH. Distinct roles in autophagy and importance in infectivity of the two ATG4 cysteine peptidases of *Leishmania major*. J Biol Chem 2013;288(5):3678-90.

[198] Barquilla A, Navarro M. Trypanosome TOR as a major regulator of cell growth and autophagy. Autophagy 2009;5(2):256-8.

[199] Madeira-da-Silva L, Beverley SM. Expansion of the target of rapamycin (TOR) kinase family and function in *Leishmania* shows that TOR3 is required for acidocalcisome biogenesis and animal infectivity. Proc Natl Acad Sci USA 2010;107(26):11965-70.

[200] Bao Y, Weiss LM, Braunstein VL, Huang H. Role of protein kinase A in *Trypanosoma cruzi*. Infect Immun 2008;76(10):4757-63.

[201] Hall BS, Gabernet-Castello C, Voak A, Goulding D, Natesan SK, Field MC. TbVps34, the trypanosome orthologue of Vps34, is required for Golgi complex segregation. J Biol Chem 2006;281(37):27600-12.

[202] Freitas-Junior LH, Chatelain E, Kim HA, Siqueira-Neto JL. Visceral leishmaniasis treatment: What do we have, what do we need and how to deliver it? Int J Parasitol Drugs Drug Resist 2012;2:11-9.

[203] Mäser P, Wittlin S, Rottmann M, Wenzler T, Kaiser M, Brun R. Antiparasitic agents: new drugs on the horizon. Curr Opin Pharmacol 2012;12(5):562-6.

[204] Bustamante JM, Tarleton RL. Potential new clinical therapies for Chagas disease. Expert Rev Clin Pharmacol 2014;7(3):317-25.

[205] Patterson S, Wyllie S. Nitro drugs for the treatment of trypanosomatid diseases: past, present, and future prospects. Trends Parasitol 2014;30(6):289-98.

[206] Ambrosio AR, De Messias-Reason IJ. *Leishmania (Viannia) braziliensis*: interaction of mannose-binding lectin with surface glycoconjugates and complement activation. An antibody-independent defence mechanism. Parasite Immunol 2005;27(9):333-40.

[207] Lambris JD, Ricklin D, Geisbrecht BV. Complement evasion by human pathogens. Nat Rev Microbiol 2008;6(2):132-42.

[208] Cestari I, Ramirez MI. Inefficient complement system clearance of *Trypanosoma cruzi* metacyclic trypomastigotes enables resistant strains to invade eukaryotic cells. PLoS One 2010;5(3):e9721.

[209] Evans-Osses I, de Messias-Reason I, Ramirez MI. The emerging role of complement lectin pathway in trypanosomatids: molecular bases in activation, genetic deficiencies, susceptibility to infection, and complement system-based therapeutics. ScientificWorldJournal 2013;2013:675898.

[210] Rudenko G. African trypanosomes: the genome and adaptations for immune evasion. Essays Biochem 2011;51:47-62.

[211] Fernandes MC, Da Silva EN, Pinto AV, De Castro SL, Menna-Barreto RF. A novel tri-azolic naphthofuranquinone induces autophagy in reservosomes and impairment of mitosis in *Trypanosoma cruzi*. Parasitology 2012;139(1):26-36.

[212] Desoti VC, Lazarin-Bidóia D, Sudatti DB, Pereira RC, Alonso A, Ueda-Nakamura T, Dias Filho BP, Nakamura CV, Silva SO. Trypanocidal action of (-)-elatol involves an oxidative stress triggered by mitochondria dysfunction. Mar Drugs 2012;10(8): 1631-46.

[213] Paris C, Loiseau PM, Bories C, Bréard J. Miltefosine induces apoptosis-like death in *Leishmania donovani* promastigotes. Antimicrob Agents Chemother 2004;48(3):852-9.

[214] Mamani-Matsuda M, Rambert J, Malvy D, Lejoly-Boisseau H, Daulouède S, Thiolat D, Coves S, Courtois P, Vincendeau P, Mossalayi MD. Quercetin induces apoptosis of *Trypanosoma brucei gambiense* and decreases the proinflammatory response of human macrophages. Antimicrob Agents Chemother 2004;48(3):924-9.

[215] Deolindo P, Teixeira-Ferreira AS, Melo EJ, Arnholdt AC, Souza Wd, Alves EW, DaMatta RA. Programmed cell death in *Trypanosoma cruzi* induced by *Bothrops jararaca* venom. Mem Inst Oswaldo Cruz 2005;100(1):33-8.

[216] Freire-de-Lima CG, Nascimento DO, Soares MB, Bozza PT, Castro-Faria-Neto HC, de Mello FG, DosReis GA, Lopes MF. Uptake of apoptotic cells drives the growth of a pathogenic trypanosome in macrophages. Nature 2000;403(6766):199-203.

[217] Ribeiro-Gomes FL, Silva MT, Dosreis GA. Neutrophils, apoptosis and phagocytic clearance: an innate sequence of cellular responses regulating intramacrophagic parasite infections. Parasitology 2006;132 Suppl:S61-8.

[218] Balanco JMF, Moreira ME, Bonomo A, Bozza PT, Amarante-Mendes G, Pirmez C, Barcinski MA. Apoptotic mimicry by an obligate intracellular parasite downregulates macrophage microbicidal activity. Curr Biol 2001;11(23):1870-3.

[219] El-Hani CN, Borges VM, Wanderley JL, Barcinski MA. Apoptosis and apoptotic mimicry in Leishmania: an evolutionary perspective. Front Cell Infect Microbiol 2012;2:96.

[220] Van Zandbergen G, Klinger M, Mueller A, Dannenberg S, Gebert A, Solbach W, Laskay T. Cutting edge: neutrophil granulocyte serves as a vector for Leishmania entry into macrophages. J Immunol 2004;173(11):6521-5.

[221] Pinheiro RO, Nunes MP, Pinheiro CS, D'Avila H, Bozza PT, Takiya CM, Côrte-Real S, Freire-de-Lima CG, DosReis GA. Induction of autophagy correlates with increased parasite load of *Leishmania amazonensis* in BALB/c but not C57BL/6 macrophages. Microbes Infect 2009;11(2):181-90.

[222] Mitroulis I, Kourtzelis I, Papadopoulos VP, Mimidis K, Speletas M, Ritis K. *In vivo* induction of the autophagic machinery in human bone marrow cells during *Leishmania donovani* complex infection. Parasitol Intern 2009;58(4):475-7.

[223] Cyrino LT, Araújo AP, Joazeiro PP, Vicente CP, Giorgio S. *In vivo* and *in vitro Leishmania amazonensis* infection induces autophagy in macrophages. Tissue Cell 2012;44(6):401-8.

[224] Romano PS, Arboit MA, Vázquez CL, Colombo MI. The autophagic pathway is a key component in the lysosomal dependent entry of *Trypanosoma cruzi* into the host cell. Autophagy 2009;5(1):6-18.

[225] Romano PS, Cueto JA, Casassa AF, Vanrell MC, Gottlieb RA, Colombo MI. Molecular and cellular mechanisms involved in the *Trypanosoma cruzi*/host cell interplay. IUBMB Life 2012;64(5):387-96.

[226] Martins RM, Alves RM, Macedo S, Yoshida N. Starvation and rapamycin differentially regulate host cell lysosome exocytosis and invasion by *Trypanosoma cruzi* metacyclic forms. Cellular Microbiology 2011;13(7): 943-54.

[227] Maeda FY, Alves RM, Cortez C, Lima FM, Yoshida N. Characterization of the infective properties of a new genetic group of *Trypanosoma cruzi* associated with bats. Acta Tropica 2011;120(3):231-7.

[228] Duque TLA, Souto XM, Andrade-Neto VV, Ennes-Vidal V, Menna-Barreto RFS. Autophagic Balance Between Mammals and Protozoa: A Molecular, Biochemical and Morphological Review of Apicomplexa and Trypanosomatidae Infections. In: *Bailly Y*, editor. Autophagy - A Double-Edged Sword - Cell Survival or Death? Rijeka, Croatia: Intéch;2013. pp. 289-317.

[229] Ameisen JC. The origin of programmed cell death. Science 1996;272(5266):1278-9.

[230] Menna-Barreto RF, De Castro SL. The double-edged sword in pathogenic trypanosomatids: the pivotal role of mitochondria in oxidative stress and bioenergetics. Biomed Res Int 2014;2014:614014.

[231] Welburn SC, Barcinski MA, Williams GT. Programmed cell death in trypanosomatids. Parasitol Today 1997;13(1):22-6.

[232] Baehrecke EH. Autophagy: dual roles in life and death? Nat Rev Mol Cell Biol 2005;6(6):505-10.

[233] Jesenberger V, Jentsch S. Deadly encounter: ubiquitin meets apoptosis. Nature Rev Mol Cell Biol 2002;3:122-21.

[234] Codogno P, Meijer AJ. Autophagy and signaling: their role in cell survival and cell death. Cell Death Differ 2005;12 Suppl 2:1509-18.

Chapter 12

Apoptosis and Infections

Yorulmaz Hatice

Additional information is available at the end of the chapter

http://dx.doi.org/10.5772/61306

Abstract

Apoptosis is a process that plays a critical role in the elimination of infected cells. Infectious diseases modulate apoptosis, and this contributes to disease pathogenesis. Apoptosis is initiated by various kinds of stimuli, including infections, radiation, etc. Increased apoptosis may assist the dissemination of intracellular pathogens or induce immunosuppression. However, apoptosis may also help eradicate pathogens from the host in many cases. Consequently, several viruses, bacteria, and parasites have evolved mechanisms to inhibit host cell by apoptosis as a strategy that may support intracellular survival and persistence of the pathogen. Bacteria are recognized by cellular receptors and elicit a multitude of signal transduction events that alter the cell's response toward apoptotic stimuli. The result of pathogenic bacteria entering into mammalian cells evokes variety of responses, including internalization or phagocytosis of the bacteria, release of cytokines, secretion of defensins, production of oxygen radicals and the triggering of apoptosis. Bacteria can trigger apoptosis through a large variety of mechanisms that include the secretion of protein synthesis inhibitors and pore forming proteins. They can also activate apoptotic proteins such as caspases, inactivate antiapoptotic proteins, or lead to upregulation of the endogenous receptor/ligand system. However, new research has shown that many bacterial pathogens can in fact prevent apoptosis during infection. As in bacteria, many viral genomes encode proteins that repress apoptosis to escape from immune attack by the host or viruses promote apoptotic death of the host cells. Virus-host interactions may determine viral persistence, extent and severity of inflammation, and pathology associated with infectious disease. The elucidation of the signaling pathways, the cellular receptors, and/or the microbial factors involved in the induction or reduction of apoptosis could reveal new therapeutic targets for blocking microbial-induced apoptosis. This chapter will summarize the most recent research on microorganisms' apoptotic and antiapoptotic strategies and the mechanisms relating to disease.

Keywords: apoptosis, infection, bacteria, viruses, parasitis

6.8. *Legionella*

Legionella pneumophila invades and replicates within alveolar macrophages, monocytes, and possibly alveolar epithelial cells and causes Legionnaires disease. The expression and/or export of apoptosis-inducing factor(s) in *L. pneumophila* is regulated by the Dot/Icm type IV-like secretion system [79,80].

L. pneumophila utilizes several strategies to ensure intracellular replication and protect itself against the host immune system. These are the following: (1) upon entry into a human phagocyte, *L. pneumophila* becomes contained in a vacuole called *Legionella*-containing phagosome that avoids the typical fusion with the lysosome. *L. pneumophila* promotes the cleavage of Rabaptin-5 by caspase 3, thus preventing the default phagosome-lysosome fusion. (2) *L. pneumophila* promotes the cleavage of Rabaptin-5 by caspase 3, thus preventing the default phagosome-lysosome fusion. (3) *L. pneumophila* does not activate caspases 1 and 7 in human monocytes consequently aborting the phagosome-lysosome fusion. (4) *L. pneumophila* inhibits host cell apoptosis by upregulating antiapoptotic genes. (5) *L. pneumophila* controls the local balance of activating cytokines (IFN-γ and/or TNF-α) that inhibit its replication and inhibiting cytokines (IL-10) that allow its survival. (6) *L. pneumophila* activates the NF-κB pathway to maintain host cell survival and/or 7-*L. pneumophila* modulates other innate immune responses to establish a replicative niche [81].

6.9. *Escherichia coli*

E. coli K1 is a leading causative agent of neonatal meningitis. OmpA of *E. coli* can directly interact with monocytes and macrophages for entry. Some *E. coli* serotypes produce a Shiga-like toxin that can bind to human intestinal epithelium and produce Shiga-like toxins, which are associated with hemorrhagic colitis and the hemolytic-uremic syndrome [82]. One pathway involving apoptosis is mediated by death receptors such as CD95 and TNF-R. Intrinsic apoptotic pathway is involved in the Shiga-like toxin-mediated apoptosis of epithelial cells [83]. *E. coli* K1 induces the expression of the antiapoptotic protein Bcl_{XL} for its own survival and that of host macrophages. In addition, the bacteria may also block the activation of caspases. Besides upregulating Bcl_{XL}, OmpA$^+$ *E. coli* interaction with macrophages may alter the signaling pathways of the host cell to use it as a protected reservoir for the time required to reach septic levels [84,85].

6.10. *Rickettsia*

Rickettsia is a genus of nonmotile, Gram-negative, non-spore-forming, and/or highly pleomorphic bacteria. *Rickettsia rickettsii* is a unicellular Gram-negative coccobacillus. *R. rickettsii* is most commonly known as the causative agent of Rocky Mountain spotted fever [86]. *R. rickettsii* prevents apoptosis by the activation of cell survival pathways. *R. rickettsii* induces NF-κB activity and the upregulation of prosurvival proteins, the downregulation of proapoptotic proteins, and a lack of cytochrome *c* release in endothelial cells. Bechelli *et al.* [87] performed a screening of pro- and antiapoptotic genes that were differentially expressed in human microvascular endothelial cells during *R. rickettsii* infection and after staurosporine challenge.

A total of 14 genes were significantly upregulated of which 8 (TRAF1, BNIP2, BCL2L1, TRAF3, BIRC2, BNIP3L, AKT1, and BIRC5) are known apoptosis suppressors while 6 are known to promote apoptosis (BOK, BCL2L13, DAPK2, TP53, ABL1, and BAK1). However, *R. rickettsii* efficiently infects neuronal cells and that the infection causes apoptotic death of neuron [88].

6.11. *Mycobacterium*

Mycobacterium tuberculosis is the causative agent of most cases of tuberculosis. *M. tuberculosis* infected macrophages die by (a) necrosis, a death modality defined by cell lysis, or (b) apoptosis, a form of death that maintains an intact plasma membrane. Necrosis is a mechanism used by bacteria to exit the macrophage, evade host defenses, and spread. *M. tuberculosis* is complex and includes both the induction of cell-death and cell-survival signals. It induces apoptosis in macrophages *in vitro* and *in vivo* [89]. Apoptosis occurs by TNF pathway. In addition, MOMP leading to the activation of the intrinsic apoptotic pathway is required. Both pathways lead to caspase 3 activation which then results in apoptosis [90]. Bacteria cell wall components connect TLR-2 molecules. *M. tuberculosis* also protects cells against apoptosis via two key pathways: first, through the induction of TLR-2-dependent activation of the NF-κB cell survival pathway, and second, by enhancing the production of soluble TNFR-2 (sTNFR2), which neutralizes the proapoptotic activity of TNF-α. *M. tuberculosis* can prevent apoptosis in alveolar epithelial cells. Danelishvili *et al.* [91] showed that *M. tuberculosis* infection of macrophages results in the downregulation of the Bcl-2 gene and the upregulation of Bax and Bad proapoptotic genes. In contrast, the increased expression of Bcl-2 and the inhibition of Bax and Bad genes were observed in alveolar epithelial cells. *M. tuberculosis* infection was associated with the repression of the Bcl-2 gene and the induction of p53 in human macrophages. In alveolar epithelial cells, the expression of p53 was unchanged during *M. tuberculosis* infection [92].

7. Viral infections

After infecting target cells, viruses replicate to produce large number of progeny virions and spread the progeny to initiate the next round of infection. Some viruses encode specific proteins to optimize their replication. Infection by viruses, however, triggers the apoptosis of the infected cell to restrict virus infection. This is done by reprogramming of the host cell apoptotic pathway to effect death of the infected host cell before the release of progeny viruses. In order to ablate host defense mechanisms, viruses have evolved proteins that are able to inhibit or delay the host protective actions by targeting strategic points in the apoptotic pathways [93]. Apoptosis can be induced by intrinsic or extrinsic signals (Table 1). Intrinsic signals may result from viral infection and include stress, cell cycle arrest, cytoplasmic calcium perturbation, and DNA damage. Extrinsic signals arise as a result of the host immune response through TNF-receptor or Fas activation or via delivery of proteases by cytotoxic lymphocytes. Once induced, apoptosis may eliminate infected cells prior to release of viral progeny [94, 95]. Many viruses have been shown to induce apoptosis, either as a mechanism for the release and dissemination of progeny virions or as a defense strategy of multicellular host organisms for the destruction of infected cells and therefore preventing the spread of the virus [96]. Innate and the acquired immune system induce apoptosis as a host defense against viral infections. The innate immune system directly activates inflammatory cells such as macrophages (e.g., granulocytes, Kupffer cells in the liver) and natural killer (NK) cells, which may directly cause death of the infected cells. On the other hand, viral RNA or proteins can bind to intracellular molecules that modulate or directly induce cell death [97]. In this immune cell-independent, virus-induced apoptosis of the host cell protein kinase R (PKR) and the cytoplasmic RNA helicase RIG-I play important roles [98]. PKR acts via the downstream transcription factor eIF-2α [99]. At the same time, acquired immune system works to eliminate the virus and the recognition of viral antigens presented by specific cells (e.g., dendritic cells). The antigen-primed CD8$^+$-T-lymphocytes cytotoxic T lymphocytes directly kill infected cells via direct cell-cell contact and release of cytotoxic and/or antiviral cytokines (e.g., interferons IFNs and/or TNF-α), whereas IFN-γ and IFN-α are also able to eliminate the virus without killing the host cell [100].

The IFNs are considered to play a critical role in innate immunity to viral infection and aside from effectively preventing intracellular viral replication can also mediate the activation and recruitment of the adaptive immune response. The IFNs can be induced by a number of stimuli, including viruses and dsRNA through mechanisms involving the activation of interferon regulatory factor (IRF-3), NF-κB, and perhaps the dsRNA-dependent PKR and the JNK2 pathway, all of which have been reported to be mediators of cell death [101,102].

7.1. Hepatitis C virus

Hepatitis C virus (HCV) causes liver cirrhosis and hepatocellular carcinoma [103]. In hepatocytes, apoptosis induction via cytotoxic T lymphocytes and macrophages largely occurs via extrinsic pathway. Ligand binding activates caspase 8 signaling cascade [104]. Another mechanism of apoptosis involves viral protein and their interactions. HCV core protein has been shown to be proapoptotic and antiapoptotic effects [105]. Machida *et al.* [106] showed that the expression of HCV proteins may directly or indirectly inhibit Fas-mediated apoptosis and death in mice by repressing the release of cytochrome *c* from mitochondria.

7.2. HIV

Human immunodeficiency virus type 1 (HIV-1)-infected individuals often suffer from neurological complications such as memory loss, mental slowing, and gait disturbance [107]. HIV infection is associated with a progressive decrease in and/or loss of CD4$^+$ and the decline of CD8$^+$ T cells and viral replication [108]. Inappropriate signaling through the binding of the HIV-1 envelope to the CD4 may induce abnormal programmed CD4$^+$ T-cell death. Viral proteins such as HIV-1 gp120 have an important role development in HIV-associated apoptosis. HIV proteins implicated in the induction of apoptosis *in vitro* include tat, nef, vpr, and protease. Cross-linking of bound gp120 on human CD4$^+$ T cells followed by signaling through the T-cell receptor for antigen was found to increase susceptibility to Fas and result in apoptosis [109,110]. In addition, deregulation in cytokine production occurs during HIV infection, perturbing the immune response. The overproduction of IL-4 and/or IL-10 cytokines is known to increase susceptibility to activation-induced cell death. IFN-α produced by HIV-1-infected dendritic cells contributes to CD4 T-cell apoptosis by the TRAIL/DR5 pathway [111]. HIV-1 protein Env triggers apoptosis by the transactivation of the p53-dependent genes Puma and Bax [112]. HIV Nef is able to induce apoptosis by extrinsic pathway [113,114].

7.3. Rabies virus

Rabies virus (RV) is a neurotropic virus and travels to the brain by following the peripheral nerves. RV, a member of the genus *Lyssavirus* of the family Rhabdoviridae, is known to cause fatal encephalomyelitis in many mammalian species. RV has developed two main mechanisms to escape the host defenses: (1) its ability to kill protective migrating T cells and (2) its ability to sneak into the nervous system without triggering the apoptosis of the infected neurons and preserving the integrity of neurites [115]. In one of the studies, Ubol *et al.* [116] showed the expression of Bax and caspase 1 activation in RV-infected neuroblastoma cells. In another study, the expression of caspase gene Nedd-2 was significantly upregulated in infected adult and suckling mice [117]. Thoulouze *et al.* [118] showed that apoptosis induced by rabies virus involves the activation of caspase 8 and disappearance of procaspases 9 and 3. In addition, AIF translocated from the cytoplasm to the nucleus, suggesting that caspase-independent pathway is also involved in RV-induced apoptosis. Sarmento *et al.* [119] showed that AIF, a caspase-independent apoptotic protein, was upregulated and translocated from the cytoplasm to the nucleus postinfection, suggesting that apoptosis induced by RV induces apoptosis by both the caspase-dependent and caspase-independent pathways.

7.4. Epstein-Barr virus

Infection with Epstein-Barr virus (EBV) is very common and usually occurs in childhood or early adulthood. In fact, up to 95% of people in the U.S. have been infected with EBV. EBV is the cause of infectious mononucleosis (also termed "mono"), an illness associated with fever, sore throat, swollen lymph nodes in the neck, and sometimes enlarged spleen. Less commonly, EBV can cause more serious disease. To establish a persistent latent infection, EBV must access the memory B-cell compartment and reside within long-lived peripheral B cells where few viral gene products are expressed in order to escape immune detection [120]. EBV encodes two viral Bcl-2 proteins, BHRF1 and BALF1, with apparently redundant functions. The viral BHRF1 gene expresses a Bcl-2 homologue protein that resembles Bcl-2 in its subcellular localization and capacity to enhance B-cell survival [121]. *In vitro*, EBV infects resting human B lymphocytes and transforms them into lymphoblastoid cell lines (LCLs). In LCLs, 11 so-called latent genes are consistently expressed. These are the EBV nuclear antigens EBNA1, EBNA2, EBNA-LP, EBNA3A, B, and/or C and the latent membrane proteins LMP1, LMP2A, and B and two noncoding RNAs [122]. LMP1 indirectly inhibits apoptosis by upregulating several cellular antiapoptotic genes presumably through the induction of the NF-κB pathway [123].

7.5. Baculovirus

Baculovirus antiapoptotic genes include p35, which encodes the most broadly acting caspase inhibitor protein known and IAP genes [124]. The baculovirus IAP blocks apoptosis induced by caspase activation. All viral IAPs (vIAPs) contain a carboxyl ring finger and a variable number of highly conserved Cys/His motifs known as baculoviral IAP repeats (BIRs). The BIR domains bind directly to caspases and inhibit their proteolytic activity and molecules that contain an IAP-binding motif (second mitochondrial-derived activator of caspases and Omi) antagonize IAP function via binding to BIR motifs displacing IAP binding to caspases or by promoting their degradation. Therefore, vIAPs act downstream of mitochondria, inhibiting the activity of procaspase 9 and effector caspases 3 and 7 [125]. The antiapoptotic protein p35 from baculovirus is thought to prevent the suicidal response of infected insect cells by inhibiting caspases [126]. Zhou *et al.* [127] showed that purified recombinant p35 inhibits human caspases 1, 3, 6, 7, 8, and 10. There may be interaction of the baculovirus antiapoptotic protein p35 with caspases. Sah *et al.* [128] demonstrated the ability of the p35 gene to inhibit oxidative stress-induced apoptosis. Oxidative damage to cellular macromolecules such as nuclear and mitochondrial DNA and proteins caused by reactive oxygen species is considered to be of key importance in the aging process. The chain of oxidative reactions initiated by ROS eventually knocks down the crucial biomolecules, thereby driving the cellular machinery to undergo apoptosis via the activation of caspases, which ultimately brings about the execution of cell death. p35 is able to directly mop out free radicals and prevent cell death by also acting in an oxidant-dependent pathway at a very upstream step in the cascade of events associated with oxidative stress-induced apoptosis [129].

7.6. Human papillomavirus (HPV)

HPVs are small DNA viruses that are known to be the most common etiological agents in cervical cancer. HPVs are implicated in the mucosal and epithelial infections that may range from a benign lesion to a malignant carcinoma [130]. Recent studies have shown that 13 different types of HPV are associated with carcinogenesis. HPVs are DNA tumor viruses whose genome is organized in three regions: the early gene (E1 to E7), the late gene (L1 and L2) regions, and the upper regulatory region (URR). The late region units, L1 and L2, encode for viral capsid proteins during the late stages of virion assembly. E1 and E2 encode proteins that are vital for extrachromosomal DNA replication and the completion of the viral life cycle. The E4 protein plays an important role for the maturation and replication of the virus. The E5 in open reading frame (ORF) interacts with various transmembrane proteins like the receptors of the epidermal growth factor, platelet-derived growth factor β, and colony stimulating factor-1. E6 and E7 ORF encode for oncoproteins that allow replication of the virus and the immortalization and transformation of the cell that hosts the HPV DNA [131]. The HPV E2 regulates the transcription of E6/E7 and facilitates apoptosis via p53-dependent pathway in HeLa cells [132]. HPV E7 is involved with cell cycling and binds to the retinoblastoma tumor suppressor protein and related proteins and induces apoptosis in mouse lens [133]. The inactivation of p53 by E6 should lead to a reduction in cellular apoptosis. Numerous studies showed that E6 could in fact sensitize cells to apoptosis. HPV E6 induce the degradation of p53. E6 expression correlated with the prolonged expression of Bcl-2 reduces the elevation of Bax and loss of p53. Several studies have shown that E2 could also induce apoptosis independent of its effects on transcription of E6 and E7 [134-136]. Tan *et al.* [137] showed that HPV16 E6 RNA interference enhances cisplatin and death receptor-mediated apoptosis in human cervical carcinoma cells. Moreover, HPV-16 E6 was shown to bind TNF R1 and protect cells from TNF-induced apoptosis in mouse fibroblasts and human histiocyte/monocyte and osteosarcoma cells. Caspase 3 and caspase 8 activation were significantly reduced in E6-expressing cells [138].

7.7. Adenovirus

Adenoviruses (Ads) were first described as the etiological agents isolated from human adenoids and respiratory secretions that cause spontaneous cytopathic effects in cultures of human cells. Adenovirus has evolved ways to commandeer host cell machinery for successful entry, viral DNA replication, and propagation of progeny virions. Adenoviral proteins interact with host-cell proteins to either exploit or inhibit cellular functions for the purpose of viral propagation. The Ad genome is a 36 kbp linear double-stranded DNA molecule that encodes five early transcription units (E1A, E1B, E2, E3, and E4), two delayed early units, and one major late unit that is processed to make five families of late mRNAs. The early genes are transcribed before viral DNA replication begins, and the late proteins are made following the onset of replication [139]. At least 51 serotypes have been distinguished based on resistance to neutralization by antibodies specific to other known serotypes. These are divided into six subgroups (A-F) based on hemagglutination patterns, oncogenicity, and genome homologies. The common subgroup C Ads, which include Ad serotypes 1, 2, 5, and 6, are endemic virtually all over the world. They cause mild upper respiratory tract infections in young children [140]. Early in infection, the expression of E1A drives the host cell into the S phase of the cell cycle in order to induce DNA synthesis that is required for viral replication. The genes in the E3 region of Ad encode several proteins that function to protect the virus-infected cell from host immune responses [141]. Table 1 describes the adenovirus immunoregulatory proteins and how they function to block or induce apoptosis of infected cells [142-150].

7.8. Human cytomegalovirus

Human cytomegalovirus (CMV), a beta herpes virus with a widespread distribution, is a major cause of morbidity and mortality in immunocompromised individuals such as organ transplant recipients and patients with AIDS. During pregnancy, CMV is a major cause of congenital disease [151]. CMV genes UL36 and UL37 encode viral inhibitor of caspase-8-induced apoptosis (vICA) and viral mitochondria inhibitor of apoptosis (vMIA), respectively. Skaletskaya *et al.* [152] identified a human cytomegalovirus cell-death suppressor denoted vICA and encoded by the viral UL36 gene. vICA inhibits Fas-mediated apoptosis by binding to the prodomain of caspase 8 and preventing its activation. vMIA blocks cytochrome *c* release and activation of downstream effector caspases in a manner analogous to Bcl-2 homologues. Like Bcl-2, vMIA localizes to mitochondria and inhibits mitochondrial permeabilization induced by apoptotic signals [153]. vMIA also counteracts serine protease HtrA2/Omi (high temperature requirement protein A2/Omi stress-regulated endonuclease)-dependent cell death and allows infected cells to survive and continuously produce a virus for several days [154]. Additional human CMV gene products, including IE1 and IE2, as well as the murine CMV UL45 homologue may influence cell susceptibility to apoptosis [155]. Transient transfection assays indicate that the IE1 and IE2 proteins regulate transcription. The IE1 and IE2 proteins each inhibit the induction of apoptosis by TNF or by the E1B 19-kDa-protein-deficient adenovirus. IE1 and IE2 proteins inhibit apoptosis in part by modulating the activity of p53[156]. In addition, IE1 and IE2 and the viral RNA beta 2.7, which bind to the mitochondrial respiratory complex I, maintain ATP production late in infection and prevent death induced by mitochondrial poison [157].

8. Apoptosis and parasitic infections

Apoptosis plays crucial roles in the interaction between the host and the parasite. This includes innate and adaptive defense mechanisms to restrict intracellular parasite replication as well as regulatory functions to modulate the host's immune response. During their evolution, parasites have developed mechanisms to induce or avoid host cell apoptosis in order to be able to survive and complete their life cycle (Table 1). Among the factors involved in that balance in infected organisms, the time of apoptosis (early or late occurrence), the cell type, and the type of parasitism (intracellular or not) are the major modulators. For example, the early apoptosis of host cells could contribute toward their fight against infection by intracellular parasites; equally, early apoptosis could favor the penetration of the parasite. The late apoptosis of cells of the defense system could be beneficial to the host clearing excess cells, thereby avoiding the detrimental effects of excessive inflammatory response in the tissue that they would cause [158].

8.1. *Toxoplasma gondii*

Toxoplasma gondii is a species of parasitic protozoa in the genus *Toxoplasma*. Humans can become infected with *T. gondii*, either through contact with soil contaminated by cat feces or by eating infected meat. Toxoplasmosis is usually asymptomatic because our immune system keeps the parasite from causing illness. The disease is more problematic for pregnant women and people who have weakened immune systems. Some results indicate a strong correlation between schizophrenia, brain cancer, and toxoplasmosis [159]. Toxoplasma promote or inhibit apoptosis. Begum-Haque *et al.* [160] demonstrated marked difference in the death of activated T cells between early (day 3 post infection) and acute (day 6 post infection) stage of *T. gondii*. The decreased production of IL-2 and augmented synthesis of IL-10 during acute stage of *T. gondii* infection may have a role in the enhanced level of apoptosis. It has been suggested that the apoptosis of T lymphocytes in *T. gondii* infection is associated with the virulence and density of the parasite in the host. In *T. gondii* infection, IFN-γ locally produced in Peyer's patches contributes to the induction of apoptosis in Peyer's patch T cells [160]. *T. gondii* inhibits the apoptosis of host cells by indirect and direct mechanisms. Granulocyte colony-stimulating factor and granulocyte-macrophage cerebrospinal fluid secreted by *T. gondii*-infected human fibroblasts increased the expression of antiapoptotic Bcl-2 family member Mcl-1 and abolished apoptosis in neutrophils *in vitro* indirectly [161]. Many studies have shown that *T. gondii* has evolved strategies to directly inhibit cell apoptosis by various mechanisms: (a) increased expression of antiapoptotic members of the Bcl-2 protein family; (b) inhibition of the cyto-chrome *c* release; (c) upregulation of IAP$_s$; (d) activation of NF-κB by *T. gondii* in distinct cell types or under distinct conditions thereby inducing the transcription of genes encoding antiapoptotic molecules, including Bfl-1 and IAPs; and/or (e) degradation of the PARP as described is involved in the inhibition of apoptosis. Although direct evidence is still lacking, it appears plausible that diminished PARP levels in *Toxoplasma*-infected cells may inhibit apoptosis in a caspase-independent fashion [162].

8.2. *Plasmodium falciparum*

Plasmodium falciparum is the agent of malaria. Enhanced levels of RBC apoptosis have been observed in clinical disorders in which anemia is a common feature such as iron and renal insufficiency, thalassemia, sickle-cell disease, and apoptosis has been associated to cerebral malaria, thrombocytopenia, and lymphocytopenia in malaria infection [163].

P. falciparum induces oxidative stress, which in turn activates the Ca^{+2} permeable cation channels followed by Ca^{+2} entry, and the stimulation of eryptosis has been coined to describe the suicidal erythrocyte death. The Ca^{+2} uptake, however, eventually triggers eryptosis of the parasitized erythrocyte, and thus the parasitized erythrocytes is doomed to be phagocytosed by macrophages [164].

P. falciparum firstly enter red blood cells. Second, parasitized red blood cell sticks endothelial cells, inducing the expression of iNOS in brain cells. The activation of caspases 8 and 9 results in apoptosis and blood-brain barrier disruption [165].

8.3. Trypanosomatids

Trypanosomatids are the causative agents of diseases such as the Chagas disease and the African sleeping sickness [166]. Trypanosomatids lack some of the key molecules contributing to apoptosis in metazoans like caspase genes, Bcl-2 family genes, and the TNF-related family of receptors. Apoptosis triggered in response to heat shock, prostaglandins, antibodies, and mutations in cell cycle regulates genes [167]. These stimuli result in loss of $\Delta\Psi m$, generation of ROS, lipid peroxidation, and increase in cytosolic Ca^{2+}. This also potentiates the release of cytochrome *c* and EndoG into the cytoplasm and the activation of proteases and nucleases to dismantle the parasites in an ordered fashion. Upon release from the mitochondrion, EndoG translocates to the nucleus to degrade DNA. These events finally lead to the execution of apoptosis [168]. *Trypanosoma brucei* causes neuronal demyelination and apoptosis after blood-brain barrier damage. This leads to apoptosis in cells of the cerebellum and brain stem. Welburn *et al.* [170] described cytoplasmic vacuolization and marginalization, extensive membrane blebbing, and condensation of nuclear chromatin in *Trypanosoma cruzi* and *T. brucei* respectively. Lectins such as ConA were among the first compounds shown to induce the expression of apoptotic markers in *T. brucei* [169]. *T. cruzi* is the etiological agent of Chagas disease. It also inhibits apoptosis through the action of parasite-derived neurotrophic factor, a parasite-derived protein in neuronal and glial cells. The parasite-derived neurotrophic factor is both a substrate and an activator of the serine-threonine kinase Akt and an antiapoptotic molecule binding to the neurotrophic surface receptor TrkA (neurotrophic tyrosine kinase receptor type 1) triggering the PI3-K/PKB pathway resulting in increased Bcl-2 expression.

This results in protection of Schwann cells from apoptosis induced by H_2O_2 and TNF-α/TGF-β (transforming growth factor b) [170-172].

8.4. *Leishmania*

Leishmanias are agents of ulcerative skin lesions (cutaneous leishmaniasis) and disseminated visceral infection (visceral leishmaniasis or kala-azar). Leishmania is able to inhibit the spontaneous apoptosis of short-lived neutrophils, increasing their life span and providing a safe place for the parasites during the first days of the infection [173]. With most apoptosis inducing stimuli, *Leishmania donovani* shows typical features of apoptotic death like cell shrinkage, nuclear condensation, and DNA fragmentation. Ca^{2+} appears to be a vital ion involved in Leishmania apoptosis. Extracellular or intracellular Ca^{2+} during oxidative stress results in the significant rescue of the fall of the mitochondrial membrane potential and consequently apoptosis [174].

9. Concluding remarks

- Apoptosis is a genetically programmed process of cellular destruction that is indispensable for the normal development and homeostasis of multicellular organisms.

- Microorganisms induce apoptosis by intrinsic and extrinsic pathway in the host cell.

- T3SS effectors have also been shown to tamper with the host's cell cycle, and some of them are able to induce apoptosis bacteria such as *Pseudomonas*, *Shigella*, *Salmonella*, and *Yersinia*.

- Microorganisms inhibit apoptosis by multiple mechanisms: protection of the mitochondria and prevention of cytochrome *c* release (i.e., *Chylamidia* sp. and/or *Neisseria* sp.), activation of cell survival pathways (i.e., *Salmonella* sp. and/or *Rickettsia* sp.), inhibition of caspases, activation of phosphoinositide 3-kinase (PI3K)-Akt/protein kinase B (PKB) pathway, and interaction with cellular caspases (i.e., *Shigella* sp. and/or *Legionella* sp.)

- Prevention of apoptosis enables microorganisms to replicate and survive in host.

- A clear understanding of the molecular basis of apoptosis inhibition or induction is needed.

- Elucidation of the mechanisms, the cellular receptors, and/or the microbial factors involved in modulating of apoptosis could reveal insights into the host-pathogen relationship and new therapeutic targets.

10. Abbreviations

AIF: apoptosis inducing factor

AP-1: activator protein 1

BH: Bcl-2 homology

CARD: caspase activation and recruitment domain

CMV: cytomegalovirus

CTL: cytotoxic T lymphocyte

DD: death domains

DED: death effector domain

$\Delta\Psi_m$: transmembrane potential

EBV: Epstein-Barr virus

FADD: Fas-associated death domain

GSK-3: Glycogen synthase kinase 3

JNK: c-Jun N-terminal kinase

HIV-1: human immunodeficiency virus type 1

IAPs: inhibitors of apoptosis proteins

IL-1β: interleukin 1β

ICE: IL-1β converting enzyme

I-KB: inhibitor of KB

IKK: I-KB kinase

IPA: invasion plasmid antigens

LlyO: listeriolysin O

MOMP: mitochondrial outer membrane permeabilization

NF-κB: nuclear factor-kappa B

TLR: toll-like receptors

TNF: tumor necrosis factor

TNFR: tumor necrosis factor receptor gene

TRAIL: tumor necrosis factor (TNF)-related apoptosis-inducing ligand

T3SS : type III secretion system

PI3K: phosphoinositide 3-kinase

PKB: protein Kinase B

PMLs: polymorphonuclear leukocytes

proIL: prointerleukin

PT: permeability transition

ROS: reactive oxygen species

RV: Rabies virus

SAPK: stress-activated protein kinase

STS: staurosporine

VDAC: voltage-dependent anionic channel

Acknowledgements

I thank Halic University and Emine Kurt for contributions.

Author details

Yorulmaz Hatice

Address all correspondence to: haticeyorulmaz@halic.edu.tr

Department of Physiology, Medical Faculty, Halic University, Istanbul, Turkey

References

[1] Degterev A, Boyce M, Yuan J (2003) A decade of caspases. Oncogene. 22:8543-8567.

[2] Ashkenazi A, Dixit VM (1998) Death receptors: signaling and modulation. Science. 281:1305-1308.

[3] Deveraux QL, Takahashi R, Salvesen GS, Reed JC (1997) X-linked IAP is a direct inhibitor of cell death proteases. Nature. 388:300-304.

[4] Du C, Fang M (2000) Smac, a mitochondrial protein that promotes cytochrome c-dependent caspase activation by eliminating IAP inhibition. Cell. 102:33-42.

[5] Wang K, Yin XM, Chao DT, Milliman CL, Korsmeyer SJ (1996) BID: a novel BH3 domain-only death agonist. Genes Dev. 10:2859-2869.

[6] Bates S, Vousden KH (1999) Mechanisms of p53-mediated apoptosis. Cell. Mol. Life Sci. 55(1):28-37.

[7] Shen Y, White E (2001) p53-dependent apoptosis pathways. Adv. Cancer Res. 82:55-84

[8]   Kam PCA, Ferch NI (2000) Apoptosis: mechanisms and clinical implications. Anesthesia. 55:1081-1093.

[9]   Fulda S, Debatin KM (2006) Extrinsic versus intrinsic apoptosis pathways in anticancer chemotherapy. Oncogene. 25(34):4798-4811.

[10]   Gross A, McDonnell JM, Korsmeyer SJ (1999) Bcl-2 family members and the mitochondria in apoptosis. Genes Dev. 13: 1899-1911.

[11]   Häcker G, Kirschnek S, Fischer SF (2006) Apoptosis in infectious disease: how bacteria interfere with the apoptotic apparatus. Med. Microbiol. Immunol. 195(1): 11-19.

[12]   Savill J, Fadok V, Henson P, Haslett C (1993) Phagocyte recognition of cells undergoing apoptosis. Immunol. Today. 14(3):131-136.

[13]   Scovassi AI, Torriglia A (2003) Activation of DNA-degrading enzymes during apoptosis. Eur J Histochem. 47(3):185-194.

[14]   Massari P, Ram S, Macleod H, Wetzler LM (2003) The role of porins in neisserial pathogenesis and immunity. Trends Microbiol. 11(2): 87-93.

[15]   Baud V, Karin M (2001) Signal transduction by tumor necrosis factor and its relatives. Trends Cell Biol. 11:372-377.

[16]   Aggarwal BB (2000) Tumor necrosis factor receptor associated signalling molecules and their role in activation of apoptosis, JNK and NF-κB. Ann. Rheum. Dis. 59:6-16.

[17]   Devin A, Lin Y, Liu ZG (2003) The role of the death-domain kinase RIP in tumour-necrosis-factor-induced activation of mitogen-activated protein kinases. EMBO Rep. 4:623-627.

[18]   Beg AA, Finco TS, Nantermet PV, Baldwin AS (1993) Tumor necrosis factor and interleukin-1 lead to phosphorylation and loss of I kappa B alpha: a mechanism for NF-kappa B activation. Mol Cell Biol. 13: 3301-3310.

[19]   Romashkova JA, Makarov SS (1999) NF-kappaB is a target of AKT in anti-apoptotic PDGF signalling. Nature. 401:86-90.

[20]   Aggarwal BB (2003) Signalling pathways of the TNF superfamily: A double-edged sword. Nat. Rev. Immunol. 3(9): 745-756

[21]   Nagata S (1999) FAS ligand-induced apoptosis. Annu Rev Genet. 33:29-55.

[22]   Chinnaiyan AM, O'rourke K, Tewari M, Dixit VM (1995) FADD, a novel death domain-containing protein, interacts with the death domain of FAS and initiates apoptosis. Cell. 81:505-512.

[23]   Scaffidi C, Fulda S, Srinivasan A, Friesen C, Li F, Tomaselli KJ, Debatin KM, Krammer PH, Peter ME (1998). Two CD95 (Apo-1/FAS) signaling pathways. EMBO J. 17:1675-1687.

[24] Fulda S, Debatin KM (2006) Extrinsic versus intrinsic apoptosis pathways in anti-cancer chemotherapy. Oncogene. 25(34):4798-4811.

[25] Ashkenazi A (2002) Targeting death and decoy receptors of the tumour-necrosis factor superfamily. Nat. Rev. Cancer. 2:420-430.

[26] Kaplanski G, Marin V, Montero-Julian F, Mantovani A, Farnarier C (2003) IL-6: a regulator of the transition from neutrophil to monocyte recruitment during inflammation. Trends Immunol. 24: 25-29.

[27] Nauseef WM, Clark RA (2000) Basic Principles in the Diagnosis and Management of Infectious Diseases. In: Mandel GL, Bennett JE, Dolin R, editors. Churchill Livingstone, New York: Vol. 1. pp. 89-112.

[28] Blomgran R, Zheng L, Stendahl O (2004) Uropathogenic *Escherichia coli* triggers oxygen-dependent apoptosis in human neutrophils through the cooperative effect of type 1 fimbriae and lipopolysaccharide. Infect. Immun. 72: 4570-4578.

[29] Weinrauch Y, Zychlinski A (1999) The induction of apoptosis by bacterial pathogens. Annu Rev Microbiol. 53: 155-187.

[30] Grassmé H, Jendrossek V, Gulbins E (2001) Molecular mechanisms of bacteria induced apoptosis. Apoptosis. 6: 441-445.

[31] Meyer TF (1999) Pathogenic neisseriae: complexity of pathogen-host cell interplay. Clin. Infect. Dis. 28: 433-441.

[32] Häcker G, Kirschnek S, Fischer SF (2006) Apoptosis in infectious disease: how bacteria interfere with the apoptotic apparatus. Med. Microbiol. Immunol. 195(1): 11-19.

[33] Faherty CS, Maurelli AT (2008) Staying alive: bacterial inhibition of apoptosis during infection. Trends Microbiol. 16(4): 173-180.

[34] Diehl JA, Cheng M, Roussel MF, Sherr CJ (1998) Glycogen synthase kinase-3b regulates cyclin D1 proteolysis and subcellular localization. Genes Dev. 12: 3499-3511.

[35] Wang HG, Rapp UR, Reed JC (1996) Bcl-2 targets the protein kinase Raf-1 to mitochondria. Cell. 87: 629-638.

[36] Aliprantis AO, Yang RB, Weiss DS, Godowski P, Zychlinsky A (2000) The apoptotic signaling pathway activated by Toll-like receptor-2. EMBO J. 19: 3325-3336.

[37] Joshi SG, Francis CW, Silverman DJ, Sahni SK (2004) NF-kappaB activation suppresses host cell apoptosis during *Rickettsia rickettsii* infection via regulatory effects on intracellular localization or levels of apoptogenic and anti-apoptotic proteins. FEMS Microbiol. Lett. 234: 333-341.

[38] Balcht AL, Smith RP (1994) *Pseudomonas aeruginosa*: Infections and Treatment. Informa Health Care. Marcel Dekker, Inc., New York, pp. 83-84.

[39] Valente E, Assis MC, Alvim IM, Pereira GM, Plotkowski MC (2000) *Pseudomonas aeruginosa* induces apoptosis in human endothelial cells. Microb. Pathog. 29: 345-356.

[40] Rudel T, van Putten JP, Gibbs CP, Haas R, Meyer TF (1992) Interaction of two variable proteins (PilE and PilC) required for pilus-mediated adherence of *Neisseria gonorrhoeae* to human epithelial cells. Mol. Microbiol. 6: 3439-3450.

[41] Makino S, van Putten JP, Meyer TF (1991) Phase variation of the opacity outer membrane protein controls invasion by *Neisseria gonorrhoeae* into human epithelial cells. EMBO J. 10: 1307-1315.

[42] McGee ZA, Stephens DS, Hoffman LH, Schlech WF, Horn RG (1983) Mechanisms of mucosal invasion by pathogenic Neisseria. Rev. Infect. Dis. 5: 708-714.

[43] Muller A, Günther D, Düx F, Naumann M, Meyer TF, Rudel T (1999) Neisserial porin (PorB) causes rapid calcium influx in target cells and induces apoptosis by the activation of caspase-3 and the Ca^{2+}-dependent protease calpain. EMBO J. 18: 339-352.

[44] Massari P, Ram S, Macleod H, Wetzler LM (2003) The role of porins in neisserial pathogenesis and immunity. Trends Microbiol. 11(2): 87-93.

[45] Massari P, King CA, Ho AY, Wetzler LM (2003) Neisserial PorB is translocated to the mitochondria of HeLa cells infected with Neisseria meningitidis and protects cells from apoptosis. Cell. Microbiol. 5(2): 99-109.

[46] Chen A, Seifert HS (2011) *Neisseria gonorrhoeae*-mediated inhibition of apoptotic signalling in polymorphonuclear leukocytes. Infect. Immun. 79(11): 4447-4458.

[47] Follows SA, Murlidharan J, Massari P, Wetzler LM, Genco CA (2009) *Neisseria gonorrhoeae* infection protects human endocervical epithelial cells from apoptosis via expression of host antiapoptotic proteins. Infect. Immun. 77(9): 3602-3610.

[48] Zychlinsky A, Thirumalai K, Arondel J, Cantey JR, Aliprantis AO, Sansonetti PJ (1996) In vivo apoptosis in *Shigella flexneri* infections. Infect. Immun. 64: 5357-5365.

[49] Suzuki T, Franchi L, Toma C, Ashida H, Ogawa M, Yoshikawa Y, Mimuro H, Inohara N, Sasakawa C, Nuñez G (2007) Differential regulation of caspase-1 activation, pyroptosis, and autophagy via Ipaf and ASC in *Shigella*-infected macrophages. PLoS Pathog. 3(8): e111.

[50] Schroeder GN, Hilbi H (2008) Molecular pathogenesis of *Shigella* spp.: controlling host cell signaling, invasion, and death by type III secretion. Clin. Microbiol. Rev. 21(1): 134-156.

[51] Zychlinsky A, Sansonetti PJ (1997) Apoptosis as a proinflammatory event: what we can learn from bacteriainduced cell death. Trends Microbiol. 5: 201-204.

[52] Zychlinsky A, Pre'vost MC, Sansonetti PJ (1992) *Shigella flexneri* induces apoptosis in infected macrophages. Nature. 358: 167-168.

[53] . Clark CS, Maurelli AT (2007) *Shigella flexneri* inhibits staurosporine-induced apoptosis in epithelial cells. Infect. Immun. 75(5): 2531-2539.

[54] Faherty C, Merrell DS, Semino-Mora C, Dubois A, Ramaswamy A, Maurelli AT (2010) Microarray analysis of *Shigella flexneri*-infected epithelial cells identifies host factors important for apoptosis inhibition. BMC Genomics. 11: 272.

[55] Kotloff KL, Winickoff JP, Ivanoff B, Clemens JD, Swerdlow DL, Sansonetti P. J, Adak GK, Levine MM (1999) Global burden of *Shigella* infections: implications for vaccine development and implementation of control strategies. Bull. WHO. 77: 651-666.

[56] Tesh VL (2010) Induction of apoptosis by Shiga toxins. Future Microbiol. 5(3): 431-453.

[57] Lindgren SW, Stojilkovic I, Heffron F (1996) Macrophage killing is an essential virulence mechanism of *Salmonella typhimurium*. Proc. Natl. Acad. Sci. 93: 4197-4201.

[58] Fink SL, Cookson BT (2006) Caspase-1-dependent pore formation during pyroptosis leads to osmotic lysis of infected host macrophages. Cell. Microbiol. 8: 1812-1825.

[59] Knodler LA, Finlay BB, Steele-Mortimer O (2005) The *Salmonella* effector protein SopB protects epithelial cells from apoptosis by sustained activation of Akt. J. Biol. Chem. 280(10): 9058-9064.

[60] Guzman CA, Domann E, Rohde M, Bruder D, Darji A, Weiss S, Wehland J, Chakraborty T, Timmis KN (1996) Apoptosis of mouse dendritic cells is triggered by listeriolysin, the major virulence determinant of *Listeria monocytogenes*. Mol. Microbiol. 20: 119-126.

[61] Rogers HW, Callery MP, Deck B, Unanue ER (1996) *Listeria monocytogenes* induces apoptosis of infected hepatocytes. J. Immunol. 156: 679-684.

[62] Stavru F, Cossart P (2011) Listeria infection modulates mitochondrial dynamics. Commun. Integr. Biol. 4(3): 364-366.

[63] Barsig J, Kaufmann SH (1997) The mechanism of cell death in *Listeria monocytogenes*-infected murine macrophages is distinct from apoptosis. Infect. Immun. 65: 4075-4081.

[64] Miller WC, Ford CA, Morris M, Handcock MS, Schmitz JL, Hobbs MM, Cohen MS, Harris KM, Udry JR (2004) Prevalence of chlamydial and gonococcal infections among young adults in the United States. JAMA. 291: 2229-2236.

[65] Fan T, Lu H, Hu H, Shi L, McClarty GA, Nance DM, Greenberg AH, Zhong G (1998) Inhibition of apoptosis in chlamydia-infected cells: blockade of mitochondrial cytochrome *c* release and caspase activation. J. Exp. Med. 187: 487-496.

[66] Fischer SF, Schwarz C, Vier J, Hacker G (2001) Characterization of antiapoptotic activities of *Chlamydia pneumoniae* in human cells. Infect. Immun. 69: 7121-7129.

[67] Bouillet P, Strasser A (2002) BH3-only proteins-evolutionarily conserved proapoptotic Bcl-2 family members essential for initiating programmed cell death. J. Cell. Sci. 115: 1567-1574.

[68] Du K, Zheng Q, Zhou M, Zhu L, Ai B, Zhou L (2011) Chlamydial antiapoptotic activity involves activation of the Raf/MEK/ERK survival pathway. Curr. Microbiol. 63(4): 341-346.

[69] Monack DM, Mecsas J, Bouley D, Falkow S (1998) *Yersinia*-induced apoptosis in vivo aids in the establishment of a systemic infection of mice. J. Exp. Med. 188: 2127-2137.

[70] Monack DM, Mecsas J, Ghori N, Falkow S (1997) *Yersinia* signals macrophages to undergo apoptosis and YopJ is necessary for this cell death. Proc. Natl. Acad. Sci. 94: 10385-10390.

[71] Cornelis GR, Boland A, Boyd AP, Geuijen C, Iriarte M (1998) The virulence plasmid of *Yersinia*, an antihost genome. Microbiol. Mol. Biol. Rev. 62: 1315-1352.

[72] Erfurth SE, Gröbner S, Kramer U, Gunst DSJ, Soldanova I, et al (2004) *Yersinia enterocolitica* induces apoptosis and inhibits surface molecule expression and cytokine production in murine dendritic cells. Infect. Immun. 72: 7045-7054.

[73] Lemaitre N, Sebbane F, Long D, Hinnebusch BJ (2006) *Yersinia pestis* YopJ suppresses tumor necrosis factor alpha induction and contributes to apoptosis of immune cells in the lymph node but is not required for virulence in a rat model of bubonic plague. Infect. Immun. 74: 5126-5131.

[74] Mittal R, Peak-Chew SY, McMahon HT (2006) Acetylation of MEK2 and I kappa B kinase (IKK) activation loop residues by YopJ inhibits signaling. Proc. Natl. Acad. Sci. 103: 18574-18579.

[75] Mittal R, Peak-Chew SY, Sade RS, Vallis Y, McMahon HT (2010) The acetyltransferase activity of the bacterial toxin YopJ of *Yersinia* is activated by eukaryotic host cell inositol hexakisphosphate. J. Biol. Chem. 285: 19927-19934.

[76] Haase R, Kirschning CJ, Sing A, Schröttner P, Fukase K, Kusumoto S, Wagner H, Ruckdeschel K (2003) A dominant role of Toll-like receptor 4 in the signaling of apoptosis in bacteria-faced macrophages. J. Immunol. 171: 4294-4303.

[77] Zheng Y, Lilo S, Brodsky IE, Zhang Y, Medzhitov R, Marcu KB, Bliska JB (2011) A *Yersinia* effector with enhanced inhibitory activity on the NF-κB pathway activates the NLRP3/ASC/caspase-1 inflammasome in macrophages. PLoS Pathog. 7(4): e1002026.

[78] Spinner JL, Seo KS, O'Loughlin JL, Cundiff JA, Minnich SA, Bohach GA, Kobayashi SD (2010) Neutrophils are resistant to *Yersinia* YopJ/P-induced apoptosis and are protected from ROS-mediated cell death by the type III secretion system. PLoS One. 5(2): e9279.

[79] Muller A, Hacker J, Brand BC (1996) Evidence for apoptosis of human macrophage-like HL-60 cells by *Legionella pneumophila* infection. Infect. Immun. 64: 4900-4906.

[80] Gao L-Y, Susa M, Ticac B, Abu Kwaik Y (1999) Heterogeneity in intracellular replication and cytopathogenicity of *Legionella pneumophila* and Legionella micdadei in mammalian and protozoan cells. Microb. Pathog. 27: 273-287.

[81] Abdelaziz DH, Gavrilin MA, Akhter A, Caution K, Kotrange S, Khweek AA, Abdulrahman BA, Grandhi J, Hassan ZA, Marsh C, Wewers MD, Amer AO (2011) Apoptosis-associated speck-like protein (ASC) controls *Legionella pneumophila* infection in human monocytes. J. Biol. Chem. 286(5):3203-3208

[82] Proulx F, Seidman EG, Karpman D (2001) Pathogenesis of Shiga toxin-associated hemolytic uremic syndrome. Pediatr. Res. 50: 163-171.

[83] Li H, Zhu H, Xu CJ, Yuan J (1998) Cleavage of BID by caspase 8 mediates the mitochondrial damage in the Fas pathway of apoptosis. Cell. 94: 491-501.

[84] Gross A, McDonnell JM, Korsmeyer SJ (1999) Bcl-2 family members and the mitochondria in apoptosis. Genes Dev. 13: 1899-1911.

[85] Sukumaran SK, Selvaraj SK, Prasadarao NV (2004) Inhibition of apoptosis by *Escherichia coli* K1 is accompanied by increased expression of BclXL and blockade of mitochondrial cytochrome *c* release in macrophages. Infect. Immun. 72(10): 6012-6022.

[86] Silverman DJ (1984) *Rickettsia rickettsii* induced cellular injury of human vascular endothelium in vitro. Infect. Immun. 44: 545-553.

[87] Bechelli JR, Rydkina E, Colonne PM, Sahni SK (2009) *Rickettsia rickettsii* infection protects human microvascular endothelial cells against staurosporine-induced apoptosis by a cIAP(2)-independent mechanism. J. Infect. Dis. 199(9): 1389-1398.

[88] Joshi SG, Kovács AD (2007) *Rickettsia rickettsii* infection causes apoptotic death of cultured cerebellar granule neurons. J. Med. Microbiol. 56(Pt 1): 138-141.

[89] Rojas M, Olivier M, Gros P, Barrera LF, García LF (1999) TNF-a and IL-10 modulate the induction of apoptosis by virulent *Mycobacterium tuberculosis* in murine macrophages. J. Immunol. 162: 6122-6131.

[90] Behar SM, Martin CJ, Booty MG, Nishimura T, Zhao X, Gan HX, Divangahi M, Remold HG (2011) Apoptosis is an innate defense function of macrophages against *Mycobacterium tuberculosis*. Mucosal Immunol. 4(3): 279-287.

[91] Danelishvili L, McGarvey J, Li YJ, Bermudez LE (2003) *Mycobacterium tuberculosis* infection causes different levels of apoptosis and necrosis in human macrophages and alveolar epithelial cells. Cell Microbiol. 5(9): 649-660.

[92] Kornfeld H, Mancino G, Colizzi V (1999) The role of macrophage cell death in tuberculosis. Cell Death Differ. 6: 71-78.

[93] Hasnain SE, Begum R, Ramaiah KVA, Sahdev S, Shajil EM, Taneja TK, Mohan M, Athar M, Sah NK, Krishnaveni M (2003) Host-pathogen interactions during apoptosis. J. Biosci. 28(3): 349-358.

[94] Smyth MJ, Kelly JM, Sutton VR, Davis JE, Browne KA, Sayers TJ, Trapani JA (2001) Unlocking the secrets of cytotoxic granule proteins. J. Leukocyte Biol. 70: 18-29.

[95] Roulston A, Marcellus RC, Branton PE (1999) Viruses and apoptosis. Annu. Rev. Microbiol. 53: 577-628.

[96] Mori I, Nishiyama Y, Yokochi T, Kimura Y (2004) Virus-induced neuronal apoptosis as pathological and protective responses of the host. Rev. Med. Virol. 14(4): 209-216.

[97] Balachandran S, Roberts PC, Kipperman T, Bhalla KN, Compans RW, Archer DR, Barber GN (2000) Alpha/beta interferons potentiate virus-induced apoptosis through activation of the FADD/Caspase-8 death signaling pathway. J. Virol. 74: 1513-1523.

[98] Herzer K, Sprinzl MF, Galle PR (2007) Hepatitis viruses: live and let die. Liver Int. 27: 293-301.

[99] Gil J, Esteban M (2000) Induction of apoptosis by the dsRNA-dependent protein kinase (PKR): mechanism of action. Apoptosis. 5: 107-114.

[100] [100] Frese M, Schwarzle V, Barth K, Krieger N, Lohmann V, Mihm S, Haller O, Bartenschlager R (2002) Interferon-gamma inhibits replication of subgenomic and genomic hepatitis C virus RNAs. Hepatology. 35: 694-703

[101] Stark GR, Kerr IM, Williams BR, Silverman RH and Schreiber RD (1998) How cells respond to interferons. Annu. Rev. Biochem. 67: 227-264)

[102] Chu WM,Ostertag D, Li ZW, Chang L, Chen Y, Hu Y, Williams B, Perrault J andKarin M (1999) JNK2 and IKKbeta are required for activating the innate response to viral infection. Immunity. 11: 721-731

[103] Lauer GM, Walker BD. (2001) Hepatitis C virus infection. N. Engl. J. Med. 345: 41-52

[104] Kumar S. (2007) Caspase function in programmed cell death. Cell Death Differ. 14: 32-43

[105] Ray RB, Lagging LM, Meyer K, Steele R, Ray R. (1995) Transcriptional regulation of cellular and viral promoters by the hepatitis C virus core protein. Virus Res. 37: 209-220

[106] Machida K, Tsukiyama-Kohara K, Seike E, Tone S, Shibasaki F, Shimizu M, Takahashi H, Hayashi Y, Funata N, Taya C, Yonekawa H, Kohara M (2001) Inhibition of cytochrome *c* release in Fas-mediated signaling pathway in transgenic mice induced to express hepatitis C viral proteins. J. Biol. Chem. 276: 12140-12146

[107] Mc Arthur JC, Brew BJ, Nath A (2005) Neurological complications of HIV infection. Lancet Neurol. 4(9):543-555.

[136] Kadaja M, Isok-Paas H, Laos T, Ustav E, Ustav M (2009) Mechanism of genomic instability in cells infected with the high-risk human papillomaviruses. PLoS Pathog. 5:e1000397-e1000412.

[137] Tan S, Hougardy BM, Meersma GJ, Schaap B, de Vries EG, van der Zee AG, de Jong S (2012) HPV16 E6 RNA interference enhances cisplatin and death receptor-mediated apoptosis in human cervical carcinoma cells. Mol. Pharmacol. 81(5):701-709.

[138] Filippova M, Song H, Connolly JL, Dermody TS, Duerksen-Hughes PJ (2002) The human papillomavirus 16 E6 protein binds to tumor necrosis factor (TNF) R1 and protects cells from TNF-induced apoptosis. J. Biol. Chem. 277: 21730-21739.

[139] Shenk T (1996) Adenoviridae: the viruses and their replication. In: B.N. Fields, D. Knipe, P. Howley (Eds.) et al., Fields Virology (3rd ed.). Lippincott-Raven, Philadelphia, pp. 2111-2148.

[140] McNees AL, Gooding LR (2002) Adenoviral inhibitors of apoptotic cell death. Virus Res. 88(1-2):87-101.

[141] Flint J, Shenk T (1997) Viral transactivating proteins. Annu. Rev. Genet. 31:177-212.

[142] Lowe SW, Ruley HE (1993) Stabilization of the p53 tumor suppressor is induced by adenovirus 5 E1A and accompanies apoptosis. Genes Dev. 7(4):535-545.

[143] Livne A, Shtrichman R, Kleinberger T (2001) Caspase activation by adenovirus e4orf4 protein is cell line specific and is mediated by the death receptor pathway. J. Virol. 75:789-798.

[144] Perez D, White E (2000) TNF-alpha signals apoptosis through a bid-dependent conformational change in Bax that is inhibited by E1B 19K. Mol. Cell. 6 (1), 53-63.

[145] Teodoro JG, Branton PE (1997) Regulation of p53-dependent apoptosis, transcriptional repression, and cell transformation by phosphorylation of the 55-kilodalton E1B protein of human adenovirus type 5. J. Virol. 71:3620-3627.

[146] Lichtenstein DL, Toth K, Doronin K, Tollefson AE, Wold WS (2004) Functions and mechanisms of action of the adenovirus E3 proteins. Int. Rev. Immunol. 23(1-2): 75-111.

[147] Benedict CA, Norris PS, Prigozy TI, Bodmer JL, Mahr JA, Garnett CT, Martinon F, Tschopp J, Gooding LR, Ware CF (2001) Three adenovirus E3 proteins cooperate to evade apoptosis by tumor necrosis factor-related apoptosis-inducing ligand receptor-1 and -2. J. Biol. Chem. 276:3270-3278.

[148] McNees AL, Garnett CT, Gooding LR (2002) The adenovirus E3 RID complex protects some cultured human T and B lymphocytes from Fas-induced apoptosis. J. Virol. 76(19):9716-9723.

[149] Korner H, Burgert HG (1994). Down-regulation of HLA antigens by the adenovirus type 2 E3/19K protein in a T lymphoma cell line. J. Virol. 68 (3):1442-1448.

[150] Querido E, Blanchette P, Yan Q, Kamura T, Morrison M, Boivin D, Kaelin WG, Conaway RC, Conaway JW, Branton PE (2001) Degradation of p53 by adenovirus E4orf6 and E1B55K proteins occurs via a novel mechanism involving a Cullin-containing complex. Genes Dev. 15(23):3104-3117.

[151] Goldmacher VS, Bartle LM, Skaletskaya A, Dionne CA, Kedersha NL, Vater C A, Han J, Lutz RJ, Watanabe S, McFarland ED, Kieff ED, Mocarski ES, Chittenden T (1999) A cytomegalovirus-encoded mitochondria-localized inhibitor of apoptosis structurally unrelated to Bcl-2. Proc. Natl. Acad. Sci. U. S. A. 96:12536-12541

[152] Skaletskaya A, Bartle LM, Chittenden T, McCormick AL, Mocarski ES, Goldmacher VS (2001) A cytomegalovirus-encoded inhibitor of apoptosis that suppresses caspase-8 activation. Proc. Natl. Acad. Sci. U. S. A. 98(14):7829-7834.

[153] Goldmacher, VS (2002) vMIA, a viral inhibitor of apoptosis targeting mitochondria. Biochimie. 84:177-185.

[154] McCormick AL, Roback L, Mocarski ES (2008) HtrA2/Omi terminates cytomegalovirus infection and is controlled by the viral mitochondrial inhibitor of apoptosis (vMIA). PLoS Pathog. 4: e1000063.

[155] Brune W, Menard C, Heesemann J, Koszinowski UH (2001) A ribonucleotide reductase homolog of cytomegalovirus and endothelial cell tropism. Science. 291(5502): 303-305.

[156] Zhu H, Shen Y, Shenk T (1995) Human cytomegalovirus IE1 and IE2 proteins block apoptosis. J. Virol. 69(12):7960-7970.

[157] McCormick AL. Control of apoptosis by human cytomegalovirus. Curr. Top. Microbiol. Immunol. 2008;325:281-95.

[158] Bienvenu AL, Gonzalez-Rey E, Picot S (2010) Apoptosis induced by parasitic diseases. Parasit Vectors. 17;3:106.

[159] Tenter AM, Heckeroth AR, Weiss LM (2000) *Toxoplasma gondii*: from animal to humans. Int. J. Parasitol. 30: 1217-1258.

[160] Begum-Haque S, Haque A, Kasper LH. Microb. Pathog. 2009 Apoptosis in *Toxoplasma gondii* activated T cells: the role of IFNgamma in enhanced alteration of Bcl-2 expression and mitochondrial membrane potential. 47(5):281-288.

[161] Channon JY, Miselis KA, Minns LA, Dutta C, Kasper LH (2002) *Toxoplasma gondii* induces granulocyte colony-stimulating factor and granulocyte-macrophage colony-stimulating factor secretion by human fibroblasts: implications for neutrophil apoptosis. Infect. Immun. 70(11):6048-6057.

[162] Lüder CG, Gross U (2005) Apoptosis and its modulation during infection with *Toxoplasma gondii*: molecular mechanisms and role in pathogenesis. Curr. Top. Microbiol. Immunol. 289:219-237

[163] Piguet PF, Kan CD, Vesin C (2002) Thrombocytopenia in an animal model of malaria is associated with an increased caspase-mediated death of thrombocytes. Apoptosis. 7(2):91-98.

[164] Brand VB, Sandu CD, Duranton C, Tanneur V, Lang KS., Huber SM, aLang F (2003) Dependence of plasmodium falciparum in vitro growth on the cation permeability of the human host erythrocyte. Cell. Physiol. Biochem. 13:347-356.

[165] Pino P, Vouldoukis I, Kolb JP, Mahmoudi N, Desportes-Livage I, Bricaire F, Danis M, Dugas B, Mazier D (2003) *Plasmodium falciparum*-infected erythrocyte adhesion induces caspase activation and apoptosis in human endothelial cells. J. Infect. Dis. 187:1283-1290.

[166] Stuart K, Brun R, Croft S, Fairlamb A, Gurtler RE, McKerrow J, Reed S, Tarleton R (2008) Kinetoplastids: related protozoan pathogens, different diseases. J. Clin. Invest. 118:1301-1310.

[167] Smirlis D, Duszenko M, Ruiz AJ, Scoulica E, Bastien P, Fasel N, Soteriadou K (2010) Targeting essential pathways in trypanosomatids gives insights into protozoan mechanisms of cell death. Parasit. Vectors. 3:107.

[168] Orrenius S, Zhivotovsky B, Nicotera P (2003) Regulation of cell death: the calcium-apoptosis link. Nat. Rev. Mol. Cell. Biol. 4:552-565.

[169] Camello-Almaraz C, Gomez-Pinilla PJ, Pozo MJ, Camello PJ (2006) Mitochondrial reactive oxygen species and Ca2+ signaling. Am. J. Physiol. Cell. Physiol. 291:1082-1088.

[170] Welburn SC, Dale C, Ellis D, Beecroft R, Pearson TW (1996) Apoptosis in procyclic *Trypanosoma brucei* rhodesiense in vitro. Cell Death Differ. 3:229-236.

[171] Pearson TW, Beecroft RP, Welburn SC, Ruepp S, Roditi I, Hwa KY, Englund PT, Wells CW, Murphy NB (2000). The major cell surface glycoprotein procyclin is a receptor for induction of a novel form of cell death in African trypanosomes in vitro. Mol. Biochem. Parasitol. 111:333-349.

[172] Chuenkova MV, PereiraPerrin M (2009) *Trypanosoma cruzi* targets Akt in host cells as an intracellular antiapoptotic strategy. Sci. Signal. 2: ra74

[173] Peters NC, Egen JG, Secundino N, Debrabant A, Kimblin N, Kamhawi S, Lawyer P, Fay MP, Germain RN, Sacks D (2008) In vivo imaging reveals an essential role for neutrophils in leishmaniasis transmitted by sand flies. Science. 321: 970-974.

[174] Shaha C (2006) Apoptosis in Leishmania species & its relevance to disease pathogenesis. Indian J. Med. Res. 123(3):233-244.

Chapter 13

Caspases as Putative Biomarkers of Cervical Cancer Development

Olga V. Kurmyshkina, Pavel I. Kovchur and Tatyana O. Volkova

Additional information is available at the end of the chapter

http://dx.doi.org/10.5772/61810

Abstract

Resistance to apoptosis is commonly accepted as the principal hallmark of a cancer cell, while caspases are recognized as the key molecular players of the apoptosis regulatory network. Since the level of caspase activity is thought to be directly coupled with aggressive features of cancer cells (such as ability to withstand immune reactions, invasiveness, drug resistance, etc.), these proteases could serve as objective diagnostic markers especially for those types of cancer where early differential diagnosis is needed. Cervical cancer develops through morphologically well-described stages—from intraepithelial lesions of 1/2/3 grade including carcinoma *in situ* to microinvasive and invasive cancer with precancerous lesions known to be potentially reversible. The percentage of cervical neoplasms diagnosed at early stages is relatively high, providing a basis for the use of cervical cancer as an *in vivo* model to investigate the mechanisms of apoptosis modulation in malignant cells. The existing diagnostic criteria, despite their usefulness, have substantial limitations with respect to cervical cancer and preneoplastic lesions, so caspases may be helpful in improving them, but there is insufficient data regarding the involvement of these enzymes in cervical cancer development. In this chapter, we report on specific patterns of activity of caspases revealed in tissue biopsies and blood lymphocytes in association with different stages of cervical cancer development. The data indicate that caspases are pivotal components of the *in vivo* molecular "portrait" of cervical cancer and have the potential of being used as biomarkers.

Keywords: Biomarkers of carcinogenesis, apoptosis, caspases, cervical cancer, human papillomavirus

1. Introduction

Carcinogenesis of solid tumors is a complex multistage process that generally develops over a long period of time through a succession of precancerous lesions. Cancer *in situ* and micro-

processing of the downstream executioner (effector) caspases 3, 6, and 7, which, in turn, destroy numerous structural, regulatory, and catalytic intracellular proteins. Enzymatic activity of caspases is tightly regulated by various endogenic inhibitors and activators, including IAP-1/-2, Bcl-2 family members, Smac/DIABLO, c-FLIP, Survivin/BIRC5, XIAP, NAIP, livin, and others. The level of activity of caspases (and thereby susceptibility to apoptotic signals) is thought to be determined by the ratio of expression levels of their endogenous modulators. Many human diseases, including neoplasms, are known to be accompanied by the repression of caspases functions; however, this repression usually not associated with abnormalities in their gene structure that distinguishes caspases from the so-called tumor suppressors. Inactivation of caspases occurs in cancer cells as a consequence of the overexpression of their inhibitors or as a result of suppression of the upstream components of the apoptotic signaling pathways, as for example death receptors [5].

Due to active mutational processes and viral etiology, cervical cancer is characterized by particularly early and rapid development of resistance to apoptosis. Numerous articles reporting on the influence of different variants of post-transcriptional E6/E7 silencing on the cell sensitivity to the apoptosis inducers provide strong evidence that hyperexpression of HPV oncogenes is the primary cause and a prerequisite for the development and maintenance of apoptotic-resistant phenotype. For example, the transfection of E6-siRNAs (small interfering RNAs) into CC cells conferred susceptibility to cisplatin-induced apoptosis [6]. Similarly, the expression of an E7-targeted RNA-aptamer disrupting the interaction between E7 and pRb resulted in the induction of apoptosis [7]. Treatment with the synthetic peptide anti-E7 antagonist was shown to suppress tumor growth in an animal xenograft model due to activation of apoptotic cell death [8]. Analysis of the published data revealed three basic groups of mechanisms that modulate apoptosis signaling pathways by engagement of HPV oncopro-teins: (1) inactivation of proapoptotic proteins resulting from direct binding to E6/E7 with subsequent ubiquitination and proteasome-mediated degradation; (2) interactions of E6/E7 proteins with the cellular transcription factors and chromatin-remodeling enzymes leading to the change of either the mRNA expression pattern of pro- and antiapoptotic factors, or the profile of microRNAs (miRNAs) targeted these factors; and (3) HPV oncogene-mediated induction of genomic instability that causes either the accumulation of inactivating mutations in the proapoptotic oncosuppressor genes, or, alternatively, the amplification of antiapoptotic genes.

1. "High-risk" HPV oncoproteins are capable of high-affinity binding to various protein components of the extrinsic or intrinsic apoptosis pathways and stimulating their degradation due to the ubiquitin-ligase activity, thereby blocking signal transduction from an apoptogenic stimulus. In HPV-positive cells, the membrane expression of CD95/ Fas, the key cell death receptor, is significantly reduced, and the DISC assembling is impaired because of accelerated destruction of the FADD adaptor protein and caspase 8, with endogenous inducers of mitochondrial apoptosis pathway (such as Bid, Bak, and Bax antagonists of Bcl-2 protein) being degraded as well (for a review, see [9, 10]). That is why the abrogation of proteasome functions (by MG132 or Bortezomib treatment, for example) potentiates the activity of caspases and sensitizes CC cells to TRAIL-/Fas-dependent apoptosis or radiation-induced cell death [11]. E5 protein, similarly to E6 and E7, can impair the mechanisms of CD95L- and TRAIL-mediated apoptosis [10, 12].

2. Epigenetic modification of apoptotic genes is one of the mechanisms of global regulation of the cell death program in cervical cancer. HPV has been shown to dramatically alter the host DNA methylation landscape, especially within the promoter regions of tumor suppressor genes and genes coding for apoptosis activator proteins, thus facilitating conversion of these genes into the heterochromatin state. For example, the promoter hypermethylation of PRDM14 gene encoding a transcription factor which is required for the expression of NOXA and PUMA proapoptotic regulators of Bcl2-family was observed in HPV16-bearing cell lines and primary tumors [13]. Histone acetylation/deacetylation is another way of modulating gene activity employed by CC cells. The treatment of CC cell lines with various histone deacetylase inhibitors results in apoptosis induction followed by activation of caspases 3, 8, and 9, PARP cleavage, and loss of mitochondrial membrane potential, thus confirming the importance of this epigenetic mechanism for the establishment of apoptotic resistance [14–16]. In a similar manner, silencing of the MLL5β histone methyltransferase has an apoptosis-inducing effect on CC cells [17]. Transcriptomic studies also suggest that among the different functional groups of genes whose expression is affected by the presence of HPV oncogenes, the apoptosis-regulatory genes (as for example, BCL2, BCLXL, and c-IAP1 [18]) constitute a substantial portion. The expression profile of pro- and antiapoptotic miRNAs in CC cells also arouses much interest among researchers. More than 100 miRNA species were documented to change their expression in the presence of the viral E6/E7 proteins, and many of their mRNA-targets were found to code for various regulators of apoptosis and, in particular, caspases, as for example survivin and Bcl-family proteins [19–22]. The novel long noncoding RNAs (lncRNAs) contributing to the development of apoptosis-resistant phenotype of CC cells have been described as well [23].

3. In cervical cancer cells, the deletions of the chromosome loci containing genes required for the apoptotic program to be implemented are found to occur at a high frequency [24, 25]. Although having sporadic nature of occurrence, these genetic abnormalities most likely confer a selective advantage to cancer cells for further expansion that probably explains why the incidence of such abnormalities increases with tumor progression [24]. In contrast, for genes encoding inhibitors of apoptosis, amplification of the corresponding genome segments is frequently observed with CC progression [26].

As evidenced by the above-stated examples, the mechanisms cervical cancer cells employ to achieve apoptotic resistance engage all the levels of intracellular regulation—genomic, transcriptomic, epigenomic, and proteomic. Although all the diversity of the known molecular pathways is ultimately directed to the suppression of caspases as the crucial mediators of cell death reactions, the experimental data showing that their proteolytic activity does undergo specific changes in the natural history of cervical neoplasms appeared to be virtually absent in literature, thus prompting us to conduct research whose results are described below.

2.2. Induction of apoptosis in immunocompetent cells as a putative factor of cervical cancer progression

Because of the viral etiology of cervical cancer, the mechanisms of its development need to be investigated in conjunction with changes occurring in the immune system. In each individual case, it is not only the properties of tumor cells (the mutation spectrum, the gene expression profile) that determine the progression of CIN to invasive metastatic state but also the survival and growth of secondary tumor foci. The abilities of the immune system to recognize and eliminate virus-infected and malignantly transformed cells are believed to be of great importance too. According to the general conception, it is due to the reactions of the immune system that both the HPV infection and dysplastic alterations of squamous epithelium (CIN1/2) are usually transient, and there is only a small percent of cases that develop to chronic or malignant form [2]. On the other hand, various mechanisms exploited by the virus and/or tumor cells for specific inhibition or avoidance of immune reactions are becoming elucidated in the last years. Furthermore, possible involvement of supplementary "environmental" factors exerting suppressive influence on the immune system of an organism is assumed. The induction of apoptosis in immunocompetent cells is supposed to be one of the mechanisms to inhibit antitumor/antiviral immunity. Although there is some evidence of increased expression of apoptosis-related markers in tumor-infiltrating or circulating lymphocytes of cancer patients, in case of cervical cancer, such information is scarce [27, 28]. CC cell lines as well as primary CC cells are known to express CD95L (FasL) [29] and, when cocultured, to induce apoptotic death in cytotoxic T-lymphocytes, the effect being abolished by anti-CD95 antibodies [30]. It was also found that treatment of peripheral blood lymphocytes taken from healthy donors with conditioned media from CC cell lines could induce apoptosis in subpopulation of CD4+ T-helpers [31]. Reasoning from these facts, we hypothesized that during its *in vivo* development cervical cancer may withstand immune reactions via promoting apoptosis in effector immune cells, and specific change of activity/expression of caspase may therefore be detected in circulating lymphocytes of CC patients.

2.3. Materials and methods

2.3.1. Patients and samples

Tissue samples and peripheral blood were obtained from 156 patients who underwent surgery in Oncological Dispensary of the Republic of Karelia: 75 women diagnosed with cervical intraepithelial neoplasia grade 3 (CIN 3, with average age at diagnosis 32.9 ± 7.4 years) and 81 women with squamous carcinoma, including 45 with stage IA (average age 31.3 ± 6.0), 21 with stage II (average age 43.6 ± 13.2), and 15 with stages III–IV (average age 46.9 ± 11.1), were examined. Stages of cancer were defined in accordance with the TNM-classification and the International Federation of Gynecology and Obstetrics system (FIGO, 1994). CIN 3 and CC diagnosis was based on comprehensive physical examination, extended colposcopy findings, cytology, and histopathology tests, in full compliance with the approved standards for the diagnosis and treatment of patients with gynecological malignancies. All women enrolled in this study were informed and gave voluntary written consent. The research was approved by the Committee on Medical Ethics of Petrozavodsk State University and the Ministry of Healthcare and Social Development of the Republic of Karelia.

Extended colposcopy was performed in each case before surgery in order to define localization and precise margins of a lesion for subsequent accurate excision of tissue fragments.

Colposcopic findings were evaluated according to the International Federation for Cervical Pathology and Colposcopy (IFCPC, 2002) terminology. Tissue samples were obtained during cervical conization or total hysterectomy. In each case, two pieces of tissue were resected from the pathologic locus, which was defined both visually and colposcopically, and one piece of morphologically normal epithelium (control) was excised from the contralateral side of the cervix outside the pathologic zone. Tissue samples were immediately submerged into RNA-stabilizing solution RNALater (Qiagen) or RPMI-1640 medium (Gibco), then frozen and stored at $-80°C$. For all patients the original diagnosis was verified by histomorphological examination.

Venous blood sampling was done right before the surgery or any other treatment. The fraction of peripheral blood mononuclear cells (PBMC) was isolated by standard procedure in Ficoll density gradient (Paneco, Russia). Forty-five samples of peripheral blood were also taken from healthy nonpregnant HPV-negative women comparable in age and anamnesis, with no pathology of the cervix (control blood group 1, age characteristics: 23.4 ± 0.9 ($n = 15$); 33.3 ± 1.7 ($n = 15$); 46.7 ± 11.1 ($n = 15$)). For patients with CIN 3 and CC stage IA, blood samples were again collected in 1 and 3 months after conization and course of immunomodulatory therapy. Control group 2 consisted of patients with CIN 3 ($n = 15$) and CC stage IA ($n = 15$) (average age 34.1 ± 7.2) who underwent only surgical treatment. Patients of the examined groups did not differ in anamnesis, virological, and histological findings.

Screening for the presence of HPV DNA and identification of HPV genotype were performed by polymerase chain reaction (PCR) using AmpliSens HPV HCR Screen kit (The Central Research Institute of Epidemiology of The Federal Service on Customers' Rights Protection and Human Well-being Surveillance, Russia) and TaqMan probes. E6/E7 oncogene mRNA was detected by the reverse transcription coupled PCR (RT-PCR), and reagents kits were purchased from DNA-Technology and Sileks companies (Russia). The PCR products were visualized by 2% agarose gel electrophoresis. The distribution of HPV genotypes is displayed in Table 1.

	HPV16	HPV16, 18, 31, 33	HPV18	HPV31	HPV33
CIN 3	52.1	34.8	13.1	-	-
CC stage IA	79.4	-	11.8	5.9	2.9
CIN 3 + CC stage IA	61.2	23.3	12.6	1.9	1.0
CIN 3 + CC stage IA (3 months after treatment)	1.1	2.2	1.1	-	-

Table 1. The distribution (%) of HPV genotypes in groups of patients diagnosed with CIN 3 or CC of stage IA.

2.3.2. Real-time PCR

Total RNA from tumor cells and PBMC was extracted with TRizol reagent (Invitrogen, USA) following the manufacturer's guidelines. The concentration and purity of the RNA template was determined by spectrophotometry (BioWave II+, Biochrom, UK). RNA nativity was determined by capillary gel electrophoresis using Experion Automated Station and RNA StdSens analysis kit (Bio-Rad, USA). The extracted RNA template was treated with DNase I (Fermentas, ThermoScientific, USA). Complementary DNA (cDNA) was synthesized from 1 μg of total RNA using random hexaprimers and ProtoScript MMLV reverse transcriptase following the protocol proposed by the manufacturer (New England BioLabs, UK). RNA and cDNA samples were stored at −80°C. Gene expression was estimated by SYBR Green real-time PCR. Amplification was performed in StepOnePlus thermal cycler (Applied Biosystems, USA) with StepOne™ Software v2.2.2 using 20 ng of cDNA per 1 reaction volume (25 μl) and qPCRmix-HS-SYBR+HighROX 5×-reaction mix (Evrogen, Russia), containing gene-specific primers at final concentration 0.5 μM. Primers for the nucleotide sequences of the investigated genes were selected from published sources (Table 2). Oligonucleotides were synthesized by the Evrogen company (Russia). The PCR protocol was 20 s at 95°C, 20 s at 60°C, and 30 s at 72°C (45 cycles). All the reactions were done in duplicates. Effectiveness of amplification (%) was monitored by standard curve approach. To determine the specificity of primer annealing, the PCR fragments were melted: 1 min at 95°C, 1 min at 60°C, and 10 s at 60°C (80 cycles, the temperature raised by 0.5°C in each cycle). To exclude the possibility of the template cDNA being contaminated by the genomic DNA, PCR was performed for each template under the same conditions with the RNA matrix (negative control). The resultant reaction products were also separated in 8% polyacrylamide gel using the Tris–borate buffer, then stained with 1% ethidium bromide solution and visualized in transmitted UV light (with GelDoc-It Imaging System, UVP, USA). The correspondence of amplicon sizes to theoretically expected ones was confirmed using the low-molecular pUC19/Msp I fragment length marker (Syntol, Russia). Gene mRNA expression was measured using the $2^{-\Delta\Delta Ct}$ method [32]. cDNA samples from normal epithelial tissue or from PBMC of healthy donor were used as calibrator.

Gene	Sequence	PCR product length	Source
GAPDH F	5′-GAAGGTGAAGGTCGGAGTC-3′	225	[33]
GAPDH R	5′-GAAGATGGTGATGGGATTTC-3′		
Caspase 6 F	5′-ACTGGCTTGTTCAAAGG-3′	181	[34]
Caspase 6 R	5′-CAGCGTGTAAACGGAG-3′		
Caspase 3 F	5′-ATGGAAGCGAATCAATGGAC-3′	240	[35]
Caspase 3 R	5′-ATCACGCATCAATTCCACAA-3′		
Caspase 9 F	5′-AACAGGCAAGCAGCAAAGTT-3′	246	[35]
Caspase 9 R	5′-CACGGCAGAAGTTCACATTG-3′		

Table 2. Primers for the nucleotide sequences of the genes under study.

2.3.3. The enzyme activity of caspases

The enzyme activity of caspases was determined by standard technique using specific substrates labeled with fluorescent marker (7-amino-4-trifluoromethylcumarin — AFC) (Bio-

Rad, USA), detected by variations in fluorescence or optical density [36]. Fifty microliters of lytic buffer prepared by mixing 920 µl of bidistilled H_2O, 40 µl of 25-fold reaction buffer, and 10 µl of each of the four inhibitors: phenylmethylsulfonyl fluoride (PMSF) (35 mg/ml), pepstatin A (1 mg/ml), aprotinin (1 mg/ml), and leupeptin (1 mg/ml), were added to the tumor sample (5 mg) or PBMC (10^6 cells). The 25-fold reaction buffer included the following components: 250 mM HEPES, pH 7.4, 50 mM EDTA, 2.5% 3-((3-chloramidopropyl)dimethylammonio)-1-propanesulfonate (CHAPS), and 125 mM dithiothreitol. After that, the cells were frozen three times in liquid nitrogen, the cell lysate then centrifuged in a microcentrifuge at 17,000g (4°C) for 30 min, and the supernatant (template) collected. The activity of caspases 3, 6, and 9 was determined in the reaction buffer by mixing the template with the corresponding specific substrate. The substrate for caspase 3 was DEVD (Asp–Glu–Val–Asp), for caspase 6 — VEID (Val–Glu–Ile–Asp), for caspase 8 — LETD (Leu–Glu–Thr–Asp), and for caspase 9 — LEHD (Leu–Glu–His–Asp). The amount of cleaved AFC was measured by spectrophotometry in FluoroMax ("Horiba-Scientific," Japan) at 395 nm 30, 60, 90, 120, 150, and 180 min after the onset of the reaction. Then, the curve of caspase activity depending on the template and substrate incubation time was plotted. Relative proteolytic activity was calculated as the slope $\Delta S/\Delta t$, where

$$\Delta S = \left[S(t_i) - B(t_i) \right] - \left[S(t_0) - B(t_0) \right], \ \Delta t = (t_i - t_0),$$

where S is the sample signal at time t, and B is the blank signal at time t, t_i is the time of measurement, and t_0 is the time of initial measurement.

2.3.4. Flow cytometry

Total leukocyte fraction was prepared via using ammonium chloride osmotic shock (155 mM NH_4Cl, 10 mM $KHCO_3$, 0.1 mM Na_2EDTA), and 100 µl of whole blood was taken for each probe. Cells were centrifuged and washed with Versene solution (0.02% Na_2EDTA in phosphate-buffered saline). To measure the surface expression of CD antigens, the cells were probed with fluorophore-conjugated monoclonal antibodies (mAbs): CD3-APC, CD4-FITC, CD8-FITC, and CD95-RPE (Dako, Denmark). One hundred microliters of cell suspension was incubated with mAbs (1:20) for 30 min at room temperature. To prevent nonspecific Fc receptor-mediated mAb binding, the FcR Blocking Reagent was used (Miltenyi Biotec, Germany) in accordance with the manufacturer's instructions. Analysis was performed with MACSQuant Analyzer flow cytometer (Miltenyi Biotec.). Figure 2 describes the scheme for discriminating cell subpopulations of interest (the gating strategy); lymphocytes were gated by forward and side scatter. The numbers of lymphocytes having the following phenotypes were evaluated by the fluorescence parameters: $CD3^+$ (T-lymphocytes), $CD3^+CD4^+$ (T-helpers), $CD3^+CD8^+$ (T-killers), $CD3^+CD95^+$, $CD3^+CD95^{high}$, $CD3^+CD4^+CD95^+$, $CD3^+CD8^+CD95^+$, $CD3^+CD4^+CD95^{high}$, and $CD3^+CD8^+CD95^{high}$. Not less than 100,000 cells were analyzed in each probe. Dead cells were excluded from analysis by propidium iodide staining.

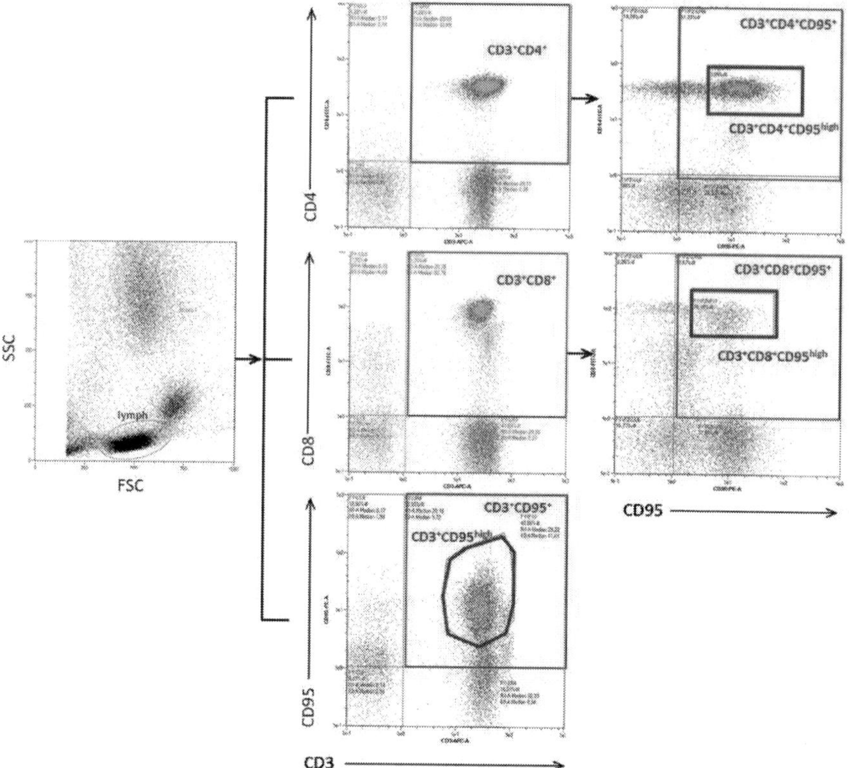

Figure 2. The scheme of the flow cytometric analysis of peripheral blood lymphocytes. Statistical reliability of the obtained results was estimated using the Student *t*-test and the nonparametric Wilcoxon–Mann–Whitney test. One-way ANOVA was also performed to compare between control/cancer groups of samples.

2.4. Results and discussion

2.4.1. Expression and activity of caspases in cervical intraepithelial neoplasia grade 3 and cervical cancer tissue samples

In the first phase of our research, we assessed the level of the relative mRNA expression of the initiator caspase 9 and executioner caspases 3 and 6 at different steps of CC development, starting from CIN 3 to advanced cancer (stages II–IV) comparing to the normal epithelium. The results are summarized in Table 3. Substantial fraction of CIN 3 samples (more than 50%) displayed the control level of caspases 3, 6, and 9 expressions, while the downregulation of caspase 3 and caspase 9 mRNAs was observed in the one-third of CIN 3 samples; caspase 6 was downregulated in 20% of samples. In the minor portion of CIN 3 cases (less than 20%), the relative amount of caspase 3, caspase 6, and caspase 9 transcripts was increased compared to the control. Microinvasive cancer (stage IA1) showed elevated mRNA levels for all three caspases in 50% of cases, with other cases displaying no difference or decrease (of various extents) relative to the control level. Analysis of stage II–IV CC samples revealed that mRNA levels of caspases 3, 6, and 9 generally did not differ from those of normal epithelium. For the rest of the advanced cancer samples, the upregulation of caspases 3 and 6 was detected, with caspase 9 showing inverse relation. In summary, cervical cancer progression was not found to be significantly correlated with mRNA expression of caspases 3, 6, and 9.

	CIN 3 (n = 25)	CC stage IA (n = 12)	CC stage II–IV (n = 10)
Caspase 3	Control, 56% ($n = 14$)	Control, 33% ($n = 4$)	Control, 60% ($n = 6$)
	↑ , 16% ($n = 4$)	↑ , 50% ($n = 6$)	↑ , 40% ($n = 4$)
	↓ , 28% ($n = 7$)	↓ , 17% ($n = 2$)	
Caspase 6	Control, 60% ($n = 15$)	Control, 25% ($n = 3$)	Control, 70% ($n = 7$)
	↑ , 20% ($n = 5$)	↑ , 50% ($n = 6$)	↑ , 30% ($n = 3$)
	↓ , 20% ($n = 5$)	↓ , 25% ($n = 3$)	
Caspase 9	Control, 52% ($n = 13$)	Control, 42% ($n = 5$)	Control, 40% ($n = 4$)
	↑ , 16% ($n = 4$)	↑ , 50% ($n = 6$)	↑ , 60% ($n = 6$)
	↓ , 32% ($n = 8$)	↓ , 8% ($n = 1$)	

Table 3. The change of mRNA levels of caspases at CIN 3 → CC progression relative to the normal epithelium. Arrows (↑ or ↓) correspond to up- or downregulation relative to the control level.

As modulation of caspases proteolytic activity is regarded as the highest level of regulation, we further explored its relative change in normal epithelium, CIN 3, and invasive CC specimens using specific tetrapeptide fluorescently labeled substrates. In 38% of CIN 3 samples tested, caspases 3 and 6 exhibited increased activity compared to the control level, with caspase 9 activity being upregulated only in a few (14%) CIN 3 cases (Figure 3). As for the rest of CIN 3 samples, the protease activity of caspases was found to diminish or correspond to the control level. In those CIN 3 samples that exhibited altered caspase 3 activity, either downregulation or upregulation was significant as compared to the normal epithelium (Figure 3A). In contrast to caspase 3, the activity of caspase 6 matched the range of control values in 50% of CIN 3 cases; the other 50% of tissue samples demonstrating significant (2.5- to 3-fold) increase in caspase 6 activity (Figure 3B). The activity of caspase 9 in CIN 3 group was generally comparable to that of the control group (Figure 3C). Stage IA was characterized by significant decrease of activity of all caspases analyzed (for caspases 3 and 9, the median activity values were significantly lower than those of the control samples); the same trend was observed for stages II–IV of CC. Note that in 23% of CC stage IA samples, the activity of caspase 6 was notably increased.

To summarize, a gradual downregulation of caspases 3, 6, and 9 occurs as invasive CC progresses, while at the stage of preinvasive cancer (CIN 3) the activity of caspases 3 and 6 may be significantly higher than that of normal epithelium. This observation may indicate, in respect of apoptosis-associated processes, a high degree of molecular heterogeneity of lesions morphologically defined as CIN 3. The activity of caspases 3 and 9 was shown to be significantly correlated with the clinical stage ($r = -0.72$, $R^2 = 0.52$, $p < 0.01$ for caspase 3; $r = -0.67$, $R^2 = 0.45$, $p < 0.01$ for caspase 9; linear regression). The correlation between caspase 6 activity and CC stage was not proved to be statistically significant. Comparing the results on the expression level of caspases 3, 6, and 9 mRNA, on the one side, and the change of their protease activity, on the other side, it can be inferred that, for the invasive forms of CC, these two regulatory levels (i.e., transcriptional and translational) are poorly correlated: as the activity of caspases decreases, their mRNA levels stay within the range of control values or increase. At the same time, mRNA expression and protease activity values were comparable with respect to CIN 3 lesions.

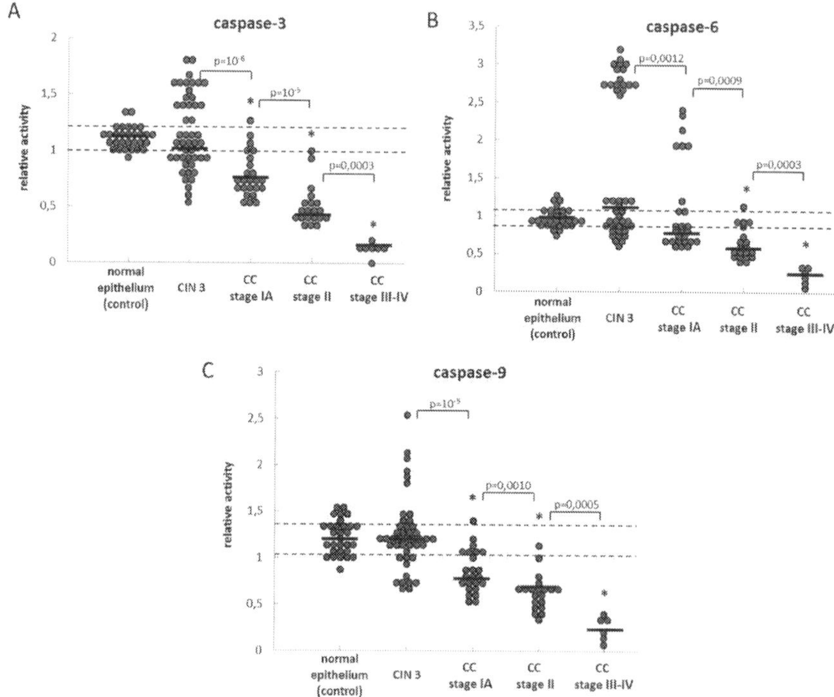

Figure 3. Scatter plots showing individual patients' changes of protease activity of caspases 3(A), 6(B), and 9(C) in pathological tissue at different stages of CC development. Horizontal bars represent the median activity level for each data set; dashed lines indicate ± standard deviation (SD) of the mean value in samples of normal epithelium (control). The results were multiplied by 10^4 for convenience. *Patient groups with median caspase activity being significantly different from that of control (the differences were considered significant if p-value was <0.05, U-test). The p-values shown on the plots were obtained by comparing successive stages of CC.

The above data allow us to speculate that the growth and development of CC is accompanied by multifaceted impairments of apoptotic processes implemented in the form of global suppression of caspases' functioning and multiple resistance of tumor cells to apoptotic stimuli. To our knowledge, this is the first systematic study investigating the changes of expression/activity of functionally different caspases at different clinical stages of cervical cancer. Caspase 3 is a universal executioner caspase of both "extrinsic" and "intrinsic" apoptotic pathways. Caspase 6 represents one of the least studied members of the family, although new evidence indicates its crucial role in amplification of apoptotic signal; as distinct from other executioner caspases (3 and 7), caspase 6 is capable of being autoactivated. However, its elevated activity does not necessarily result in apoptosis [37]. It was also revealed that the range of caspase 6 targets does not overlap with that of other effector caspases [38]. Caspase 9 is a crucial intracellular sensor of apoptogenic stimuli such as hypoxia, genome destabilization, redox dysregulation, deficiency of prosurvival, and growth factors—all these stimuli inevitably affect the process of tumor development as a consequence of rapid accumulation of cell mass and overexpression of HPV oncogenes. Therefore, the suppression of caspase 9 is required to escape their influence. It is worthy of notice that the inhibition of caspase activity is an early event in CC progression since it was observed in the majority of stage IA1 samples (i.e., invasive microcarcinoma) and apparently reflects the establishment of an aggressive phenotype of CC cells.

The principal conclusion that can be made on the basis of our study and data available from literature is that caspases occupy the key position in the multistep model of natural CC development and have prospects of being used as biomarkers for this type of oncopathology. The importance of studying the profile of expression/protease activity of caspases with the use of cervical tissue samples is emphasized in the work of Arechaga-Ocampo et al. [39], where the authors come to a conclusion that the activity of caspases upon CC development is determined by the individual spectrum of genomic aberrations and the microenvironment. Therefore, the information on caspases expression patterns derived by the use of model cell systems describes isolated cases rather than reflects the *in vivo* situation. In view of this aspect, the elevated levels of activity of caspases 3 and 6 that we observed in CIN 3 samples seem to be of a particular interest. Indeed, there are some published reports arguing for frequent upregulation of apoptotic markers (caspases, internucleosomal DNA fragmentation) in clinical tissue specimens of CIN 3 [40–44], but nevertheless, the data pointing at the absence of such alterations should also be taken into consideration [45]. At this point, it is difficult to judge objectively whether the increase of caspase activity is linked to the higher risk of a subsequent fast progression of CIN to invasive cancer or, in contrast, the probability of long-term disease persistence without signs of invasive growth. To understand the biological significance of certain changes in activity of caspases and the driving forces underlying these changes upon CC development, it is necessary to continue research with inclusion of other regulatory molecular factors. Caspases as being multifunctional proteinases may participate in some processes that rule the HPV life cycle progression [46]. In particular, there is evidence showing that production of E-proteins is coupled with the increase of stability and activity level of caspases [47]. It was also discovered that HPV oncoproteins contain specific sites for caspase-dependent proteolysis [48]. The study of Moody et al. [46], carried on with the use of organotypic raft model of stratified epithelium, demonstrates the potential role for the proteolytic activity of caspases in the HPV life cycle. According to Moody et al., in raft cultures generated from normal keratinocytes, the active (processed) form of caspase 3 was present in small amounts and was concentrated predominantly in the basal cell layer. In similar cultures

Chapter 14

The Apoptotic Microtubule Network During the Execution Phase of Apoptosis

Manuel Oropesa Ávila, Alejandro Fernández Vega, Juan Garrido Maraver, Marina Villanueva Paz, Isabel De Lavera, Mario De La Mata, Mario D. Cordero, Elizabet Alcocer Gómez, Ana Delgado Pavón, Mónica Álvarez Córdoba, David Cotán and José Antonio Sánchez-Alcázar

Additional information is available at the end of the chapter

http://dx.doi.org/10.5772/61481

Abstract

Apoptosis is a regulated energy-dependent process of cell death characterized by specific morphological and biochemical features in which caspase activation has a central role. During apoptosis, cells undergo characteristic morphological rearrangements in which the cytoskeleton participates actively. From a historical point of view, this reorganization has been assigned mainly to actinomyosin ring contraction with microtubule and intermediate filaments, both reported to be depolymerized at early stages of apoptosis. However, recent results have shown that the microtubule cytoskeleton is reformed during the execution phase of apoptosis, forming an apoptotic microtubule network (AMN). AMN is closely associated with the plasma membrane, forming a cortical ring or cellular "cocoon." Apoptotic microtubules' reorganization has been reported in many cell types and under many apoptotic inducers. Recently, it has been proposed that AMN is essential for preserving plasma membrane permeability and cell morphology during the execution phase of apoptosis. Apoptotic microtubules' depolymerization leads cells to secondary necrosis and the release of toxic intracellular contents that can harm surrounding cells and initiate inflammation. Therefore, microtubules' reorganization in physiological apoptosis during development and in the adult organism or in pathological apoptosis induced by anticancer treatments or chronic inflammation is essential for tissue homeostasis, preventing cell damage and inflammation.

Keywords: apoptosis, microtubules, cytoskeleton

1. Introduction

1.1. Apoptosis

Apoptosis is an intracellular signaling pathway conserved across evolution dependent on a caspase-mediated proteolytic cascade that leads to programmed cell death through a series of cellular changes distinct of cell necrosis. Apoptosis plays a critical role in tissue remodeling during development, tissue homeostasis, cleaning of senescent cells, and removal of the cells with severe DNA damage. Given that cell necrosis causes the release of toxic molecules and causes inflammation, an important function of apoptosis is to isolate specific cells and prepare them for disposal by phagocytosis. This program of cell death is carried out by organelle-directed regulators including the Bcl-2 proteins and ultimately executed by proteases of the caspase family.

The apoptotic process can be divided into three functionally distinct phases : (a) induction – cellular environmental changes that result in the activation of intracellular apoptotic mechanisms (entry into the execution phase); (b) execution – processes that result in the degradation of intracellular components by a family of proteases called caspases which cleave a variety of important structural and regulatory proteins at conserved aspartic acid residues to alter their function irreversibly; and (c) cleaning – those events associated with the removal of apoptotic cells and cellular debris by "professional" phagocytes such as macrophages, immature dendritic cells, and neutrophils.

2. Apoptosis and secondary necrosis

In contrast to apoptotic cells *in vivo*, cells that perform apoptosis *in vitro* cell cultures are not removed by phagocytes and suffer a late process of secondary necrosis, which is defined as the loss of membrane integrity and release of cell contents into the extracellular space. Secondary necrosis *in vivo* will take place in the presence of massive apoptosis that overloads the ability of available macrophages as well as deficiencies in the number of macrophages and/ or saturation of the immune system as seen in chronic inflammation. Secondary necrosis is the natural result of apoptosis in unicellular eukaryotes but can also occur in multicellular organisms and is involved in the genesis of many human pathologies.

Elimination of apoptotic cells by phagocytosis instead of progressing to necrosis secondary has major advantages to multicellular organisms. These advantages include increased speed for cell elimination, increased degradation of cellular component and a most cell-energy-efficient reuse of the components of engulfed apoptotic cells, allowing proper recycling of molecules. Furthermore, and most significantly, apoptotic cell elimination by phagocytosis when the apoptotic cell is still wrapped by a plasma membrane integrated allows that cell disintegration including membrane permeabilization takes place in the secure compartment of the macrophage phagolysosome. This mechanism prevents secondary necrosis, which is potentially pathological through the release of partially degraded cellular components such

as "damage-associated molecular patterns" (DAMPs) that are pro-inflammatory and immu‐
nogenic. Among DAMPs are proteases, nucleosomes, proteolytically processed autoantigens,
calcium-binding protein (calgranulin), high mobility group box 1 (HMGB-1), and urate
crystals. The cells in secondary necrosis may also be phagocytosed by macrophages *in vivo*
following different mechanisms that have been reviewed recently. The removal process of
apoptotic cells and cells undergoing secondary necrosis are not identical and generally they
have several consequences in terms of the inflammatory and immunogenic responses.
Apoptotic clearance *in vivo* includes the sensing of corpses via "find me" signals, the recognition
of corpses via "eat me" signals and their cognate receptors, the signaling pathways that regulate
cytoskeletal rearrangement necessary for engulfment and the responses of the phagocytes.

While typically the removal of apoptotic cells is anti-inflammatory and "immunologically
silent", phagocytosis of cells in secondary necrosis is pro-inflammatory and immunogenic and
thus represents another mechanism by which secondary necrosis generates pathogenic
consequences. Secondary necrosis, therefore, can produce acute and chronic diseases and it
has been recently implicated in a growing number of clinical situations that occur with acute
and chronic inflammation including many autoimmune alterations, ischemia, atherosclerosis,
chronic obstructive pulmonary disease (COPD), pulmonary inflammation associated with
oxidative stress in smokers, cystic fibrosis, asthma, bronquioestasis, and infections. On the
other hand, secondary necrosis that affects tumors has recently gained prominence because of
its recognition as a process with beneficial implications in anticancer therapies by promoting
the activation of the immune system response and consequently the disposal of tumor cells.
Chemotherapy and radiotherapy generally act by the induction of apoptosis in tumor cells. It
has been demonstrated that after treatment with ionizing radiation or chemotherapeutic
agents, cancer cells became highly immunogenic when administrated into immunocompetent
mice. Several observations suggest that such immunogenicity is associated with progression
of apoptotic cells to secondary necrosis and release of DAMPs molecules.

3. Reorganization of the cytoskeleton during apoptosis

The execution phase of apoptosis lasts approximately one hour and is characterized by typical
morphological features: cell shrinkage, plasma membrane "blebbing," chromatin condensa‐
tion, and DNA and cells fragmentation. To perform these dramatic morphological changes
that accompany the execution phase of apoptosis, apoptotic cells make a series of profound
changes (breaks and rearrangements) in the cell cytoskeleton.

The cytoskeleton is made up of three main types of filamentous proteins, actin filaments,
intermediate filaments, and microtubules, that assemble into higher-order polymers in healthy
cells and coordinately act to increase tensile strength, allow cell motility, maintain cell
morphology, participate in cell division, and provide platforms for positioning and transport
of cellular components.

Previous experiments have shown that actinomyosin cytoskeleton plays an essential role in
cellular remodeling during the early events of the execution phase of apoptosis while micro-

tubules and intermediate filaments are disorganized. However, some researchers have demonstrated the reorganization of microtubules in apoptotic cells at late stages. This rearrangement occurs during the execution phase of apoptosis, playing a key role in maintaining the integrity of the plasma membrane and dispersion of cellular and nuclear fragments.

Basically, cytoskeletal rearrangements during apoptosis can be summarized as in Figure 1. Initially, actinomyosin ring contraction is activated via phosphorylation of myosin light chain (MLC) II. MLC phosphorylation is under the control of the Myosin Light Chain kinase (MLCK), ROCK (Rho-associated coiled-coil-forming protein) kinases and is regulated by MLC phosphatase (MLCP). It has been discovered that the ROCK kinases actively phosphorylate a large cohort of actin-binding proteins and intermediate filament proteins to modulate their functions. ROCK kinases activates MLC by direct phosphorylation and by inhibiting the activity of MLCP. This movement of the actinomyosin contractile ring is facilitated by the early disruption of microtubules and intermediate filaments, which allows full contraction of the ring and the formation of protrusions of plasma membrane ("blebs"). Subsequently, actin cytoskeleton is depolymerized and, coinciding with the absence of an organized structure of the various elements of the cell cytoskeleton, apoptotic microtubules are repolymerized close to the cytosolic side of the plasma membrane.

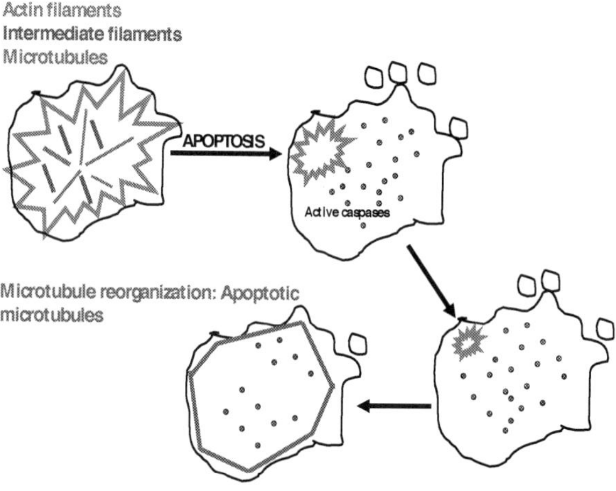

Figure 1. Reorganization of the cytoskeleton during apoptosis.

During the execution phase of apoptosis, ROCK I is cleaved by caspase-3 at a conserved DETD1113/G sequence and its carboxy-terminal inhibitory domain is removed, resulting in deregulated and constitutive kinase activity that is necessary and sufficient for actinomyosin ring contraction.

MLC phosphorylation is also regulated by the calcium-dependent MLCK calmodulin. Thus, an increase in calcium may also activate actinomyosin ring contraction. In the ring contraction activated by ROCK kinases not only MLCK is involved but also LIM kinase through phosphorylation and inactivation of cofilin, which stabilizes actin polymers.

After actinomyosin ring contraction, actin filaments are depolymerized by the action of active caspases. The Rho effector protein kinase C-related kinase (PRK) 1 is cut by active caspases leading to a constitutively active kinase fragment and after that, PRK1 can induce depolymerization of actin filaments. Likewise, the kinase p21-activated kinase (PAK) 2, a Rac effector, can be activated by caspases and active PAK has been demonstrated to induce stress fibers disassembly. Furthermore, caspases can cleave gelsolin which depolymerizes the actin cytoskeleton in a Ca^{2+}-independent manner.

Intermediate filaments that help maintain the integrity of tissues and cells are disrupted at the onset of apoptosis by the action of caspases. The intermediate filament cleavage causes fragmentation and aggregation and the breaking of the nuclear lamins facilitates nuclear disintegration. Cleavage occurs at a conserved location within the rod domain, causing loss of filament integrity and disorganization of the nuclear and cytoplasmic intermediate filaments networks. Caspase cleavage of intermediate filaments is important for the timely execution of apoptosis as evidenced by the delay incurred when caspase-insensitive forms of lamins and desmin are overexpressed. Equally interesting, the caspase 6-mediated cleavage of lamin A/C is required for complete chromatin condensation during apoptosis. The early onset and efficient cleavage of intermediate filaments proteins may be fostered by physical proximity as key effectors of the apoptotic machinery, including procaspase 3, bind cytoplasmic intermediate filaments.

After microtubules, actin, and intermediate filaments' depolymerization, apoptotic cells are devoid of the structured elements of the cytoskeleton. Then, microtubules are reorganized leading to the formation of the apoptotic microtubule network (AMN), which becomes the only element of the cytoskeleton during the execution phase of apoptosis (Figure 2).

The molecular mechanism involved in the early microtubule depolymerization during the execution phase of apoptosis is unknown although several hypotheses have been postulated. Microtubule dynamics is governed by several effectors, microtubule-associated proteins (MAPs), motor proteins such as kinesin, gradients Ran-GTP+ends proteins and proteins that bind to tubulin. These microtubule-associated proteins are in turn under the control of phosphatases and kinases. One of these regulatory kinases CDK1 is associated with cyclin B, a key enzyme for the entry into mitosis and essential for mitotic spindle formation. One of the activities ascribed to CDK1 is the interphase microtubule depolymerization at the beginning of mitosis. Among CDK1 substrates are numerous microtubular effectors such as MAP4 and XMAP215 that when are phosphorylated their ability to stabilize microtubules may be reduced. Furthermore, CDK1 can directly catalyze tubulin β phosphorylation, preventing its incorporation to microtubules. During apoptosis, increased CDK1 activity and other CDKs have been detected, suggesting that they could act as important regulators in the modifications of the microtubule cytoskeleton during the apoptotic process and the formation of AMN. CDK1 activity may also be responsible for actinomyosin ring contraction at the onset of apoptosis since previous studies have shown it can activate MLC kinase by phosphorylation. However, in PC12 cells, tubulin depolymerization at the onset of apoptosis has been associated

of caspase inhibitors. However, it has been postulated that active caspases may cleave the C-terminal regulatory regions of tubulins, which increases their ability to polymerize and thus facilitate the formation of apoptotic microtubules.

Jon Lane's group has reported that active, GTP-bound Ran is indeed required to support apoptotic microtubule assembly and that release of RanGTP into the apoptotic cytoplasm serves as a trigger for microtubule nucleation. They showed that the RanGTP-activated spindle-assembly factor, TPX2 (targeting protein for Xklp2), escapes from the nucleus during the execution phase and associates with apoptotic microtubule bundles. Consequently, silencing TPX2 expression by siRNA abrogates apoptotic microtubule assembly. They propose that formation of the apoptotic microtubule array shares several features in common with mitotic and meiotic spindle assembly with a particular dependence upon RanGTP and the microtubule-binding protein TPX2. Together, these observations suggest that, like mitotic and meiotic cells, apoptotic cells utilize the RanGTPase pathway to stimulate the coordinated assembly of a specialized microtubule network. Although AMN lacks the morphological and functional precision of the spindle apparatus, it nevertheless represents an important example of regulated, non-centrosomal microtubule assembly and organization and further highlights that the apoptotic execution phase should be recognized as a dynamic, tightly controlled process of cellular demise.

In another approach, the study of apoptotic microtubules components has revealed that in addition to the expected alpha and beta tubulin subunits, they intensively recruit other microtubule-associated proteins (MAP) such as MAP-4. These findings may be interesting for elucidating the role of MAPs in AMN nucleation. Given previous evidence linking MAP4 with microtubule nucleation, bundling, and stabilization, this protein could play an important role in AMN formation and maintenance.

7. Intracellular calcium chelation disrupts AMN and increases the permeability of plasma membrane

The elevation of intracellular calcium plays a pivotal role in the induction of the biochemical processes that characterize the execution phase of apoptosis. Among these changes, an early translocation of phosphatidylserine to the outer leaflet of the cellular membrane seems to be a key step in apoptosis, which has been shown to depend on caspase-3 activity and cytosolic Ca^{2+} concentration. The resulting exposure of phosphatidylserine in the plasma membrane surface may serve as a "eat me" signal that triggers phagocytosis by macrophages.

Intracellular calcium levels are slightly elevated in genuine apoptotic cells with AMN. This slight elevation of intracellular calcium may play a role in the nucleation of the AMN because treatment with EGTA-AM that causes intracellular calcium chelation impairs AMN formation. These observations are consistent with a model whereby a slight increase in intracellular calcium, favored by the absence of actin and intermediate filament networks, triggers the reorganization of apoptotic microtubules in the execution phase of apoptosis.

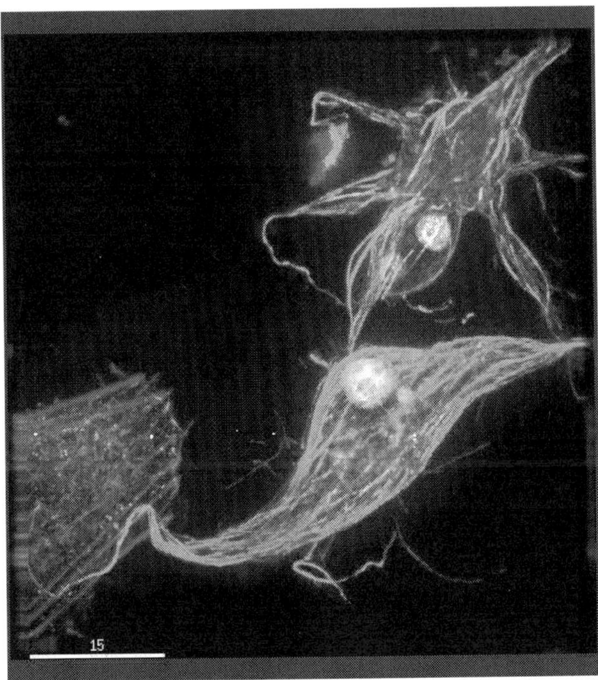

Figure 2. Human lung cancer apoptotic cells, H460: Green: anti-tubulin, AMN. Blue: actin. Red: Mitotracker, mitochondria; White: Hoechst, nuclei.

8. AMN disruption increases the permeability of plasma membrane

Given the critical role of apoptotic microtubules to maintain apoptotic cell integrity, any physical or chemical interference with AMN formation or stability might influence the process of genuine apoptosis and can induce a sort of derailed apoptosis with the release of toxic intracellular compounds that can have many pathological consequences in the context of multicellular organisms. Thus, as previously mentioned, the incubation of apoptotic cells with colchicine, an agent that depolymerizes microtubules, causes AMN depolymerization and increases plasma membrane permeability. In contrast, microtubules stabilization by taxol prevents both AMN disorganization and plasma membrane permeability.

Hypothermia or cold storage is widely used to protect cells and tissues against injurious processes. However, cold exposure can be deleterious for apoptotic cells given that microtubules which are essential to prevent plasma membrane permeability and secondary necrosis are cold labile. In contrast, AMN stabilization may preserve apoptotic cell integrity and prevents secondary necrosis.

Microtubules result from the polymerization of tubulin dimers in protofilaments that associate through lateral contacts. Microtubules are dynamic structures alternating growing and shrinking phases ended by catastrophes and rescues, respectively. *In vitro*, microtubule dynamics are under the control of the tubulin concentration and numerous other physico-chemical parameters. Among them, temperature plays a crucial role as microtubules depolymerize upon a temperature shift from 37°C to 4°C. This could be due to the modification of different dynamics parameters, especially the increase of catastrophe and the disappearance of rescue events at such temperatures. Because low temperatures depolymerize microtubules, cold exposition of apoptotic cells and re-warming to 37ºC leads to AMN disorganization and secondary necrosis. In contrast, apoptotic microtubules' stabilization by taxol prevents AMN disruption and secondary necrosis after cold/warming exposition.

Furthermore, apoptotic microtubules' depolymerization by cold/warming exposure was associated with less-efficient removal of dead cells and increased production of pro-inflammatory cytokines by macrophages. Taken together, these results indicate that temperature has an essential role for the correct execution of the apoptotic process. Furthermore, cold exposition of apoptotic cells impairs the proper phosphatidylserine externalization and interaction with macrophages, indicating that temperature is also critical for the efficient clearance of apoptotic cells. Consistent with a role of AMN for proper phosphatidylserine exposure, phagocytosis of apoptotic cells with stabilized microtubules by taxol coincided with high phosphatidylserine externalization while it was reduced when apoptotic microtubules were absent by cold/warming exposure and phosphatidylserine externalization was low, confirming previous results published by Moss *et al.* (2006). These data also suggest that the induction of secondary necrosis in apoptotic cells by cold/warming exposure is capable of inducing inflammation through the increased production of pro-inflammatory cytokines such as IL-1β and TNF-α.

Therefore, cold or hypothermia although often used as a means of reducing cellular alterations or injury during periods of storage or transport of biological material, may induce secondary necrosis in cells already undergoing apoptosis.

These findings can be relevant in order to preserve apoptotic cells by cold storage and avoid the toxic and pro-inflammatory events induced by secondary necrosis.

9. Apoptotic microtubules delimit an active-caspase free area in the cellular cortex

AMN indeed may work as physical barrier impeding active caspases to access and cleave critical proteins in the cellular cortex and plasma membrane, which are essential for plasma membrane integrity (Figure 3). AMN disorganization in apoptotic cells by a short incubation with colchicine allowed caspase-mediated cleavage of cell cortex and plasma membrane proteins such as α-spectrin, paxilin, Focal Adhesion Kinase (FAK), E-cadherin, plasma membrane Ca^{2+} ATPase-4 (PMCA-4), Na^+/Ca^{2+} exchanger (NCX), integrin β4, and Na^+/K^+ pump subunit β. These events were associated with increased cell permeability, calcium and sodium influx, and bioenergetics collapse, which in turn precipitate secondary necrosis. The essential role of caspase-mediated cleavage of cortical and plasma membrane proteins after AMN disassembly was confirmed because the addition of both colchicine and Z-VAD (a pan-caspase inhibitor) blocked protein cleavage and significantly prevented plasma membrane permeability, cell detachment, LDH release, calcium and sodium overload, and bioenergetics failure.

Under physiological conditions, plasma membrane cytoskeleton supports plasma membrane. The significance of this supporting cytoskeleton network has been mainly demonstrated in erythrocytes in which deficiencies or defects in the cytoskeletal proteins spectrin or spectrin-associated proteins were associated with increased fragility and lysis of plasma membrane. Changes in the cytoskeletal network beneath the plasma membrane (as α-spectrin cleavage by caspases) may contribute to increase membrane permeability during the transition to secondary necrosis. Thus, spectrin-deficient spherocytes had decreased membrane mechanical stability, which probably contributes to cell lysis. It is thought that the spectrin skeleton acts universally to support the otherwise mechanically vulnerable cell surface bilayer. One way that membrane skeleton/bilayer interactions have been demonstrated is through the physiology of mechano-susceptible ion channels (channel whose gating is altered by abnormally high bilayer tension). These initially unresponsive channels become progressively more mechano-responsive under stretch and chemical reagents damage of the plasma membrane skeleton. The conclusion of these studies is that the intact membrane skeleton is mechano-protective and its perturbation may increase the ion permeability of the plasma membrane.

Likewise, focal adhesions are large, dynamic protein complexes through which the cytoskeleton of a cell interacts with the extracellular matrix. When cells adhere to the extracellular matrix, integrin receptors initiate signals to recruit more integrins and other cytoskeleton proteins (such as talin, tensin, vinculin, zyxin, and actinin), adapters (such as paxillin, Crk-associate substrate (p130CAS)1, and Crk), and kinases (such as FAK and Src) to their cytoplasmic tails, forming a "focal adhesion complex". Focal adhesions provide not only mechanical support to cells through the connection with the actin cytoskeleton and mechanically couple the cell with the extracellular matrix but also signals necessary for anchorage-dependent cellular responses such as proliferation, migration, and inhibition of anoikis, a type of apoptosis induced by cell detachment. Integrins, heterodimers formed by one beta and one alpha subunits, interact with extracellular proteins via short amino acid sequences, such as the RGD (found in proteins such as fibronectin, laminin, or vitronectin) or the DGEA and GFOGER found in collagen.

Hydrolysis of "focal adhesion complex" proteins by caspases after AMN depolymerization could break the membrane-cytoskeleton linkage and decrease the physical support leading to cell detachment. It has been reported that FAK, integrin β4, and paxilin are cleaved in apoptotic cells when AMN was disorganized by colchicine treatment. Furthermore, this cleavage is a caspase-dependent process because it was blocked by z-VAD. These results suggest that disruption of FAK, integrin β4, and paxilin may contribute to cell detachment and the morphological changes observed in apoptotic cells undergoing secondary necrosis.

In agreement with these results, it has been demonstrated that E-cadherins are also cellular cortex proteins targeted by caspases when AMN is depolymerized. Cadherins are transmembrane glycoproteins involved in cell–cell adherence. Recent developments indicate that classical cadherins may act as adherence-activated signaling receptors. Previously, it has been showed that cadherins are also targeted during apoptosis. Specific cell–cell and cell–matrix contacts regulate cell growth in epithelial cells and disruption of these contacts induces apoptotic cell death. According to these observations, caspase-mediated cleavage of E-cadherins and the loss of cell–cell contacts are likely to represent an important process during the extrusion of apoptotic cells undergoing secondary necrosis.

The Na^+/K^+-ATPase (Na^+/K^+-pump) acts as an electrogenic ion transporter in the plasma membrane whose primary role is to maintain high intracellular K^+ and low intracellular Na^+ concentrations. Each cycle of Na^+/K^+- ATPase enzyme pumps three Na+ ions out of the cell, moves two K^+ ions into the cell, and uses 1 ATP. Dysfunction of the Na^+/K^+-pump results in depletion of intracellular K^+, accumulation of intracellular Na^+ and, consequently, leads to membrane depolarization. Secondary, Na^+/K^+-pump failure increases intracellular free Ca^{2+} ($[Ca^{2+}]i$) due to activation of voltage-gated Ca^{2+} channels and reversed operation of the Na^+-Ca^{2+} exchanger (NCX). Previous studies have shown that the intracellular Na^+ concentration increases prior to a loss of plasma membrane integrity and that cell shrinkage in apoptotic Jurkat cells is accompanied by a net efflux of ions due to an inactivation of Na^+/K^+-ATPase. The latter authors also observed that Na^+/K^+-ATPase subunits were degraded in populations with reduced volume during apoptosis. Because active cellular volume regulation requires Na$^+$/K$^+$-ATPase activity, both events may act synergistically in the induction of cell shrinkage. The Na^+-K^+-ATPase is composed of two subunits. The α-subunit (~113 kD) is the catalytic subunit that binds ATP and both sodium and potassium ions and also contains the phosphorylation site. On the other hand, the smaller β-subunit (~35 kDa glycoprotein) is absolutely essential in facilitating the plasma membrane localization and α-subunit activation. Studies on purified enzyme also suggested that both subunits were essential for activity because any efforts to separate α and β resulted in inactive enzyme. It has been shown that Na^+/K^+-ATPase β-subunits but not α-subunits were cleaved by caspases after AMN depolymerization, which may contribute to ionic imbalance and increase plasma membrane permeability. Degradation of Na^+/K^+-ATPase β-subunit associated with mitochondria and plasma membrane depolarization during apoptosis has been previously reported. Consistent with this hypothesis, sodium levels were notably increased after AMN depolymerization by colchicine and partially restored when AMN was disorganized but caspases were blocked by z-VAD.

Changes in cytosolic calcium have an essential role during apoptosis by triggering the activation of Ca^{2+}-dependent processes thereby inducing global intracellular and morphological modifications including phosphatidylserine externalization. However, AMN disruption by colchicine allows the caspase-mediated cleavage of key proteins involved in calcium extrusion such as PMCA-4 and NCX and, consequently, provokes calcium overload. PMCAs are vital caspase substrates for the regulated subprogram leading to secondary necrosis. Cells expressing PMCA-4 mutants that lack the caspase cleavage site(s) prevent calcium influx during apoptosis and notably delay secondary necrosis. Furthermore, the Na^+/Ca^{2+} transporter (NCX) also participates in calcium efflux in addition to PMCAs. NCX has a low calcium affinity but high calcium transporting activity, which is required to rapidly eject large amounts of calcium. While the NCX contribution in regulating resting cytosolic calcium may be less important than that of PMCAs, its function may avoid calcium overload in cells undergoing apoptosis. It has been reported that both PMC-4 and NCX are cleaved by caspases when AMN is depolymerized by colchicine treatment and suggest that inactivation of plasma membrane calcium transporters is a relevant process leading apoptotic cells to secondary necrosis.

Like many other insults, increased cytosolic calcium can trigger either apoptosis or necrosis. The final result of cell death is probably controlled by the concentration of cytoplasmic calcium. Whereas low to moderate calcium levels (200–400 nM) induces apoptosis, higher concentration of calcium (>1 μM) is associated with necrosis. This may help to understand why an initial slight calcium increase is pro-apoptotic and favors AMN formation whereas the late calcium influx through the plasma membrane that cannot be expelled out of the cell is associated with secondary necrosis. Calcium overload can induce: 1) activation of calpains leading to more extensive disruption of the protein components of cytoskeleton, 2) activation of calcium-dependent phospholipases causing liberation of arachidonic acid and formation of lysophosphatides that alter membrane structures and disruption of membrane permeability with ensuing secondary necrosis and/or 3) Mitochondrial permeabilization and bioenergetics collapse.

10. Apoptotic cells with AMN enhance phosphatidylserine exposure and interactions with macrophages

Phagocytic clearance of apoptotic cells or efferocytosis consists of four main distinct steps: accumulation of professional phagocytes at the site where apoptotic cells are located, recognition of apoptotic cells through a number of binding molecules and receptors, engulfment by a unique uptake process and digestion of engulfed cells within phagocytes. The efficient phagocytosis of apoptotic cells by macrophages reduces the potential for an inflammatory response by ensuring that the dying cells are eliminated before their intracellular contents are released to the extracellular medium. Early apoptotic cells are targeted for phagocytosis through the translocation of phosphatidylserine from the inner to the outer leaflet of the plasma membrane. The externalization of phosphatidylserine is an early event of apoptosis occurring while the plasma membrane remains intact and cells exclude membrane-impermeant dyes. Phosphatidylserine exposure has been reported to be a process that needs energy and depends on caspase activation but its mechanism is still not clearly understood. A combined effect of down-regulation of a phospholipid translocase activity and activation of a lipid scramblase may participate in phosphatidylserine translocation.

Consistent with a role of AMN for proper phosphatidylserine exposure, it has been shown that the phagocytosis of apoptotic cells with AMN is associated with high appearance of phosphatidylserine on the cell surface while it was reduced when AMN was depolymerized by colchicine treatment. Phosphatidylserine externalization and phagocytosis of apoptotic cells were restored when AMN was depolymerized in the presence of Z-VAD, suggesting that caspase-dependent plasma membrane permeabilization impairs proper phosphatidylserine externalization.

analysis. Interpretation of these changes can also be more difficult if the study is not made in a homogeneous synchronized population because examination of a population of apoptotic cells can show a mixed population of cells with mitochondrial hyperpolarization or depolarization. Another interesting conclusion of these works is that AMN disorganization in apoptotic cells seems to be a consequence of mitochondria depolarization rather that increased calcium influx.

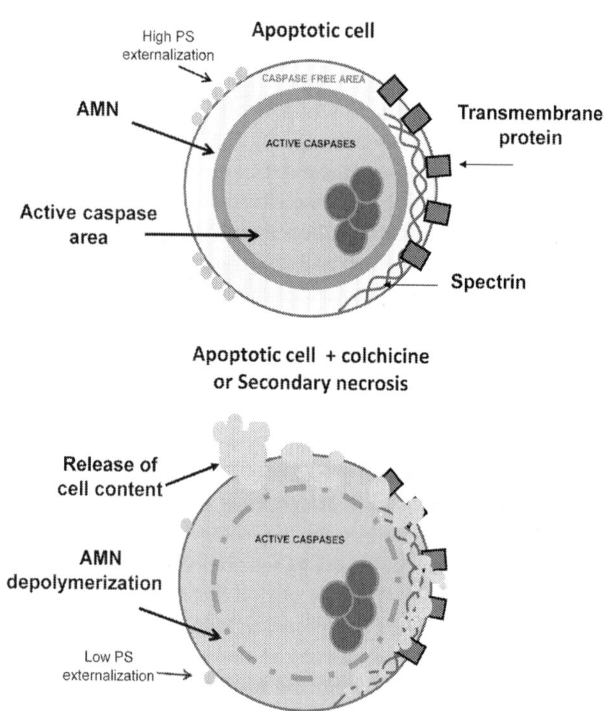

Figure 3. Scheme summarizing the main findings on AMN during the execution phase of apoptosis.

Furthermore, the observation of mitochondrial membrane potential fluctuations with cycles of hyperpolarization in the living cell imaging studies could reflect changes in the bioenergetic status of the apoptotic cell. When ATP levels are high, mitochondria are hyperpolarized by reverse flow of FoF1-ATPase that reduces ATP levels and permits the forward operation of FoF1-ATPase, which increases ATP levels again.

In summary, examining mitochondrial membrane potential during apoptosis, two features are striking: the cyclic changes of mitochondrial membrane potential associated with high ATP levels and the presence of AMN and the final mitochondrial depolarization coinciding with AMN disassembly and increased plasma permeability in the transition to secondary necrosis. These findings also support the hypothesis that AMN is essential for preserving plasma membrane permeability and, therefore, cell bioenergetics during the execution phase of apoptosis.

12. Apoptotic cells can be stabilized

Given that apoptotic cells maintain the integrity of the plasma membrane and cellular cortex, an innovative method aimed at the long-term stabilization and preservation of apoptotic cells has been developed. This method consists of the combined treatment of apoptotic cells with taxol, Zn^{2+}, and coenzyme Q_{10} (CoQ). This experimental approach guarantees apoptotic cell integrity by preventing plasma membrane permeability and secondary necrosis for at least 96 hours in cell cultures.

The rationale for using this stabilizing combination is: a) the use of taxol, which as a microtubule stabilizing agent, prevents AMN depolymerization and subsequently the access of active caspases to the cellular cortex; b) the use of Zn^{2+}, which as a caspase inhibitor, avoids excessive degradation of cellular components and caspase-dependent cleavage of cellular cortex and plasma membrane proteins; and c) the use of CoQ, which as an antioxidant, protects against oxidative membrane damage which is increased in apoptotic cells.

Stabilized apoptotic cells can be seen as dying cells in which the cellular cortex and plasma membrane are maintained intact or alive. In a metaphorical sense, they can be considered as "living dead" or "zombie cells." Stabilized apoptotic cells have many of the typical hallmarks of genuine apoptosis such as plasma membrane impermeability, integrity of plasma membrane and cellular cortex proteins, low intracellular calcium levels, plasma membrane potential, phosphatidylserine exposure, and the ability of being engulfed by phagocytes.

Recently, interest in apoptosis research has increased considerably for a number of reasons including development of new treatments, cell culture technology, metabolic engineering of mammalian cells, and gene therapy. Furthermore, apoptotic cells play important roles in biomedical research because due to their characteristics, they are being widely used to evaluate the cytotoxic effects of various drugs by quantifying apoptotic cells by flow cytometry. However, this determination is often affected by the process of cell manipulation (cell harvesting, cell centrifugation, cell pipetting) required for flow cytometry assays, which may disrupt plasma membrane and lead apoptotic cells to secondary necrosis. Thus, accurate and reliable apoptosis quantifications are particularly difficult in adherent cell cultures. Stabilization of apoptotic cells before cell harvesting permits a more accurate and reliable quantification of the actual number of apoptotic cells or the correct measurement of biochemical parameters of genuine apoptotic cells such as mitochondrial membrane potential, intracellular calcium concentration, pH, caspase activity, and many others without the interference of plasma membrane disruption.

In addition, apoptotic cells are currently utilized in various forms of clinical treatments, primarily with the objective of inducing immunological tolerance in the recipient individual. The stabilization of apoptotic cells before administration may help to guarantee that the inoculated apoptotic cells retain their characteristic features until they are engulfed by phagocytes. The inoculation of stabilized apoptotic cells can be additionally utilized for delivering compounds of interest such as therapeutic proteins (for protein replacement therapy) or drugs to recipient macrophages.

There are processes of cell death which by their nature hinder the correct AMN formation (e.g., mitochondrial toxics and cold exposure) and subsequently apoptotic cells are not able to keep up plasma membrane integrity and cells undergo secondary necrosis that may cause serious adverse effects. Apoptotic cells' stabilization may permit the development of therapies for the correct formation and stabilization of AMN and thus promote a more physiological and controlled sort of cell death. Taking into account the above arguments, there is a need to find drugs and protocols for apoptotic cell stabilization.

13. Conclusion

In summary, AMN plays an essential role in the preservation of plasma membrane integrity during the execution phase of apoptosis. In the context of multicellular organisms, the most critical aspect of apoptosis is that cell death occurs without the release of potential pathogenic or harmful intracellular molecules, and without inflammation or injury to adjacent cells. In this manner, the primary role of AMN could be to guarantee that the dying cell is confined to prevent damage to the surrounding tissues and the potentially devastating consequences to processes such as tissue homeostasis and elimination of damaged cells. It is therefore essential to know the effects of different physical or chemical agents on the correct formation of this structure allowing us to better understand their effects on the immune system response and the potential adverse reactions.

Furthermore, apoptotic cells can be stabilized for accurate detection and quantification of apoptosis in cultured cells. Stabilization of apoptotic cells might likewise permit a more secure administration of apoptotic cells in clinical applications and open new alternatives for the functional reconstruction of apoptotic cells for longer preservation.

Acknowledgements

This work was supported by the FIS PI13/00129 Grant, Ministerio de Sanidad, Spain and Fondo Europeo de Desarrollo Regional (FEDER-Unión Europea), Proyecto de Investigación de Excelencia de la Junta de Andalucía CTS-5725, and by AEPMI (Asociación de Enfermos de Patología Mitocondrial) and ENACH (Asociación de Enfermedades Neurodegenerativas por Acumulación Cerebral de Hierro).

The authors declare that they have no conflict of interest.

Author details

Manuel Oropesa Ávila[1], Alejandro Fernández Vega[1], Juan Garrido Maraver[1],
Marina Villanueva Paz[1], Isabel De Lavera[1], Mario De La Mata[1], Mario D. Cordero[2],
Elizabet Alcocer Gómez[1], Ana Delgado Pavón[1], Mónica Álvarez Córdoba[1], David Cotán[1] and
José Antonio Sánchez-Alcázar[1]

1 Andalusian Centre for Developmental Biology-CSIC-University Pablo de Olavide, Seville,
Spain

2 Research Laboratory, Oral Medicine Department, University of Seville, Seville, Spain

References

[1] Kerr JF, Wyllie AH, Currie AR. Apoptosis: a basic biological phenomenon with wide-ranging implications in tissue kinetics. *Br J Cancer*. 1972;26(4):239-57.

[2] Savill J, Dransfield I, Gregory C, Haslett C. A blast from the past: clearance of apoptotic cells regulates immune responses. *Natur Rev*. 2002;2(12):965-75.

[3] Aslan JE, Thomas G. Death by committee: organellar trafficking and communication in apoptosis. *Traffic* (Copenhagen, Denmark). 2009;10(10):1390-404.

[4] Wilson MR. Apoptosis: unmasking the executioner. *Cell Death Differ*. 1998;5(8):646-52.

[5] Afford S, Randhawa S. Demystified...: Apoptosis. *Mol Pathol*. 2000;53(2):55-63.

[6] Silva MT, do Vale A, dos Santos NM. Secondary necrosis in multicellular animals: an outcome of apoptosis with pathogenic implications. *Apoptosis*. 2008;13(4):463-82. Epub 2008/03/07.

[7] Silva MT. Secondary necrosis: the natural outcome of the complete apoptotic program. *FEBS Lett*. 2010;584(22):4491-9. Epub 2010/10/27.

[8] Poon IK, Hulett MD, Parish CR. Molecular mechanisms of late apoptotic/necrotic cell clearance. *Cell Death Differ*. 2010;17(3):381-97. Epub 2009/12/19.

[9] Hochreiter-Hufford A, Ravichandran KS. Clearing the dead: apoptotic cell sensing, recognition, engulfment, and digestion. *Cold Spring Harbor Persp Biol*. 2013;5(1):a008748.

[10] Fadok VA, Bratton DL, Konowal A, Freed PW, Westcott JY, Henson PM. Macrophages that have ingested apoptotic cells in vitro inhibit proinflammatory cytokine production through autocrine/paracrine mechanisms involving TGF-beta, PGE2, and PAF. *J Clin Investig*. 1998;101(4):890-8. Epub 1998/03/21.

cortex during the execution phase of apoptosis. *Cell Death Dis.* 2013;4:e527. Epub 2013/03/09.

[90] Akhmanova A, Steinmetz MO. Tracking the ends: a dynamic protein network controls the fate of microtubule tips. *Nat Rev Mol Cell Biol.* 2008;9(4):309-22.

[91] Wade RH. On and around microtubules: an overview. *Mol Biotechnol.* 2009;43(2): 177-91.

[92] Lee JC, Timasheff SN. In vitro reconstitution of calf brain microtubules: effects of solution variables. *Biochemistry.* 1977;16(8):1754-64.

[93] Olmsted JB, Borisy GG. Ionic and nucleotide requirements for microtubule polymerization in vitro. *Biochemistry.* 1975;14(13):2996-3005.

[94] Regula CS, Pfeiffer JR, Berlin RD. Microtubule assembly and disassembly at alkaline pH. *J Cell Biol.* 1981;89(1):45-53.

[95] Suprenant KA, Marsh JC. Temperature and pH govern the self-assembly of microtubules from unfertilized sea-urchin egg extracts. *J Cell Sci.* 1987;87 (Pt 1):71-84.

[96] Fygenson DK, Braun E, Libchaber A. Phase diagram of microtubules. *Physic Rev.* 1994;50(2):1579-88.

[97] Modig C, Wallin M, Olsson PE. Expression of cold-adapted beta-tubulins confer cold-tolerance to human cellular microtubules. *Biochemic Biophysic Res Commun.* 2000;269(3):787-91.

[98] Palek J, Sahr KE. Mutations of the red blood cell membrane proteins: from clinical evaluation to detection of the underlying genetic defect. *Blood.* 1992;80(2):308-30. Epub 1992/07/15.

[99] Perrotta S, Gallagher PG, Mohandas N. Hereditary spherocytosis. *Lancet.* 2008;372(9647):1411-26. Epub 2008/10/23.

[100] Morris CE. Mechanoprotection of the plasma membrane in neurons and other non-erythroid cells by the spectrin-based membrane skeleton. *Cell Mole Biol Lett.* 2001;6(3):703-20. Epub 2001/10/13.

[101] Peters LL, Jindel HK, Gwynn B, Korsgren C, John KM, Lux SE, et al. Mild spherocytosis and altered red cell ion transport in protein 4. 2-null mice. *J Clin Investig.* 1999;103(11):1527-37. Epub 1999/06/08.

[102] Chen CS, Alonso JL, Ostuni E, Whitesides GM, Ingber DE. Cell shape provides global control of focal adhesion assembly. *Biochemic Biophysic Res Commun.* 2003;307(2): 355-61. Epub 2003/07/16.

[103] Critchley DR. Focal adhesions – the cytoskeletal connection. *Curr Opin Cell Biol.* 2000;12(1):133-9. Epub 2000/02/19.

[104] Gilmore AP, Burridge K. Molecular mechanisms for focal adhesion assembly through regulation of protein-protein interactions. *Structure*. 1996;4(6):647-51. Epub 1996/06/15.

[105] Miyamoto S, Katz BZ, Lafrenie RM, Yamada KM. Fibronectin and integrins in cell adhesion, signaling, and morphogenesis. *Ann N Y Acad Sci*. 1998;857:119-29. Epub 1999/01/26.

[106] Clark EA, Brugge JS. Integrins and signal transduction pathways: the road taken. *Science*. New York, NY. 1995;268(5208):233-9. Epub 1995/04/14.

[107] Giancotti FG, Ruoslahti E. Integrin signaling. *Science*. New York, NY. 1999;285(5430): 1028-32. Epub 1999/08/14.

[108] Wen LP, Fahrni JA, Troie S, Guan JL, Orth K, Rosen GD. Cleavage of focal adhesion kinase by caspases during apoptosis. *J Biologic Chem*. 1997;272(41):26056-61. Epub 1997/11/05.

[109] Fouquet S, Lugo-Martinez VH, Chambaz J, Cardot P, Pincon-Raymond M, Thenet S. [Control of the survival/apoptosis balance by E-cadherin: role in enterocyte anoikis]. *J de la Societe de biologie*. 2004;198(4):379-83. Epub 2005/06/23. Le controle de la balance survie/apoptose par la E-cadherine: implication dans l'anoikis des enterocytes.

[110] Herren B, Levkau B, Raines EW, Ross R. Cleavage of beta-catenin and plakoglobin and shedding of VE-cadherin during endothelial apoptosis: evidence for a role for caspases and metalloproteinases. *Mol Biol Cell*. 1998;9(6):1589-601. Epub 1998/06/17.

[111] Schmeiser K, Grand RJ. The fate of E- and P-cadherin during the early stages of apoptosis. *Cell Death Differ*. 1999;6(4):377-86. Epub 1999/06/25.

[112] Steinhusen U, Weiske J, Badock V, Tauber R, Bommert K, Huber O. Cleavage and shedding of E-cadherin after induction of apoptosis. *J Biologic Chem*. 2001;276(7): 4972-80. Epub 2000/11/15.

[113] Hermiston ML, Gordon JI. In vivo analysis of cadherin function in the mouse intestinal epithelium: essential roles in adhesion, maintenance of differentiation, and regulation of programmed cell death. *J Cell Biol*. 1995;129(2):489-506. Epub 1995/04/01.

[114] Ruoslahti E, Reed JC. Anchorage dependence, integrins, and apoptosis. *Cell*. 1994;77(4):477-8. Epub 1994/05/20.

[115] Rakowski RF, Gadsby DC, De Weer P. Stoichiometry and voltage dependence of the sodium pump in voltage-clamped, internally dialyzed squid giant axon. *J Gen Physiol*. 1989;93(5):903-41.

[116] Xiao AY, Wei L, Xia S, Rothman S, Yu SP. Ionic mechanism of ouabain-induced concurrent apoptosis and necrosis in individual cultured cortical neurons. *J Neurosci*. 2002;22(4):1350-62.

[117] Bortner CD, Gomez-Angelats M, Cidlowski JA. Plasma membrane depolarization without repolarization is an early molecular event in anti-Fas-induced apoptosis. *J Biologic Chem*. 2001;276(6):4304-14.

[118] Mann CL, Bortner CD, Jewell CM, Cidlowski JA. Glucocorticoid-induced plasma membrane depolarization during thymocyte apoptosis: association with cell shrinkage and degradation of the Na(+)/K(+)-adenosine triphosphatase. *Endocrinology*. 2001;142(12):5059-68.

[119] Jorgensen PL, Hakansson KO, Karlish SJ. Structure and mechanism of Na,K-ATPase: functional sites and their interactions. *Annu Rev Physiol*. 2003;65:817-49. Epub 2003/01/14.

[120] Kaplan JH. Biochemistry of Na,K-ATPase. *Annu Rev Biochem*. 2002;71:511-35.

[121] Dussmann H, Rehm M, Kogel D, Prehn JH. Outer mitochondrial membrane permeabilization during apoptosis triggers caspase-independent mitochondrial and caspase-dependent plasma membrane potential depolarization: a single-cell analysis. *J Cell Sci*. 2003;116(Pt 3):525-36. Epub 2003/01/01.

[122] Carafoli E, Santella L, Branca D, Brini M. Generation, control, and processing of cellular calcium signals. *Critic Rev Biochem Mol Biol*. 2001;36(2):107-260. Epub 2001/05/24.

[123] Linck B, Qiu Z, He Z, Tong Q, Hilgemann DW, Philipson KD. Functional comparison of the three isoforms of the Na+/Ca2+ exchanger (NCX1, NCX2, NCX3). *Am J Physiol*. 1998;274(2 Pt 1):C415-23. Epub 1998/03/05.

[124] McConkey DJ, Orrenius S. The role of calcium in the regulation of apoptosis. *J Leukocyte Biol*. 1996;59(6):775-83. Epub 1996/06/01.

[125] Orrenius S, Nicotera P, Zhivotovsky B. Cell death mechanisms and their implications in toxicology. *Toxicologic Sci*. : an official journal of the Society of Toxicology. 2011;119(1):3-19. Epub 2010/09/11.

[126] Zhivotovsky B, Orrenius S. Calcium and cell death mechanisms: a perspective from the cell death community. *Cell Calc*. 2011;50(3):211-21. Epub 2011/04/05.

[127] Erwig LP, Henson PM. Clearance of apoptotic cells by phagocytes. *Cell Death Differ*. 2008;15(2):243-50. Epub 2007/06/16.

[128] Savill J, Fadok V. Corpse clearance defines the meaning of cell death. *Nature*. 2000;407(6805):784-8.

[129] Voll RE, Herrmann M, Roth EA, Stach C, Kalden JR, Girkontaite I. Immunosuppressive effects of apoptotic cells. *Nature*. 1997;390(6658):350-1. Epub 1997/12/06.

[130] Fadok VA, Bratton DL, Frasch SC, Warner ML, Henson PM. The role of phosphatidylserine in recognition of apoptotic cells by phagocytes. *Cell Death Differ*. 1998;5(7): 551-62.

[131] Fink SL, Cookson BT. Apoptosis, pyroptosis, and necrosis: mechanistic description of dead and dying eukaryotic cells. *Infect Immun*. 2005;73(4):1907-16.

[132] Castedo M, Hirsch T, Susin SA, Zamzami N, Marchetti P, Macho A, et al. Sequential acquisition of mitochondrial and plasma membrane alterations during early lymphocyte apoptosis. *J Immunol*. 1996;157(2):512-21.

[133] Nicotera P, Leist M, Ferrando-May E. Intracellular ATP, a switch in the decision between apoptosis and necrosis. *Toxicol Lett*. 1998;102-103:139-42.

[134] Moreira ME, Barcinski MA. Apoptotic cell and phagocyte interplay: recognition and consequences in different cell systems. *Anais da Academia Brasileira de Ciencias*. 2004;76(1):93-115.

[135] Wyllie AH. Glucocorticoid-induced thymocyte apoptosis is associated with endogenous endonuclease activation. *Nature*. 1980;284(5756):555-6. Epub 1980/04/10.

[136] Duanmu C, Lin CM, Hamel E. Tubulin polymerization with ATP is mediated through the exchangeable GTP site. *Biochim Biophys Acta*. 1986;881(1):113-23.

[137] Leist M, Single B, Castoldi AF, Kuhnle S, Nicotera P. Intracellular adenosine triphosphate (ATP) concentration: a switch in the decision between apoptosis and necrosis. *J Exper Med*. 1997;185(8):1481-6.

[138] Richter C, Schweizer M, Cossarizza A, Franceschi C. Control of apoptosis by the cellular ATP level. *FEBS Lett*. 1996;378(2):107-10.

[139] Eguchi Y, Shimizu S, Tsujimoto Y. Intracellular ATP levels determine cell death fate by apoptosis or necrosis. *Cancer Res*. 1997;57(10):1835-40.

[140] Leist M, Single B, Naumann H, Fava E, Simon B, Kuhnle S, et al. Inhibition of mitochondrial ATP generation by nitric oxide switches apoptosis to necrosis. *Exper Cell Res*. 1999;249(2):396-403.

[141] Atlante A, Giannattasio S, Bobba A, Gagliardi S, Petragallo V, Calissano P, et al. An increase in the ATP levels occurs in cerebellar granule cells en route to apoptosis in which ATP derives from both oxidative phosphorylation and anaerobic glycolysis. *Biochim Biophys Acta*. 2005;1708(1):50-62.

[142] Zamaraeva MV, Sabirov RZ, Maeno E, Ando-Akatsuka Y, Bessonova SV, Okada Y. Cells die with increased cytosolic ATP during apoptosis: a bioluminescence study with intracellular luciferase. *Cell Death Differ*. 2005;12(11):1390-7.

[143] Sanchez-Alcazar JA, Ruiz-Cabello J, Hernandez-Munoz I, Pobre PS, de la Torre P, Siles-Rivas E, et al. Tumor necrosis factor-alpha increases ATP content in metabolically inhibited L929 cells preceding cell death. *J Biologic Chem*. 1997;272(48):30167-77.

[144] Green DR, Reed JC. Mitochondria and apoptosis. *Science*. New York, NY. 1998;281(5381):1309-12.

[145] Kim JM, Bae HR, Park BS, Lee JM, Ahn HB, Rho JH, et al. Early mitochondrial hyperpolarization and intracellular alkalinization in lactacystin-induced apoptosis of retinal pigment epithelial cells. *J Pharmacol Exper Therapeut*. 2003;305(2):474-81.

[146] Li PF, Dietz R, von Harsdorf R. p53 regulates mitochondrial membrane potential through reactive oxygen species and induces cytochrome c-independent apoptosis blocked by Bcl-2. *EMBO J*. 1999;18(21):6027-36.

[147] Marzo I, Brenner C, Zamzami N, Jurgensmeier JM, Susin SA, Vieira HL, et al. Bax and adenine nucleotide translocator cooperate in the mitochondrial control of apoptosis. *Science* (New York, NY. 1998;281(5385):2027-31.

[148] Matarrese P, Falzano L, Fabbri A, Gambardella L, Frank C, Geny B, et al. Clostridium difficile toxin B causes apoptosis in epithelial cells by thrilling mitochondria. Involvement of ATP-sensitive mitochondrial potassium channels. *J Biologic Chem*. 2007;282(12):9029-41.

[149] Matarrese P, Gambardella L, Cassone A, Vella S, Cauda R, Malorni W. Mitochondrial membrane hyperpolarization hijacks activated T lymphocytes toward the apoptotic-prone phenotype: homeostatic mechanisms of HIV protease inhibitors. *J Immunol*. 2003;170(12):6006-15.

[150] Perl A, Gergely P, Jr., Nagy G, Koncz A, Banki K. Mitochondrial hyperpolarization: a checkpoint of T-cell life, death and autoimmunity. *Trends Immunol*. 2004;25(7):360-7.

[151] Vander Heiden MG, Chandel NS, Williamson EK, Schumacker PT, Thompson CB. Bcl-xL regulates the membrane potential and volume homeostasis of mitochondria. *Cell*. 1997;91(5):627-37.

[152] Khaled AR, Reynolds DA, Young HA, Thompson CB, Muegge K, Durum SK. Interleukin-3 withdrawal induces an early increase in mitochondrial membrane potential unrelated to the Bcl-2 family. Roles of intracellular pH, ADP transport, and F(0)F(1)-ATPase. *J Biologic Chem*. 2001;276(9):6453-62.

[153] Nicholls DG, Budd SL. Mitochondria and neuronal survival. *Physiol Rev*. 2000;80(1): 315-60.

[154] Matsuyama S, Llopis J, Deveraux QL, Tsien RY, Reed JC. Changes in intramitochondrial and cytosolic pH: early events that modulate caspase activation during apoptosis. *Natur Cell Biol*. 2000;2(6):318-25.

[155] Santamaria G, Martinez-Diez M, Fabregat I, Cuezva JM. Efficient execution of cell death in non-glycolytic cells requires the generation of ROS controlled by the activity of mitochondrial H+-ATP synthase. *Carcinogenesis*. 2006;27(5):925-35.

[156] Chinopoulos C, Adam-Vizi V. Mitochondria as ATP consumers in cellular pathology. *Biochim Biophys Acta*. 2010;1802(1):221-7.

[157] Peachman KK, Lyles DS, Bass DA. Mitochondria in eosinophils: functional role in apoptosis but not respiration. *Proc Natl Acad Sci U S A*. 2001;98(4):1717-22.

[158] Kroemer G, Zamzami N, Susin SA. Mitochondrial control of apoptosis. Immunology today. 1997;18(1):44-51.

[159] Schiff PB, Fant J, Horwitz SB. Promotion of microtubule assembly in vitro by taxol. *Nature*. 1979;277(5698):665-7.

[160] Schiff PB, Horwitz SB. Taxol stabilizes microtubules in mouse fibroblast cells. *Proc Natl Acad Sci U S A*. 1980;77(3):1561-5.

[161] Huber KL, Hardy JA. Mechanism of zinc-mediated inhibition of caspase-9. *Protein Sci*: a publication of the Protein Society. 2012;21(7):1056-65. Epub 2012/05/11.

[162] Perry DK, Smyth MJ, Stennicke HR, Salvesen GS, Duriez P, Poirier GG, et al. Zinc is a potent inhibitor of the apoptotic protease, caspase-3. A novel target for zinc in the inhibition of apoptosis. *J Biologic Chem*. 1997;272(30):18530-3.

[163] Smith AF, Longpre J, Loo G. Inhibition by zinc of deoxycholate-induced apoptosis in HCT-116 cells. *J Cell Biochem*. 2012;113(2):650-7.

[164] Stennicke HR, Salvesen GS. Biochemical characteristics of caspases-3, -6, -7, and -8. *J Biologic Chem*. 1997;272(41):25719-23. Epub 1997/11/05.

[165] Bentinger M, Brismar K, Dallner G. The antioxidant role of coenzyme Q. *Mitochondrion*. 2007;7 Suppl:S41-50.

[166] Darzynkiewicz Z, Bruno S, Del Bino G, Gorczyca W, Hotz MA, Lassota P, et al. Features of apoptotic cells measured by flow cytometry. *Cytometry*. 1992;13(8):795-808.

[167] van Engeland M, Ramaekers FC, Schutte B, Reutelingsperger CP. A novel assay to measure loss of plasma membrane asymmetry during apoptosis of adherent cells in culture. *Cytometry*. 1996;24(2):131-9.

[168] Saas P, Kaminski S, Perruche S. Prospects of apoptotic cell-based therapies for transplantation and inflammatory diseases. *Immunotherapy*. 2013;5(10):1055-73.

[169] Perez B, Paquette N, Paidassi H, Zhai B, White K, Skvirsky R, et al. Apoptotic cells can deliver chemotherapeutics to engulfing macrophages and suppress inflammatory cytokine production. *J Biologic Chem*. 2012;287(19):16029-36. Epub 2012/03/22.

Chapter 15

Steroidal Saponins and Cell Death in Cancer

María L. Escobar-Sánchez, Luis Sánchez-Sánchez and
Jesús Sandoval-Ramírez

Additional information is available at the end of the chapter

http://dx.doi.org/10.5772/61438

Abstract

Steroidal saponins are natural glycosidic compounds of amphiphilic character. Their diverse biological activities are directly related to the variability of their structural constitutive frameworks, aglycones, and sugars. Several studies have demonstrated the therapeutic potential of steroidal saponins by their capacity to induce programmed cell death in different tumor cell lines. The process of cell death is required to maintain cellular and tissular homeostasis; it has been established that disturbances in the balance between cellular proliferation and cell death lead to several pathologies, including cancer. The antitumor activity of steroidal saponins has been intensely studied allowing elucidation of their different molecular mechanisms of action; this knowledge is crucial to the establishment of new therapeutic strategies against cancer.

Keywords: Steroidal saponins, cancer, cell death, apoptosis, cytotoxicity

1. Introduction

Saponins are a broad group of glycosides widely distributed in higher order terrestrial plants, and in lower marine organisms. They include a diverse group of compounds containing a steroidal or triterpenoid aglycone and one or more sugar chains [1]. Steroidal saponins are present almost exclusively in monocotyledonous angiosperms, not only in the families of *Dioscoreaceae, Asparagaceae, Liliaceae,* and *Amaryllidaceae* but also in the dicotyledonous *Solanaceae*. Triterpenoid saponins are more common in dicotyledonous angiosperms (e.g., the families *Caryophyllaceae, Quillajaceae, Sapindaceae*) [2]. Steroidal saponins are less common than triterpenoid saponins. Usually, glycosteroidal alkaloids are included in the very large alkaloid group.

Mankind has used, for thousands of years, many saponin-containing plants as soaps. Saponins have an amphiphilic character and as soaps, they are surface-active compounds and produce micelles. They have a wide spectrum of uses; in ancient folk medicine, they have been used as venoms, hemolytes, antimicrobials, and anti-inflammatories. Saponins are responsible for numerous biological effects in traditional Chinese and Japanese medicines. New uses are in the cosmetic and pharmaceutical industries, as starting materials in the semisynthesis of many high-cost products. The latter are difficult to produce through total synthesis due to their great structural complexity and numerous chiral centers. The foaming property of saponins in water resulted in the coining of the word saponin (from Latin *sapo*, soap). Properties and pharma-cological activities of saponins were described in great detail in 1927, before a single saponin had been fully characterized [1].

Steroidal saponins have a wide range of pharmacological applications, including use as expectorants and to inhibit platelet aggregation, and also have hemolytic, insecticidal, anti-inflammatory, antitumor, antidiabetic, antifungal/antiyeast, antibacterial, antiparasitic, antihyperlipidemic, and anti-oxidative properties, among others [3]. Taking into account the above applications, the physiological role of saponins in animals and plants has been related to their defense systems. One of the first uses in the health field was made in the immune system, since they activate the immune response to antigens, functioning as adjuvants that improve the effectiveness of orally administered vaccines by facilitating the absorption of large molecules [4]. Later studies have allowed identifying saponins as inductors of cell death by means of several molecular mechanisms.

2. Chemical characteristics of saponins

Structurally, saponins are composed of a lipid-soluble aglycone that consists of a steroidal or triterpenoid skeleton and a water-soluble moiety, composed of sugar residues. The latter can differ in the type and amount of cyclic carbohydrates. The natural properties of saponins allow them to be dissolved in water where they form colloidal solutions that foam upon shaking [5]. The structure of saponins derived from plant sources are different from those found in animals. The same structural difference is observed in steroidal or triterpenoid saponins. In general, their water solubility depends on their sugar moiety number [6].

The triterpenoid aglycone consists of a skeleton of 30 carbon atoms, showing in general a pentacyclic structure. In triterpene saponins, ten main classes are found: dammaranes, tirucallanes, curcubitanes, lanostanes (all with a four six-membered ring skeleton), cycloar-tanes (possessing a cyclopropane attached to a four six-membered ring skeleton), lupanes and hopanes (in which a cyclopentane ring is attached to a four six-membered ring skeleton), oleananes, taraxasteranes, and ursanes (composed by a five six-membered ring skeleton) [7]. In all cases, several skeletons have been found to undergo ring cleavage (seco-skeletons), homologation (homo-skeletons), degradation (nor-skeletons), or rearrangements (abeo-skeletons).

All steroidal saponins contain a 27 carbon atom aglycone skeleton and are classified in three main subclasses: spirostan, furostan, and cholestane saponins [8]. Spirostan saponins contain an aglycone that is composed of four six-membered and two five-membered rings (named as A, B, C, D, E, and F-rings, Figure 1); aglycones of furostan saponins possess only A, B, C, D, and E rings (three six-membered and two five-membered rings), while aglycones of cholestane saponins have only the tetracyclic A, B, C, and D system (three six-membered and one five-membered rings). Biosynthetically, spirostans and furostans derive from a cholestane skeleton through selective oxidation pathways.

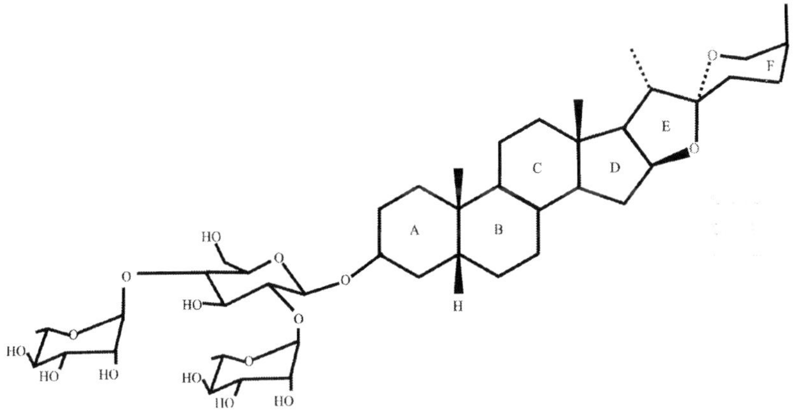

Figure 1. Structure of a spirostan saponin. Typical hexacyclic ABCDEF-ring system.

The steroidal saponin dioscin (Figure 7) had a huge importance as the favorite starting material in the steroid industry. A first transformation, an enzymatic or acidic hydrolysis, produced its aglycone diosgenin, and then modification of the diosgenin homoallylic enol and the spiroketal moieties gave progestagens, androstagens, corticosteroids, and some other important biological compounds. It is also possible to obtain a partial hydrolysis working under smooth-controlled conditions. Dioscin, and its chemically related saponins polyphyllin D and balanitins have a remarkable anticancer activity. These monodesmosidic saponins present oligosaccharide chains in which the first sugar, β-D-glucopyranose, is attached to the diosgenin C-3 position, and this in turn is substituted via its 2-OH and 4-OH positions. Commonly, α-L-rhamnopyranose, α-L-arabinofuranose, and other sugars constitute their oligosugar chains [9].

3. Diverse biological activities of saponins

As previously mentioned, steroidal saponins have an extensive variety of biological activities (Figure 2), including the absorption of cholesterol from the small intestine [10]. Mice treated with saponins from the plant *Tribulus terrestris* L. showed total cholesterol reduction in the liver and total serum [11], and hyperlipidemia was prevented. This control of cholesterol occurs through interaction with saponins, producing insoluble complexes that are excreted in bile, thus inhibiting entero-hepatic cholesterol recycling and reducing blood cholesterol levels (reviewed in [12]).

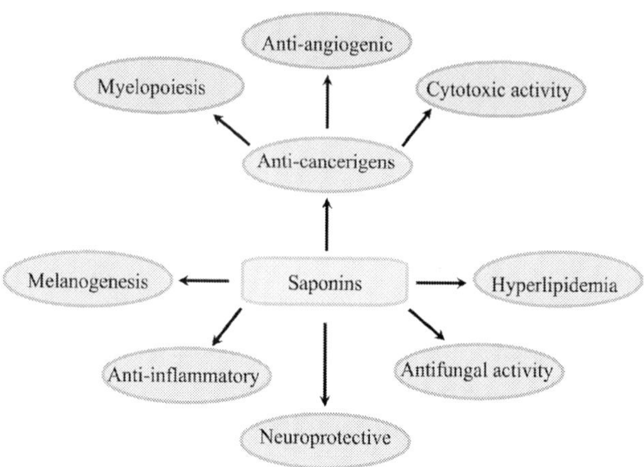

Figure 2. Biological activities of saponins.

3.1. Steroidal saponins from ginseng

Steroidal saponins are present in different types of plants. Ginseng (the root of *Panax ginseng*, C.A. Mey.) contains a series of ginsenosides that belong to the family of steroidal saponins, and these exhibit several biological properties (Figure 3). Ginsenosides are chemically structured by a skeleton consisting of four trans/anti-fusioned rings with modifications related to the type and number of sugar moieties and the attachment sites of the hydroxyl groups [13]. The two major components of the ginsenoside family are protopanaxadiol and protopanaxatriol [14]. The sugar moieties in the protopanaxadiol and protopanaxatriol are attached to the 3-position and 6-position of a dammarane-type triterpene, respectively (Figure 4). The protopanaxadiols include ginsenoside Rb1, Rb2, Rc, and Rd, while protopanaxatriols include ginsenoside Re, Rf, and Rg1.

Several reports indicate that each ginsenoside has distinct biological effects; it has been shown that purified ginsenoside protopanaxadiol Rb1 has a neuroprotective effect on PC12 (rat adrenal pheochromocytoma cell line) cells inhibiting the cell death by decreasing both the amount of active caspase-3 as well as DNA fragmentation, and increasing the amount of the anti-apoptotic Bcl-xL protein [15]. Besides acting as a neuroprotective, Rb1 has anti-angiogenic function inhibiting the process of new blood vessel formation [16], as well as an anti-inflammatory function [17].

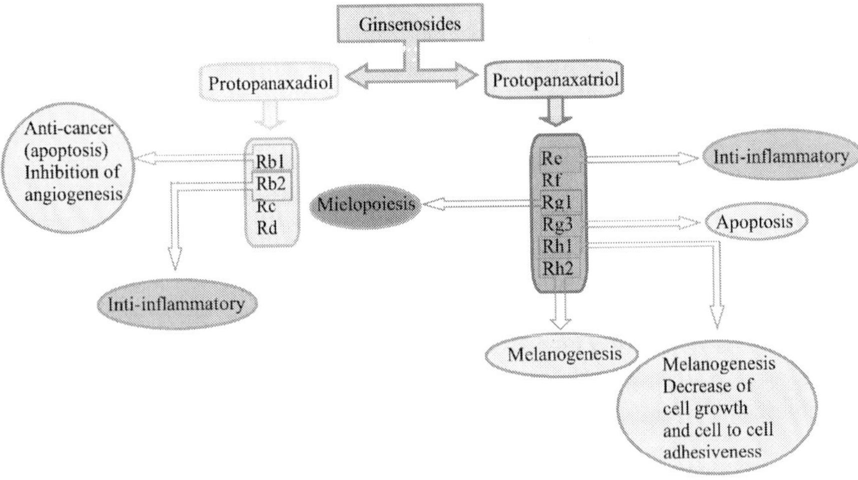

Figure 3. The two major groups of ginsenosides and their specific biological functions.

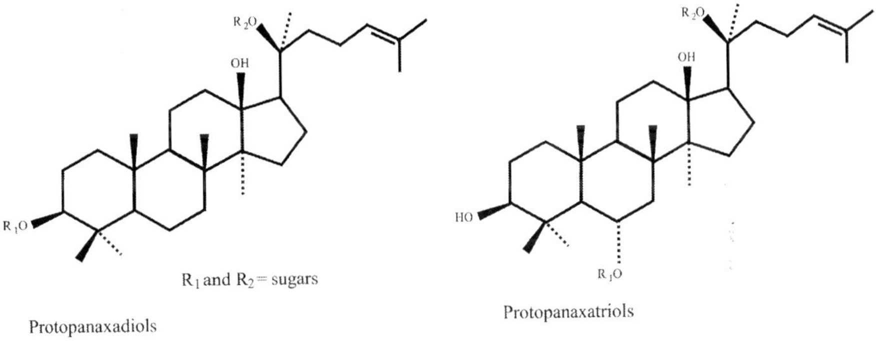

Figure 4. Structure of ginsenosides protopanaxadiols and protopanaxatriols.

The ginsenosides of the protopanaxatriol group have been shown to have several functions, some of which are similar to those exhibited by the protopanaxadiols. An anti-inflammatory effect has also has been shown by using the Re member [18]. With respect to the promotion of cell death by this group, it has been observed that Rg3 induces cell death in hepatocellular carcinoma cells in a selective form, since it does not affect normal cells [19]. The treatment of several types of tumors implies the use of chemotherapeutic agents that possess secondary reactions such as myelosupression. It has been shown that the protopanaxadiol Rg1 enhances myelopoiesis in vitro and reconstitutes bone marrow after myelosuppression treatment in mice [20].

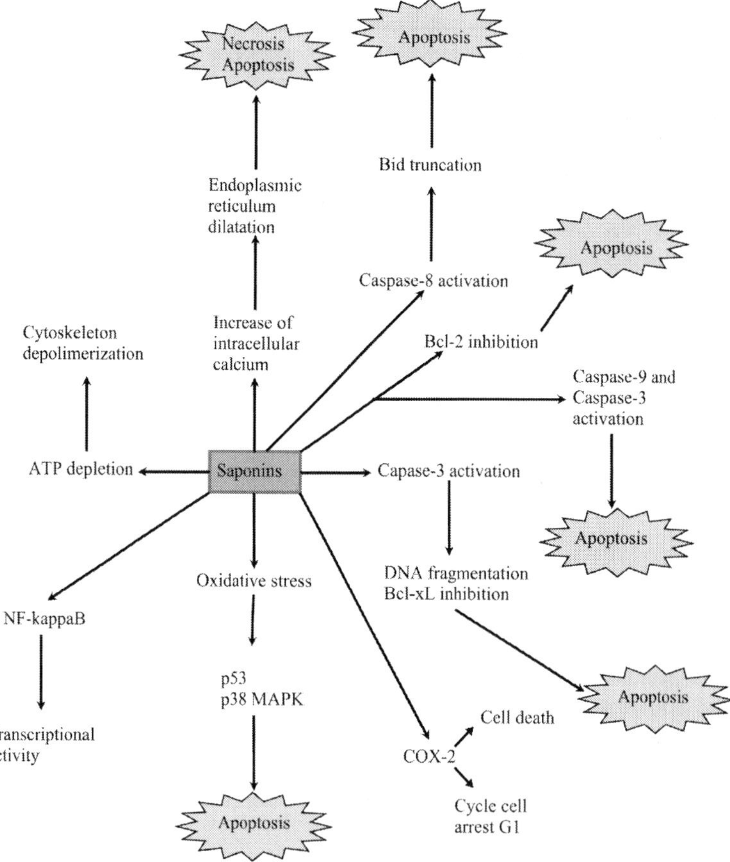

Figure 9. Different molecular cell death pathways activated by saponins. Saponins are able to induce the intrinsic and the extrinsic pathway of activation of the apoptosis cell death. In the same form, it influences inside different molecular levels including inside transcription factors as NF-kappaB as well as in the route of signaling of the MAPK. The correct polymerization of the cytoskeleton components is affected by the saponins, since this provokes a depletion of ATP inhibiting the correct polymerization of actin. In several occasions, the steroidal saponins not only induce the cell death process but also can inhibit the cell cycle progression.

6. Conclusions

Steroidal saponins are compounds that manifest antiproliferative activity and necrotic induction, and promote apoptotic or autophagic cell death in tumor cells. The important biological property of these compounds is their capacity to induce programmed cell death (apoptosis) in different tumor cell lines. In view of the fact that the compounds used in anticancer treatments are unspecific and inefficient in terminal patients and may have side effects stemming from their cytotoxic activity, research groups are looking for new compounds with antiproliferative activity that are noncytotoxic and have selective action. This aspect is relevant because it implies that the side effects related to cytotoxic activity could be reduced quite significantly. The knowledge of different molecular mechanisms of cell death triggered by saponins is of great importance because these compounds have been shown to have significant potential as antitumor agents, and may be apt for use in treating cancers, with important cost-benefit advantages and reduced side effects.

Acknowledgements

MLES thanks CONACyT for grant 180526. LSS thanks PAPIIT IN222114 for academic and financial support. J.S.R. thanks for Grant 176858. This chapter is partial fulfillment of the Doctorado en Ciencias Médicas y Biológicas de la Universidad Autónoma Benito Juárez de Oaxaca, México. The authors kindly thank Allen J. Coombes (BUAP Botanic Garden) for checking the English in the manuscript.

Author details

María L. Escobar-Sánchez[1*], Luis Sánchez-Sánchez[2] and Jesús Sandoval-Ramírez[3]

*Address all correspondence to: escobarluisa@ciencias.unam.mx

1 Lab. de Microscopía Electrónica. Depto. de Biología Celular. Facultad de Ciencias, UN-AM., México

2 Facultad de Estudios Superiores Zaragoza, Universidad Nacional Autónoma de México, México

3 Facultad de Ciencias Químicas, Benemérita Universidad Autónoma de Puebla, Ciudad Universitaria, Puebla, México

References

[1] Hostettmann K, Marston A. Saponins. Cambridge University Press, Cambridge, pp. 10–121. 1995. ISBN: 0-521-32970-1

[2] Bruneton J. Pharmacognosy, Phytochemistry, Medicinal Plants. Lavoisier Publishing, Paris, pp. 538–544 (ISBN 2-4730-0028-7). 1995.

[3]   Sparg SG, Light ME, van Staden J. Biological activities and distribution of plant saponins. J Ethnopharmacology 2004;94:219–43. DOI: 10.1016/j.jep.2004.05.016

[4]   Bomford R, Stapleton M, Winsor S, Beesley JE, Jessup EA, Price KR, Fenwick GR. Adjuvanticity and ISCOM formation by structurally diverse saponins. Vaccine 1992;10:572–7. DOI: 10.1016/0264-410X(92)90435M

[5]   Tyler VE, Brady LR, Robbers JE. Pharmacognosy, (9th edn.) Lea & Ferbiger, Philadelphia, 1988. ISBN 0-8121-1071-4

[6]   Zhu L, Tan J, Wang B, Guan L, Liu Y, Zheng C. In-vitro antitumor activity and antifungal activity of pennogenin steroidal saponins from Paris polyphylla var. yunnanensis. Iran J Pharm Res 2011;10(2):279–86.

[7]   Vincken JP, Heng L, de Groot A, Gruppen H. Saponins, classification and occurrence in the plant kingdom. Phytochemistry 2007;68:275–97. DOI: 10.1016/j.phytochem.2006.10.008

[8]   Challinor VL, De Voss JJ. Open-chain steroidal glycosides, a diverse class of plant saponins. Nat Prod Rep 2013;30:429–54. DOI: 10.1039/c3np20105h

[9]   Deng S, Yu B, Hui Y, Yu H, Han X. Synthesis of three diosgenyl saponins: dioscin, polyphyllin D and balanitin-7.Carbohydrate Res 1999;317:53--62. DOI: 10.1016/S0008-6215(99)00066-X

[10]   Oakenfull D, Sidhu GS. Could saponins be a useful treatment for hypercholesterolaemia? Eur J Clin Nutr 1990;44:79–88. http://www.ncbi.nlm.nih.gov/pubmed/2191861

[11]   Chu S, Qu W, Pang X, Sun B, Huang X. Effect of saponin from *Tribulus terrestris* on hyperlipidemia. Zhong Yao Cai 2003;26:341–4. http://www.ncbi.nlm.nih.gov/pubmed/14535016

[12]   Cheeke PR, Piacente S, Oleszek W. Anti-inflammatory and anti-arthritic effects of *Yucca schidigera*: a review. J Inflamm 2006;3:6. DOI: 10.1186/1476-9255-3-6.

[13]   Jang SI, Lee YW, Cho CK, Yoo HS, Jang JH. Identification of target genes involved in the antiproliferative effect of enzyme-modified ginseng extract in HepG2 hepatocarcinoma cell. Evid Based Complement Alternat Med 2013;2013:502568. DOI: 10.1155/2013/502568.

[14]   Qi LW, Wang CZ, Yuan CS. American Ginseng: potential structure-function relationship in cancer chemoprevention. Biochem Pharmacol 2010;80:947–54. DOI: 10.1016/j.bcp.2010.06.023.

[15]   Hashimoto R, Yu J, Koizumi H, Ouchi Y, Okabe T. Ginsenoside Rb1 prevents MPP+-Induced apoptosis in PC12 cells by stimulating estrogen receptors with consequent activation of ERK1/2, Akt and Inhibition of SAPK/JNK, p38 MAPK. Evid Based Complement Alternat Med 2012;2012:693717. http://dx.DOI.org/10.1155/2012/693717

[16]   Leung KW, Cheung LWT, Pon YL, Wong RNS, Mak NK, Fan TPD, Au SCL, Tombran-Tink J, Wong AST. Ginsenoside Rb1 inhibits tube-like structure formation of en-

dothelial cells by regulating pigment epithelium-derived factor through the oestrogen β receptor Br J Pharmacol 2007;152(2):207–15. DOI: 10.1038/sj.bjp.0707359

[17] Joh EH, Lee IA, Jung IH, Kim DH. Ginsenoside Rb1 and its metabolite compound K inhibit IRAK-1 activation. The key step of inflammation. Biochem Pharmacol 2011;82:278–86. DOI: 10.1016/j.bcp.2011.05.003.

[18] Lee IA, Hyam SR, Jang SE, Han MJ, Kim DH. Ginsenoside Re ameliorates inflammation by inhibiting the binding of lipopolysaccharide to TLR4 on macrophages. J Agric Food Chem 2012;60(38):9595–602. DOI: 10.1016/j.lfs.2007.05.009

[19] Lee JY, Jung KH, Morgan MJ, Kang YR, Lee HS, Koo GB, Hong SS, Kwon SW, Kim YS. Sensitization of TRAIL-induced cell death by 20(S)-ginsenoside Rg3 via CHOP-mediated DR5 upregulation in human hepatocellular carcinoma cells. Mol Cancer Ther 2013;12(3):274–85. DOI: 10.1158/1535-7163.MCT-12-0054.

[20] Raghavendran HR, Sathyanath R, Shin J, Kim HK, Han JM, Cho J, Son CG. *Panax ginseng* modulates cytokines in bone marrow toxicity and myelopoiesis: ginsenoside Rg1 partially supports myelopoiesis. PLoS One 2012;7(4):e33733. DOI: 10.1371/journal.pone.0033733.

[21] Chen CF, Chiou WF, Zhang JT. Comparison of the pharmacological effects of Panax ginseng and *Panax quinquefolium*. Acta Pharmacol Sin 2008;29(9):1103–8. DOI: 10.1111/j.1745-7254.2008.00868.x.

[22] Odashima S, Ota T, Kohno H, Matsuda T, Kitagawa I, Abe H, Arichi S. Control of phenotypic expression of cultured B16 melanoma cells by plant glycosides. Cancer Res 1985;45:2781–4. http://www.ncbi.nlm.nih.gov/pubmed/3986809

[23] Odashima S, Nakayabe Y, Honjo N, Abe H, Arichi S. Induction of phenotypic reverse transformation by ginsenosides in cultured Morris hepatoma cells. Eur J Cancer 1979;15:885–92. http://www.ncbi.nlm.nih.gov/pubmed/227695

[24] Tilwari A, Shukla NP, Devi U. Effect of five medicinal plants used in Indian system of medicines on immune function in Wistar rats. Afr J Biotechnol 2011;10:16637–45. DOI: 10.5897/AJB10.2168

[25] Álvarez L, Pérez MC, González JL, Navarro V, Villarreal ML, Olson JO. SC-1, an antimycotic spirostan saponin from *Solanum chrysotrichum*. Planta Medica 2001;67(4):372–374. DOI: 10.1055/s-2001-14332.

[26] Kataria H, Shah N, Kaul SC, Wadhwa R, Kaur G. Water extract of ashwagandha leaves limits proliferation and migration, and induces differentiation in glioma cells. Evid Based Complement Alternat Med 2011;2011:267614. DOI: 10.1093/ecam/nep188. Epub 2011 Feb 14

[27] Coleman JJ, Okoli I, Tegos GP, Holson EB, Wagner FF, Hamblin MR, Mylonakis E. Characterization of plant-derived saponin natural products against *Candida albicans*. ACS Chem Biol 2010;5(3):321–32. DOI: 10.1021/cb900243b.

[49] Zhang C, Feng S, Zhang L, Ren Z. A new cytotoxic steroidal saponin from the rhizomes and roots of Smilax scobinicaulis. Nat Prod Res 2013;27(14):1255–60. DOI: 10.1080/14786419.2012.725396

[50] Pan ZH, Li Y, Liu JL, Ning DS, Li DP, Wu XD, Wen YX. A cytotoxic cardenolide and a saponin from the rhizomes of Tupistra chinensis. Fitoterapia 2012;83:1489–93. DOI: 10.1016/j.fitote.2012.08.015.

[51] Gnoula C1, Mégalizzi V, De Nève N, Sauvage S, Ribaucour F, Guissou P, Duez P, Dubois J, Ingrassia L, Lefranc F, Kiss R, Mijatovic T. Balanitin-6 and -7: diosgenyl saponins isolated from Balanites aegyptiaca Del. display significant anti-tumor activity in vitro and in vivo. Int J Oncol 2008;32(1):5–15. DOI: 10.3892/ijo.32.1.5.

[52] Gao LL, Li FR, Jiao P, Yang MF, Zhou XJ, Si YH, Jiang WJ, Zheng TT. *Paris chinensis* dioscin induces G2/M cell cycle arrest and apoptosis in human gastric cancer SGC-7901 cells. World J Gastroenterol 2011;17(39):4389–95. DOI: 10.3748/wjg.v17.i39.4389.

[53] Itoh Y, Nagase H. Matrix metalloproteinases in cancer. Essays Biochem 2002;38:21–36. http://www.ncbi.nlm.nih.gov/pubmed/12463159

[54] Chen PS, Shih YW, Huang HC, Cheng HW. Diosgenin, a steroidal saponin, inhibits migration and invasion of human prostate cancer PC-3 cells by reducing matrix metalloproteinases expression. PLoS ONE 2011;6(5):e20164. DOI: 10.1371/journal.pone.0020164.

[55] Lee J, Jung K, Kim YS. Diosgenin inhibits melanogenesis through the activation of phosphatidylinositol-3-kinase pathway (PI3K) signaling. Life Sci 2007;81:249–54. DOI: 10.1016/j.lfs.2007.05.009.

[56] da Silva BP, De Sousa AC, Silva GM, Mendes TP, Parente JP. A new bioactive steroidal saponin from *Agave attenuata*. Z Naturforsch C 2002;57:423–8. http://www.ncbi.nlm.nih.gov/pubmed/12132678

[57] Chen YS, He Y, Chen C, Zeng Y, Xue D, Wen FY, Wang L, Zhang H, Du JR. Growth inhibition by pennogenyl saponins from *Rhizoma paridis* on hepatoma xenografts in nude mice. Steroids 2014;83:39–44. DOI: 10.1016/j.steroids.2014.01.014.

[58] Tait SW, Green DR. Mitochondria and cell death: outer membrane permeabilization and beyond. Nat Rev Mol Cell Biol 2010;11(9):621–32. DOI: 10.1038/nrm2952.

[59] Wu J, Kaufman RJ. From acute ER stress to physiological roles of the Unfolded Protein Response. Cell Death Differ 2006;13:374–84.

[60] Kubo S, Mimaki Y, Terao M, Sashida Y, Nikaido T, Ohmoto T. Acylated cholestase glycosides from the bulbs of *Ornithogalum saundersiae*. Phytochemistry 1992;31:3969 – 73. DOI: 10.1016/S0031-9422(00)97565-4

[61] Garcia-Prieto C, Riaz KB, Chen Z, Zhou Y, Hammoudi N, Kang Y, Lou C, Mei Y, Jin Z, Huang P. Effective killing of leukemia cells by the natural product OSW-1 through

disruption of cellular calcium homeostasis. Biol Chem 2013;288(5):3240–50. DOI: 10.1074/jbc.M112.384776.

[62] Sy LK, Yan SC, Lok CN, Man RY, Che CM. Timosaponin A-III induces autophagy preceding mitochondria-mediated apoptosis in HeLa cancer cells. Cancer Res 2008;68:10229–37. DOI: 10.1158/0008-5472.CAN-08-1983.

[63] Diaz-Troya S, Perez-Perez ME, Florencio FJ, Crespo JL. The role of TOR in autophagy regulation from yeast to plants and mammals. Autophagy 2008;4(7):851–65. http://www.ncbi.nlm.nih.gov/pubmed/18670193

[64] King FW, Fong S, Griffin C, Shoemaker M, Staub R, Zhang YL, Cohen I, Shtivelman E. Timosaponin AIII is preferentially cytotoxic to tumor cells through inhibition of mTOR and induction of ER stress. PLoS One 2009;4(9):e7283. DOI: 10.1371/journal.pone.0007283.

[65] Cai J, Liu M, Wang Z, Ju Y. Apoptosis induced by dioscin in Hela cells. Biol Pharm Bull 2002;25:193–6. DOI.org/10.1248/bpb.25.193.

[66] Tor YS, Yazan LS, Foo JB, Armania N, Cheah YK, Abdullah R, Imam MU, Ismail N, Ismail M. Induction of apoptosis through oxidative stress-related pathways in MCF-7, human breast cancer cells, by ethyl acetate extract of *Dillenia suffruticosa*. BMC Complement Altern Med 2014;14:55. DOI: 10.1186/1472-6882-14-55.

[67] Shishodia S, Aggarwal BB. Diosgenin inhibits osteoclastogenesis, invasion, and proliferation through the downregulation of Akt, IkappaB kinase activation and NF-kappaB-regulated gene expression. Oncogene 2006;25:1463–7. DOI:10.1038/sj.onc.1209194

[68] Hsieh MJ, Tsai TL, Hsieh YS, Wang CJ, Chiou HL. Dioscin-induced autophagy mitigates cell apoptosis through modulation of PI3K/Akt and ERK and JNK signaling pathways in human lung cancer cell lines. Arch Toxicol 2013;87:1927–37. DOI: 10.1007/s00204-013-1047-z. Epub 2013 Apr 4.

[69] Atkinson SJ, Hosford MA, Molitoris BA. Mechanism of actin polymerization in cellular ATP depletion. J Biol Chem 2004;279:5194–9. DOI:10.1074/jbc.M306973200.

[70] Moalic S, Liagre B, Corbière C, Bianchi A, Dauça M, Bordji K, Beneytout JL. A plant steroid, diosgenin, induces apoptosis, cell cycle arrest and COX activity in osteosarcoma cells. FEBS Lett 2001;506(3):225–30. DOI: 10.1016/S0014-5793(01)02924-6.

Chapter 16

Cell Death Induction by Targeting Tumor Metabolism

Karin von Schwarzenberg

Additional information is available at the end of the chapter

http://dx.doi.org/10.5772/61480

Abstract

Over the last century, a broader interest in the topic of tumor metabolism has emerged. From the 1920s onward, when Otto Warburg proposed increased aerobic glycolysis of tumor cells, a deeper understanding has established that tumor cells have an altered metabolism which is directly linked to cancer progression. It was soon discovered that not only do environmental changes lead to alterations in metabolism but that oncogenes have a profound influence in these alterations. They not only induce nutrient uptake and synthesis of proteins and DNA but can lead to a switch toward glycolysis, which identifies them as a major player in tumor metabolism. These observations have raised the interest to target metabolic pathways for cancer therapy and, interestingly, some of the first discovered chemotherapeutics target metabolic pathways and are still in clinic. Concerns that these targets will also affect normal cells has intensified research to understand how changes in tumor metabolism promote tumor growth and which enzymes and signaling pathways are involved. These observations led to the discovery of new targets and drugs that specifically affect tumor metabolism and can exploit the dependence of tumor cells on the metabolic changes.

Keywords: Tumor metabolism, glycolysis, AMPK, lipid metabolism, p53

1. Introduction

Already in the 1920s, Otto Warburg described that tumor cells typically use aerobic glycolysis for energy production rather than oxidative phosphorylation (OXPHOS) despite sufficient oxygen and the lesser yield of ATP. By now, altered tumor metabolism is recognized as a hallmark of cancer cells, which allows them to escape from the typical regulatory constraints that prevent normal cells from uncontrolled growth and proliferation. A number of theories have been proposed to explain this phenomenon, amongst them independence of oxygen especially in hypoxic areas, which often occur in tumors. Furthermore, tumor cells often have mutations in mitochondria, which could explain a shift toward glycolysis. Glucose and

glutamine uptake also have the advantage that they can not only be used as ATP source but also as building blocks for essential metabolites required for uncontrolled growth (e.g., amino acids, nucleotide triphosphates, NADPH) [1, 2]. Glucose is used for the formation of nucleic acids via the pentose phosphate pathway and glycolytic intermediates are used for fatty acid biosynthesis. Therefore, highly proliferating tumor cells need to change different aspects of their metabolism to meet the high demand of energy in the form of ATP and secure the supply of the major classes of macromolecules: carbohydrates, proteins, lipids, and nucleic acids (Figure 1).

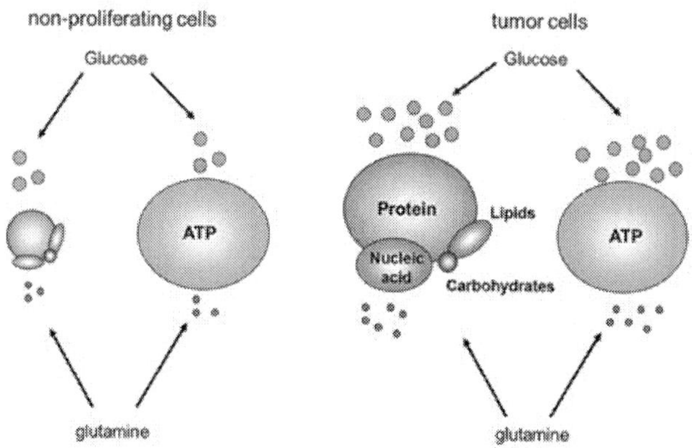

Figure 1. Tumor cells use glucose and glutamine not only for ATP production but also as source of essential metabolites. To ensure uncontrolled growth and proliferation, they have a much higher demand of nutrients, which is provided by aerobic glycolysis.

In recent years, it has also been revealed that the metabolic reprogramming is not necessarily induced by different metabolic requirements but also regulated by oncogenes and tumor suppressors. The PI3K/Akt pathway is often deregulated in tumor cells. Akt is one of the most important proteins in cells which have elevated glycolysis as it promotes the increased expression and membrane localization of the glucose receptor-1 (GLUT-1) and stimulates phosphofructokinase activity. It thus promotes augmented glucose uptake and increased glycolytic activity. The tumor suppressor p53, on the other hand, can shift tumor cells from glycolysis to oxidative phosphorylation [3, 4]. The major inducer and regulator of glycolysis is the hypoxia-inducible factor-1α (HIF1α), a transcription factor that is often found in highly metastatic and neoplastic tumor cells and strongly promotes glycolysis.

Alterations in tumor metabolism are quite heterogeneous, which impedes the finding of a generalized target. The changes depend on the availability of nutrients, oxygen, and the pH, and in turn depend on the different tumor vasculature. Proliferation requires nutrient uptake, metabolite and DNA synthesis, and energy production. Thus, genetic alterations in signaling pathways that drive the cells to proliferate are often involved in changes of the tumor metabolism. But drugs that target, e.g., DNA synthesis to inhibit proliferation are also directed against normal proliferating cells and therefore lack selectivity. Therefore, the systematic characterization of the metabolic pathways that differ in cancer cells is an ongoing challenge which must lead to the discovery of drugs that specifically target proteins or enzymes altered in tumors. Targeting tumor metabolism has become a promising field in cancer therapy but requires an in-depth understanding of the metabolic regulation and signaling pathways involved, which will be reviewed here.

2. Targeting tumor metabolism

Understanding the metabolic pathways altered in tumor cells targeting the resources or specific pathways that fuel deregulated tumor metabolism has shown to be an attractive strategy for cancer therapeutics. Hexokinase inhibitors like 2-deoxy-D-glucose have already shown promising results in preclinical studies but also dichloracetate, which targets pyruvate dehydrogenase kinase (PDK1); gemcitabine, which inhibits nucleic acid synthesis; or metformin, which induces the AMP-activated protein kinase (AMPK) are interesting compounds for targeting tumor metabolism. Interestingly, tumors with high glucose uptake detected by the 2-[18F]fluoro-2-deoxy-D-glucose–positron emission tomography (FDG–PET) scan show a worsened outcome [5], confirming the importance for drugs targeting tumor metabolism.

The complexity and tight regulation of the tumor metabolism raises the possibility to target multiple pathways, enzymes, and proteins, some of the most important ones of which will be addressed here.

2.1. Glycolysis

2.1.1. Hexokinase

Hexokinase catalyzes the first and rate-limiting step in glycolysis (see Figure 2) by the ATP-dependent phosphorylation of glucose to yield glucose-6-phosphate (G6P). There are four different hexokinase isoforms (HK1-4), which show different tissue distribution and enzyme activity. The high affinity kinases HK1 and HK2 are inhibited by excess G6P, are associated with the mitochondria, and implicated in cell survival [6]. Hexokinase-2 was shown to be highly expressed in tumor cells but only in a limited number of normal tissues and is partly responsible for the increased glycolytic activity of tumor cells. Hexokinase level could also be correlated with tumor stage and patient survival [7].

Due to the importance of hexokinase for the glycolytic flux, there exist a variety of inhibitors such as 2-deoxy-D-glucose (2DG), 3-bromopyruvate, and lonidamine, which showed promising effects in preclinical studies [8].

2DG is a glucose analogue that is phosphorylated by hexokinase to form 2DG phosphate, which cannot be further metabolized. Treatment of tumor cells with 2DG inhibits glycolysis, leading to ATP depletion, cell growth inhibition, and apoptosis [9]. 2DG has extensive metabolic effects and not only affects glycolysis but also OXPHOS. In normoxia, it can interfere with N-linked glycosylation and induce the unfolded protein response [10].

Despite promising activities *in vitro*, the effects of 2DG as a single agent *in vivo* and in clinical trials were a disappointment. This might be due to off-target effects like the activation of the PI3K pathway or the induction of pro-survival autophagy [11]. Furthermore, the dose needed for complete inhibition of glycolysis in patients showed severe side effects. In combination therapy, however, lower doses of 2DG, which are better tolerated, could improve the efficiency [12].

2.1.2. Hypoxia-inducible factor-1α

Tumor growth is associated with intratumoral hypoxia due to the lack of sufficient vascularization. The physiological consequence for the tumor to survive in this hostile environment is to increase angiogenesis to achieve adequate oxygen delivery, to adapt the metabolism by increasing glucose uptake, and switch to glycolysis for energy supply. All these actions are initiated by the transcription factor hypoxia-inducible factor-1α (HIF1α), which is the major regulator of glycolysis and induces more than a hundred genes involved in metabolism [13]. Increased activity of HIF1α is known to correlate with poor patient outcome and, interestingly, some highly neoplastic tumors like renal cell carcinoma frequently carry a mutation that leads to a constitutive active HIF1α. The best known action of HIF1α is the induction of VEGF (vasculature endothelial growth factor), which is required for angiogenesis [14, 15]. It also induces the expression of the glucose transporters GLUT-1 and -3, and activates a number of glycolytic enzymes like aldolase, phosphofructokinase, enolase, and lactate dehydrogenase. It furthermore activates pyruvate dehydrogenase-1 (PDK1) and, thereby, reduces the flow of pyruvate used by the TCA cycle, decreasing OXPHOS and further increasing glycolysis, as shown in Figure 2.

HIF1 is a heterodimer composed of the constitutive expressed HIF1β and the oxygen-sensitive HIF1α. In the presence of oxygen, HIF1α is hydroxylated by prolyl hydroxylases on prolyl-residues, which are required for the binding of the ubiquitin E3 ligase Von-Hippel-Lindau (VHL). Binding of VHL results in the degradation of HIF1α by the proteasome. Under some circumstances like overexpression of the PI3K/Akt pathway or the induction of reactive oxygen and nitrogen species, HIF can be stabilized even under normoxia [16].

Due to the importance of HIF1α in tumor growth and survival and the poor outcome of patients with high levels of HIF1α, targeting this protein seemed to be a promising option in tumor therapy. Interestingly, until now no specific inhibitors of HIF1α exist in the clinic, although there have been various efforts, and some experimental drugs exist, which inhibit transcription, translation, or DNA binding of HIF1α [17]. The most promising drugs inhibit HIF1α as a side effect; like rapamycin, which inhibits mTOR activity. The PI3K/Akt/mTOR pathway plays an important role for HIF1α translation and rapamycin as well as the PI3K inhibitor LY294002 have shown to be able to reduce HIF protein expression as well as expression of its target genes [18, 19].

The proteasome inhibitor bortezomib is another drug that was reported to inhibit adaption to hypoxia of tumors and to functionally inhibit HIF1α. Bortezomib is approved for the treatment of multiple myeloma and, therefore, an interesting candidate for analyzing effects on HIF. It inhibits the expression of the HIF target genes VEGF and erythropoietin and the recruitment of p300 coactivator [20].

2.2. Signaling proteins and growth control elements

2.2.1. PI3K/Akt

The PI3K pathway is one of the most common altered signaling pathways in cancer. The main actor is the serine/threonine kinase Akt, which is able to activate many downstream targets involved in cell growth, survival, and cell cycle progression. A constitutive activation of Akt not only leads to strong pro-survival signals but also has a strong impact on tumor metabolism. Increased Akt signaling has been shown to directly correlate with increased glucose metabolism in a variety of tumor cells [21]. It supports aerobic glycolysis even in untransformed cells when overexpressed. Akt increases glucose uptake through promoting the translocation of the glucose receptor GLUT1 to the plasma membrane and also induces several glycolytic enzymes like hexokinase and phosphofructokinase. Furthermore, it mediates the increase of a variety of fatty acid- and cholesterol-synthesis enzymes. It promotes binding of the hexokinase to the mitochondrial membrane and thereby, like Bcl-2 or Bcl-X, increases mitochondrial membrane integrity and inhibits apoptosis [22]. Finally, Akt strongly induces signaling via mTOR by phosphorylating and inhibiting its negative regulator tuberous sclerosis complex 2 (TSC2). mTOR is a key metabolic checkpoint which leads to induction of protein and lipid biosynthesis when activated and, therefore, promotes cell growth. The importance of Akt in glucose metabolism is also supported by the fact that targeting the PI3K pathway in animal models and the use of kinase inhibitors in patients lead to a decrease in glucose uptake as measured by FDG-PET uptake [23, 24].

These observations lead to the assumption that inhibition of Akt would lead to inhibition of glycolysis and especially kill tumor cells that are dependent on a permanent supply of glucose. Using Akt inhibitors on tumor cells with aberrant PI3K signaling would, on the one hand, sensitize them to glucose-starvation-induced apoptosis as shown by Elstrom *et al.* (2004) [21] and, on the other hand, would open a new therapeutic window for PI3K inhibitors in cells dependent on aerobic glycolysis.

2.2.2. AMPK

The AMP-activated protein kinase (AMPK) belongs to a family of serine/threonine kinases and is highly conserved from yeast to mammals. It consists of a catalytic α subunit and regulatory β and γ subunits. It is the most important energy-sensing protein in the cell and activated by a number of metabolic or oncogenic stresses like glucose and nutrient deprivation or hypoxia. AMPK senses the levels of AMP and ATP and is activated by an increased AMP/ATP ratio. Upon activation, it increases catabolic processes that generate ATP-like fatty acid oxidation and glycolysis and inhibits anabolic processes that consume ATP-like protein and lipid synthesis [25] (shown in Figure 3). The dominant upstream kinase that regulates AMPK activity is the liver kinase B1 (LKB1), which is a known tumor suppressor. Loss-of-function mutations of this kinase were first discovered in Peutz-Jeghers syndrome, an autosomal

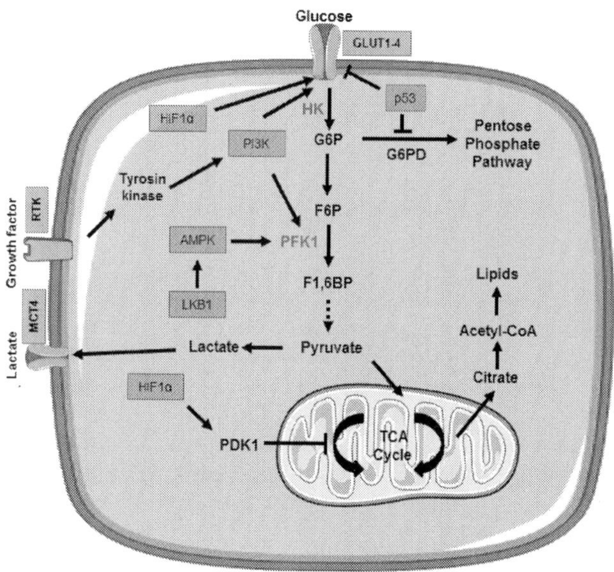

Figure 2. Simplified schema of tumor metabolism showing glycolysis, TCA cycle, lipid metabolism, and pentose phosphate pathway. Shown are proteins that are overexpressed or mutated to guarantee an increased glycolysis and decreased oxidative phosphorylation (HIF1α, PI3K/Akt, p53). HK (hexokinase), PFK1 (phosphofructokinase-1), GLUT (glucose receptor), G6P (glucose-6-phosphate), PDK1 (pyruvate dehydrogenase kinase-1), TCA cycle (tricarboxylic acid cycle), RTK (receptor tyrosine kinase), MCT4 (monocarboxylate transporter-4).

dominant genetic disorder characterized amongst others by an increased risk of gastrointestinal adenocarcinoma [26]. Mutations of LKB1 were also found in several tumors like sporadic lung adenocarcinoma or cervical carcinoma. One of the major downstream targets of AMPK is mTOR, which induces protein synthesis. During energy stress, AMPK leads to an inhibition of mTOR on the level of TSC2 and consequently to an inhibition of protein synthesis. Interestingly, ablation of LKB1 abolishes this inhibitory effect [27].

Although decreased activity of AMPK was shown to promote tumor growth, in recent years, it has become evident that AMPK activation can also prevent tumor formation, especially under conditions of hypoxia and glucose deprivation. Deletion of AMPKα1 synergizes with myc to promote lymphangiogenesis of B cell lymphoma, which supports the role as a tumor suppressor. In breast carcinoma, reduced AMPK phosphorylation inversely correlated with histological grade and axillary node metastasis [28]. Furthermore, AMPK was shown to inhibit cell proliferation by stabilization of p53 or regulation of cyclin-dependent kinase (CDK) inhibitors p21 and p27. But although activated AMPK can inhibit tumor cell proliferation, loss of AMPK is not sufficient to allow proliferation in the absence of nutrients. There are even reports that tumor cells lacking AMPK undergo apoptosis during metabolic stress and are

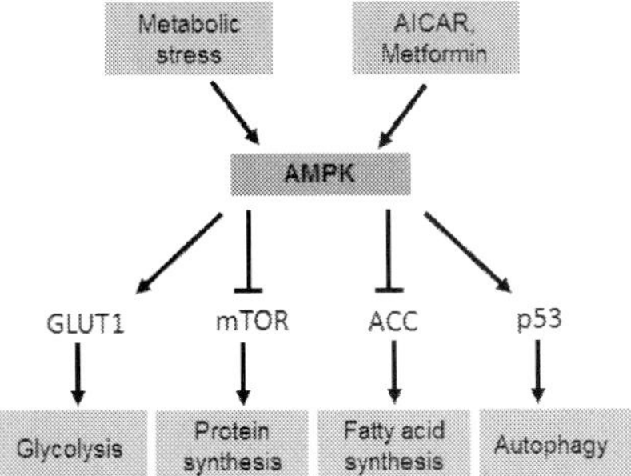

Figure 3. Major functions of AMPK activation. AMPK activation leads to an induction of glycolysis, fatty acid oxidation, and autophagy, and to an inhibition of protein synthesis.

resistant to oncogenic transformation [29]. On the other hand, pharmacological activators of AMPK like metformin, AICAR, or A769662 inhibited or delayed tumor formation in animals [30]. Faubert *et al.* (2013) could show that silencing AMPK promotes a metabolic shift to the Warburg effect with increased glucose uptake, glycolytic flux, and flow of carbon to the tricarboxylic cycle to fuel pathways of ATP production and biosynthesis [31]. The pro-apoptotic mechanisms of AMPK might be mainly associated with the inhibition mTOR. In line with this assumption, it was shown that AMPK activation by metformin or the AMP analogue AICAR correlated with mTOR inhibition in renal cell carcinoma [32].

Pro-survival mechanisms of AMPK can be partly explained by autophagy induction, which is mediated by p53, and maintenance of proliferative quiescence [33]. Therefore, the role of AMPK as an oncogene or tumor suppressor seems to be dependent on the degree of AMPK activation and duration of nutrient deprivation. Despite the promotion of AMPK agonists like metformin as anticancer treatment, there needs to be a deeper understanding of AMPK regulation in tumorigenesis to implement AMPK-targeting drugs into clinic.

2.2.3. p53

The tumor suppressor protein p53 is the most frequent mutated gene in cancer. It is well known for its role in DNA damage repair and apoptosis induction after cellular stress. Depending on the stress signals, it either leads to growth arrest by inducing p21, Gadd45 or p48 or in response to severe stresses it induces genes involved in apoptosis (Puma, Bax, Fas) or senescence (p21). But it is emerging that it also plays a pivotal role in tumor metabolism by inhibiting glycolysis and switching tumor metabolism to OXPHOS and thus inducing apoptosis.

p53 has multiple functions in tumor metabolism and considering its role as tumor suppressor, it is not surprising that p53 counteracts metabolic changes associated with cancer growth: it transcriptionally represses the expression of the glucose transporters GLUT1 and GLUT4 and indirectly represses expression of GLUT3 via NFκB inhibition and, therefore, inhibits the first rate-limiting step in glycolysis [34]. At the third step of glycolysis, phosphofructokinase-1

(PFK1) is inhibited by various metabolites that indicate sufficient supply of energy like ATP, citrate, and lactate. AMP and fructose-2,6-bisphosphate (F26B), which indicate low energy, activate PFK1 and, thus, increase glycolytic flux. TP53-inducible glycolysis and apoptosis regulator (TIGAR) is an important protein activated by p53. It acts as a phosphatase and degrades F26B decreasing the activity of PFK1 and consequently lowers the glycolytic rate [4, 35]. There are also some other glycolytic players that are inhibited by p53-like expression of pyruvate dehydrogenase kinase 2 (PDK2), which leads to a decreased conversion from pyruvate to lactate. Inhibition of the glycolytic pathway on multiple levels would assume that glucose would be shuttled into the pentose phosphate pathway (PPP) but it was recently shown that p53 also inhibits PPP by binding and inhibiting glucose-6-phosphate dehydro-genase (G6PD), the enzyme that catalyzes the first step of the PPP [36]. Therefore, p53 can reduce the production of NADPH and ribose-5-phosphate, which are important for reactive oxygen defense and DNA synthesis.

p53 not only inhibits glycolysis but also enhances OXPHOS: it transcriptionally activates cytochrome c oxidase 2 (SCO2), which is required for the assembly of the cytochrome c oxidase complex (complex IV in the mitochondrial electron transport chain) [37] and induces the expression of AIF (apoptosis-inducing factor), which is important for OXPHOS, most likely by ensuring the proper assembly of mitochondrial respiratory complex I. Furthermore, p53 regulates mitochondrial DNA copy number and mitochondrial quality control by removing damaged mitochondria [38]. By these actions, p53 directly counteracts the Warburg effect and therefore opposes metabolic changes that are essential for malignant transformation.

The metabolic functions of p53 are emerging as critical for tumor suppression and apoptosis induction via targeting tumor metabolism. Nevertheless, the many different stress signals that activate different p53 responses are still not fully understood. Therefore, future studies to understand the molecular mechanisms that activate p53 and mediate the responses to metabolic stress need to be performed to finally exploit and manipulate them for tumor therapy.

2.3. Pathways

2.3.1. Pentose phosphate pathway

Glucose is the main energy source not only for tumor cells but for all organisms. It enters the cell via glucose transporters and is then phosphorylated by hexokinase to form glucose-6-phosphate (G6P). From this point, G6P can either enter glycolysis to produce energy in form of ATP or it is shuttled to the pentose phosphate pathway (PPP). There, it is either hydrogen-ated by G6PD (glucose-6-phosphate dehydrogenase) and further converted to yield ribulose 5-phosphate in the oxidative branch. Or it enters the nonoxidative branch, which is catalyzed by transketolase, and generates ribose-5-phosphate, which is a precursor of biomolecules like nucleotides and, therefore, important for DNA synthesis (see Figure 4). During the oxidative phase, $NADP^+$ acts as electron acceptor during the oxidative reactions and 2 molecules of NADPH are yielded. NADPH plays an important role in the protection of the cell from oxidative stress. Therefore, the PPP plays a pivotal role in reductive biosynthesis like lipid and nucleotide synthesis and is essential for antioxidant defense. It is strongly connected with glycolysis and glucose is shuttled to the pathway where it is needed most. Historically, most attention was paid to glycolysis as it provides the energy for biosynthesis, but highly proliferating cells also need large amounts of lipids as energy storage and building blocks for membranes and nucleotides for DNA replication [39]. These needs are fulfilled by the PPP. In recent years, more attention was paid to changes in the pathway and effort is taken to find ways to target the PPP for tumor therapy.

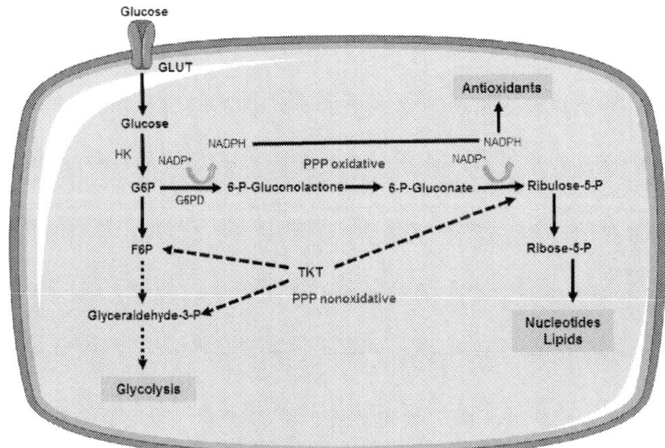

Figure 4. Schematic representation of the PPP and its connection with glycolysis. The oxidative branch of the PPP yields NADPH, which can be used for antioxidant defense and biosynthetic reactions. The nonoxidative branch produces ribulose-5-phosphate and glycolytic intermediates. G6PD – glucose-6-phosphate dehydrogenase; TKT – transketolase.

Considering its importance for proliferation, it is not astonishing that increased PPP flux and overexpression of G6PD have been found in several tumors like large B-cell lymphoma or lung adenocarcinoma. It was also shown that it plays a role in promoting malignant cell growth and inducing anchorage-independent growth in NIH3T3 (mouse embryonic fibroblast cell line) cells overexpressing G6PD [40, 41]. These findings suggest G6PD as an oncogene and interesting target for tumor therapy, which is also supported by the fact that the tumor suppressor p53 can bind to G6PD and inhibit its dimerization leading to a decreased G6PD activity, reduced glucose consumption and NADPH production, and, therefore, decreased tumor cell growth [36]. Furthermore, combinations with G6PD inhibitors like DHEA or 6-aminonicotinamide (6AN) with chemotherapeutics like 2-deoxy-D-glucose or oxythiamine, a TKT inhibitor, are shown to increase the inhibition of cancer growth and enhance the radiosensitivity of human gliomas and squamous carcinoma cell lines [42, 43]. Also, many chemotherapeutics like 5-fluoracil or gemcitabine led to the induction of ROS via DNA damage, which could be enhanced by the use of a G6PD inhibitor, and strengthened the importance of the PPP as a target for chemotherapy.

2.3.2. Lipid metabolism

Not only glucose metabolism is altered in cancer cells but also lipid biosynthesis is enhanced to meet the requirements of fast-proliferating cells. Normal cells obtain lipids via the uptake of free fatty acids (FAs) or low density lipoproteins (LDL) from the bloodstream. New synthesis of fatty acids and cholesterol is restricted to a few specialized tissues like the liver, adipose tissue, or the lactating breast. In cancer cells, however, these restrictions are interrupted and new synthesis of lipids is observed. Lipids are needed for the new synthesis of phospholipid membranes, lipid modified signaling molecules, and as energy storage to survive times of nutrient deprivation. FAs consist of a terminal carboxyl group and a hydrocarbon chain, mostly occurring in even numbers of carbons that can be either saturated or unsaturated. The acetyl groups needed for fatty acid biosynthesis are mostly provided by citrate, which is generated in the TCA cycle. The rate-limiting step of lipid synthesis is the conversion of acetyl-CoA to malonyl-CoA by the acetyl-CoA-carboxylase (ACC), which is then further processed by fatty acid synthase (FAS) to yield saturated and unsaturated fatty acids as shown in Figure 5. Several enzymes in fatty acid biosynthesis have been explored as potential targets for tumor therapy. Especially FAS and ACC are in focus as their inhibition by siRNA or chemical inhibitors led to growth arrest of tumor cells [44].

FAS is a prominent target as inhibition preferentially kills tumor cells and many tumors show increased FAS expression and dependence on *de novo* fatty acid synthesis, whereas nontumor cells rely on exogenous FAs. Inhibition of FAS, for example, leads to apoptosis induction of cells derived from lymph node metastasis of prostate carcinoma LNCaP cells and to an inhibition of HER2 expression in breast cancer cells [45, 46].

ACC is the most regulated enzyme in FA-synthesis. It is positively regulated by citrate and glutamate and inactivated by AMPK. Inhibition of ACC by siRNA or the chemical inhibitor soraphen A induced apoptosis in breast cancer and prostate cancer cells [47, 48]. Another ACC inhibitor, TOFA, showed growth inhibition of ovarian cancer cells and ovarian tumor mouse xenografts [49].

Apart from fatty acids, cholesterol plays a critical role in tumor growth and survival. Cholesterol is either obtained by uptake of LDL from the extracellular environment by the LDL receptor (LDLR) or it is newly synthesized in the mevalonate pathway (Figure 5). The first steps of the mevalonate pathway include the condensation of acetyl-CoA with acetoacetyl-CoA to form 3-hydroxy-3-methylglutaryl-CoA (HMG-CoA). The reduction of HMG-CoA to mevalonate by the HMG-CoA reductase is the rate-limiting step in cholesterol synthesis. Cholesterol is an important component of biological membranes as it modulates fluidity of the lipid bilayer and is part of the detergent-resistant lipid rafts that are membrane parts with high lipid content, which coordinate activation of a variety of signal-transduction pathways. Intermediates of the mevalonate pathway like geranylgeranyl or farnesyl are responsible for isoprenylation of small GTPases like Ras or Rho. Small GTPases play an important role in various cellular events (e.g., intracellular signal transduction, proliferation) and are anchored in the membrane. Accumulation of cholesterol has been associated especially with prostate cancer and an aberrant mevalonate pathway has been linked to several cancers [50].

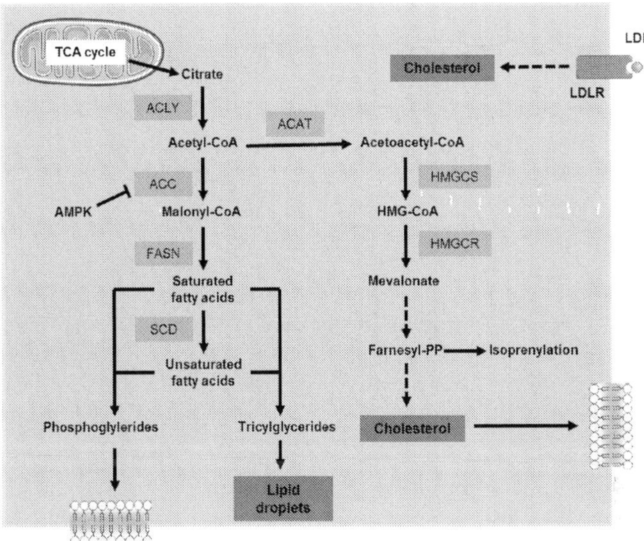

Figure 5. Schematic overview of the lipid and cholesterol metabolism. Phosphoglycerides, which are needed for membrane components, and triglycerides, which are stored in lipid droplets, are generated via multiple steps from fatty acids. Cholesterol is produced either by uptake of LDL or by synthesis via the mevalonate pathway and is an important component of the plasma membrane. Intermediates are used for isoprenylation of small GTPases. ACLY – ATP citrate lyase; ACC – acetyl-CoA-carboxylase; FASN – fatty acid synthase; SCD – steaoryl CoA desaturase; ACAT – acetyl-CoA-acetyltransferase, LDLR – low density lipoprotein receptor; HMGCS – 3-hydroxy-3-methylglutaryl-CoA synthase; HMGCR – HMG-CoA reductase.

The HMG-CoA reductase is the target for a group of cholesterol lowering drugs, the statins. Interestingly, there are some reports that statins reduce the risk for some cancers like colorectal cancer or hepatocellular cancer. But other studies found no connections. The effect seems highly dependent on the tumor type. Statins show antiproliferative activities and apoptosis induction in a variety of cancers. The cytostatic properties of statins are mainly due to, on the one hand, inhibition of cholesterol synthesis and, on the other hand, inhibiting the formation of isoprenoids. This could be confirmed by the addition of mevalonate or geranylgeranyl pyrophosphate, which could overcome the antiproliferative effects of statins [51]. Blocking the activity of small GTPases like the oncogene Ras leads to inhibition of pro-survival signaling cascades.

The induction of apoptosis by statins has been widely studied and several mechanisms of action have been revealed. In breast cancer cells, it was shown that apoptosis is induced by the activation of the c-Jun terminal kinase JNK and by downregulation of Bcl-2. In chronic myeloid leukemia, simvastatin induces apoptosis by inhibiting the NFκB pathway [52, 53].

These promising preclinical data lead to several phase I and II studies with the result that success of statins in tumor therapy is very tumor-specific and dependent on how important the mevalonate pathway is for the tumor.

2.4. pH regulation

2.4.1. Extra-and intracellular pH

The intracellular and extracellular pH of cells is one of the major factors that influence molecular processes involved in cell cycle progression and proliferation. Therefore, tumor formation and response of tumors to chemotherapeutics is also highly affected by the environmental acidity. Fast-growing tumors require a complementary vasculature to fulfill the extensive need of nutrients and oxygen. But solid tumors often develop faster than the blood supply, which results in a hypoxic environment. Due to the high glycolytic rate, tumor cells produce increased amounts of H^+ (lactate, carbonic acids) and therefore create a hostile microenvironment. The resulting acidic environment is toxic for normal cells and establishes an advantage for growth of tumors, which evolve adaptive mechanisms that allow them to survive [54]. By these adaptive mechanisms, tumor cells transport protons and lactic acid into the extracellular space via acid-base regulators like monocarboxylate transporters and Na^+/H^+ exchangers, leading to an acidic microenvironment and to a slightly alkaline cytosol. These pH conditions have shown to be beneficial for metastasis, apoptosis resistance, and increased proliferation. An acidic extracellular space promotes the activation of certain proteases (cathepsins, metalloproteinases) that degrade components of the basement membrane and extracellular matrix (ECM) and, therefore, create the prerequisite for metastasis by enabling tumor cells to get to the bloodstream. Cathepsin family members, especially cathepsin B and K, are overexpressed in metastatic tumors and silencing of these cathepsins was shown to reduce tumor cell invasion [55, 56].

Matrix metalloproteinases (MMP) are also more active at acidic pH. They have been considered as prognostic biomarkers for some metastatic cancers and a number of MMP inhibitors have been developed and tested in clinical trials [57].

The slightly alkaline intracellular pH created by tumor cells was shown to promote proliferation and inhibit apoptosis. Caspases, for example, need an acidic pH to be activated and it was shown in a variety of tumor cells that chemotherapeutics that induce apoptosis lead to an acidification of the cytosol. An alkaline pHi also plays a role in the uptake of chemotherapeutics as the most common drugs are weak bases with intracellular targets that are protonated at lower pH and neutral at higher pH. Therefore, permeation through the plasma membrane and accumulation in the cell is impaired [58].

The importance of pH regulation for tumor cells drove the development of drugs that disrupt tumor pH-regulating systems. Na^+/H^+ exchangers like NHE1 are the predominant regulators of pHi. Alkalization of the cytosol triggered by NHE1 is linked to malignant transformation. NHE1 inhibition leads to apoptosis induction of different tumor cell lines and xenograft models and showed a decrease in tumor formation with oncogene-transformed fibroblast lacking NHE1 or KRAS-transformed tumor xenografts. But so far there have not been promising effects for NHE1 inhibitors in monotherapy, especially due to immense toxic side effects [54].

The increased glycolytic flux of tumor cells requires a system to export lactic acid from the cell, which is done by monocarboxylate transporters (MCTs). Among the 14 family members, MCT1 and 4 are specialized for the cotransport of lactate and H^+, and a high expression of especially MCT4 was found in rapidly growing tumors such as triple-negative breast cancer [59]. Owing to their pivotal role in pHi regulation by securing a slight alkaline cytosolic pH despite high production of lactate, there was ongoing research for the development of small molecule inhibitors. And indeed, MCT inhibition impaired glioblastoma cell proliferation, migration, and survival [60]. Treatment with metformin could sensitize tumor cells to MCT inhibition, which indicates the possibility that combination with other metabolic or pH regulating drugs could target tumors dependent on glycolysis.

2.4.2. V-ATPase

The vacuolar H^+-ATPase (V-ATPase) is a multisubunit enzyme, which is located in the membranes of almost all eukaryotes. It is responsible for acidifying intracellular organelles like endosomes, lysosomes, golgi-derived vesicles, or secretory vesicles and is therefore responsible for the pH homeostasis in the cell. Inhibition of the V-ATPase leads to apoptosis induction and inhibition of migration in a variety of tumor cells and has therefore become an interesting target in tumor therapy [61, 62]. As reported above, regulation of pH is important for tumor cell survival and metastasis as an acidic extracellular microenvironment facilitates metastasis of tumor cells by activating proteases like MMPs. The V-ATPase is expressed on the plasma membrane of metastatic and chemoresistant tumor cells and pumps protons across the membrane creating an acidic microenvironment [63].

Apart from regulating pH we could show that V-ATPase inhibition leads to an increased transcription of genes involved in glycolysis, fatty acid, and cholesterol synthesis. Furthermore, V-ATPase inhibition leads to an increased glucose consumption and especially a strong induction of the hypoxia-inducible factor-1α (HIF1α) occurs [64]. As described above, HIF1α is the major inducer of glycolysis and metabolic stress and induction by the V-ATPase inhibitor archazolid strongly indicates that archazolid leads to changes in the tumor metabolism, which could be exploited for cancer therapy.

3. Conclusion

Although in recent years much work has been done in understanding the regulation of tumor metabolism and the discovery of drugs that specifically target tumor-related pathways, the complex interplay between oncogenic signaling pathways and tumor metabolism demands further research. Metabolic reprogramming may render cancer cells highly dependent on specific enzymes or processes that could be exploited for cancer therapy and induce apoptosis. But the search for relevant targets may be complicated due to the high diversity of tumor metabolism and the possibility of compensatory mechanisms. Another challenge is to identify those metabolic pathways that are essential targets and understanding the mechanism of apoptosis induction is inevitable in tumor therapy. It is important to understand that there is not one single tumor-specific metabolism but several programs that differ from tumor to tumor and are adjusted to the special requirements of different tumors in different tissues. Therefore, targeting different enzymes in glycolysis might be beneficial in some tumors but may have no effect in others that still have the ability to use oxidative phosphorylation for energy production. Also, normal cells need to produce energy, nucleotides, or other metabolites, as well and it is important to investigate why there exists a therapeutic window that does not harm normal cells. The complexity of the tumor metabolism furthermore demands *in vivo* models to study the effect of drugs on a whole organism.

Only by further understanding the regulation of tumor metabolism, we will be able to translate this knowledge into the discovery of drugs that will lead to tumor growth inhibition and a better clinical outcome of cancer patients.

Author details

Karin von Schwarzenberg

Address all correspondence to: karin.von.schwarzenberg@cup.uni-muenchen.de

Department of Pharmacy, Pharmaceutical Biology, Ludwig-Maximilians-University Munich, Munich, Germany

References

[1] Munoz-Pinedo C, El Mjiyad N, Ricci JE. Cancer metabolism: current perspectives and future directions. *Cell Death Dis*. 2012;3:e248.

[2] Butler EB, Zhao Y, Munoz-Pinedo C, Lu J, Tan M. Stalling the engine of resistance: targeting cancer metabolism to overcome therapeutic resistance. *Canc Res*. 2013;73:2709-2717.

[3] Vander Heiden MG. Targeting cancer metabolism: a therapeutic window opens. *Nat Rev Drug Discov*. 2011;10:671-684.

[4] Berkers CR, Maddocks OD, Cheung EC, Mor I, Vousden KH. Metabolic regulation by p53 family members. *Cell Metab*. 2013;18:617-633.

[5] Downey RJ, Akhurst T, Gonen M, Vincent A, Bains MS, Larson S, et al. Preoperative F-18 fluorodeoxyglucose-positron emission tomography maximal standardized uptake value predicts survival after lung cancer resection. *J Clin Oncol*. 2004;22:3255-3260.

[6] Wilson JE. Isozymes of mammalian hexokinase: structure, subcellular localization and metabolic function. *J Exp Biol*. 2003;206:2049-2057.

[7] Mathupala SP, Rempel A, Pedersen PL. Glucose catabolism in cancer cells: identification and characterization of a marked activation response of the type II hexokinase gene to hypoxic conditions. *J Biol Chem*. 2001;276:43407-43412.

[8] Maher JC, Krishan A, Lampidis TJ. Greater cell cycle inhibition and cytotoxicity induced by 2-deoxy-D-glucose in tumor cells treated under hypoxic vs aerobic conditions. *Canc Chemother Pharmacol*. 2004;53:116-122.

[9] Robinson GL, Dinsdale D, Macfarlane M, Cain K. Switching from aerobic glycolysis to oxidative phosphorylation modulates the sensitivity of mantle cell lymphoma cells to TRAIL. *Oncogene*. 2012;31:4996-5006.

[10] Kurtoglu M, Gao N, Shang J, Maher JC, Lehrman MA, Wangpaichitr M, et al. Under normoxia, 2-deoxy-D-glucose elicits cell death in select tumor types not by inhibition of glycolysis but by interfering with N-linked glycosylation. *Mol Canc Ther*. 2007;6:3049-3058.

[11] Xi H, Kurtoglu M, Liu H, Wangpaichitr M, You M, Liu X, et al. 2-Deoxy-D-glucose activates autophagy via endoplasmic reticulum stress rather than ATP depletion. *Canc Chemother Pharmacol*. 2011;67:899-910.

[12] Raez LE, Papadopoulos K, Ricart AD, Chiorean EG, Dipaola RS, Stein MN, et al. A phase I dose-escalation trial of 2-deoxy-D-glucose alone or combined with docetaxel in patients with advanced solid tumors. *Canc Chemother Pharmacol*. 2013;71:523-530.

[13] Semenza GL. HIF-1 mediates metabolic responses to intratumoral hypoxia and oncogenic mutations. *J Clin Invest*. 2013;123:3664-3671.

[14] Forsythe JA, Jiang BH, Iyer NV, Agani F, Leung SW, Koos RD, et al. Activation of vascular endothelial growth factor gene transcription by hypoxia-inducible factor 1. *Mol Cell Biol*. 1996;16:4604-4613.

[15] Baldewijns MM, van Vlodrop IJ, Vermeulen PB, Soetekouw PM, van Engeland M, de Bruine AP. VHL and HIF signalling in renal cell carcinogenesis. *J Pathol*. 2010;221:125-138.

[16] Semenza GL. Defining the role of hypoxia-inducible factor 1 in cancer biology and therapeutics. *Oncogene*. 2010;29:625-634.

[17] Onnis B, Rapisarda A, Melillo G. Development of HIF-1 inhibitors for cancer therapy. *J Cell Mol Med*. 2009;13:2780-2786.

[18] Thomas GV, Tran C, Mellinghoff IK, Welsbie DS, Chan E, Fueger B, et al. Hypoxia-inducible factor determines sensitivity to inhibitors of mTOR in kidney cancer. *Nat Med*. 2006;12:122-127.

[19] Hudson CC, Liu M, Chiang GG, Otterness DM, Loomis DC, Kaper F, et al. Regulation of hypoxia-inducible factor 1alpha expression and function by the mammalian target of rapamycin. *Mol Cell Biol*. 2002;22:7004-7014.

[20] Shin DH, Chun YS, Lee DS, Huang LE, Park JW. Bortezomib inhibits tumor adaptation to hypoxia by stimulating the FIH-mediated repression of hypoxia-inducible factor-1. *Blood*. 2008;111:3131-3136.

[21] Elstrom RL, Bauer DE, Buzzai M, Karnauskas R, Harris MH, Plas DR, et al. Akt stimulates aerobic glycolysis in cancer cells. *Canc Res*. 2004;64:3892-3899.

[22] Majewski N, Nogueira V, Bhaskar P, Coy PE, Skeen JE, Gottlob K, et al. Hexokinase-mitochondria interaction mediated by Akt is required to inhibit apoptosis in the presence or absence of Bax and Bak. *Mol Cell*. 2004;16:819-830.

[23] Engelman JA, Chen L, Tan X, Crosby K, Guimaraes AR, Upadhyay R, et al. Effective use of PI3K and MEK inhibitors to treat mutant Kras G12D and PIK3CA H1047R murine lung cancers. *Nat Med*. 2008;14:1351-1356.

[24] Holdsworth CH, Badawi RD, Manola JB, Kijewski MF, Israel DA, Demetri GD, et al. CT and PET: early prognostic indicators of response to imatinib mesylate in patients with gastrointestinal stromal tumor. *AJR. Am J Roentgenol*. 2007;189:W324-330.

[25] Liang J, Mills GB. AMPK: a contextual oncogene or tumor suppressor? *Canc Res*. 2013;73:2929-2935.

[26] Hemminki A. The molecular basis and clinical aspects of Peutz-Jeghers syndrome. *Cell Mol Life Sci*. 1999;55:735-750.

[27] Shaw RJ, Bardeesy N, Manning BD, Lopez L, Kosmatka M, DePinho RA, et al. The LKB1 tumor suppressor negatively regulates mTOR signaling. *Canc Cell*. 2004;6:91-99.

[28] Hadad SM, Baker L, Quinlan PR, Robertson KE, Bray SE, Thomson G, et al. Histological evaluation of AMPK signalling in primary breast cancer. *BMC Cancer*. 2009;9:307.

[29] Laderoute KR, Amin K, Calaoagan JM, Knapp M, Le T, Orduna J, et al. 5'-AMP-activated protein kinase (AMPK) is induced by low-oxygen and glucose deprivation conditions found in solid-tumor microenvironments. *Mol Cell Biol*. 2006;26:5336-5347.

[30] Luo Z, Zang M, Guo W. AMPK as a metabolic tumor suppressor: control of metabolism and cell growth. *Future Oncol*. 2010;6:457-470.

[31] Faubert B, Boily G, Izreig S, Griss T, Samborska B, Dong Z, et al. AMPK is a negative regulator of the Warburg effect and suppresses tumor growth in vivo. *Cell Metab*. 2013;17:113-124.

[32] Woodard J, Joshi S, Viollet B, Hay N, Platanias LC. AMPK as a therapeutic target in renal cell carcinoma. *Canc Biol Ther*. 2010;10:1168-1177.

[33] Liang J, Shao SH, Xu ZX, Hennessy B, Ding Z, Larrea M, et al. The energy sensing LKB1-AMPK pathway regulates p27(kip1) phosphorylation mediating the decision to enter autophagy or apoptosis. *Nat Cell Biol*. 2007;9:218-224.

[34] Schwartzenberg-Bar-Yoseph F, Armoni M, Karnieli E. The tumor suppressor p53 down-regulates glucose transporters GLUT1 and GLUT4 gene expression. *Canc Res*. 2004;64:2627-2633.

[35] Vousden KH, Ryan KM. p53 and metabolism. *Nat Rev Canc*. 2009;9:691-700.

[36] Jiang P, Du W, Wang X, Mancuso A, Gao X, Wu M, et al. p53 regulates biosynthesis through direct inactivation of glucose-6-phosphate dehydrogenase. *Nat Cell Biol*. 2011;13:310-316.

[37] Matoba S, Kang JG, Patino WD, Wragg A, Boehm M, Gavrilova O, et al. p53 regulates mitochondrial respiration. *Science*. 2006;312:1650-1653.

[38] Lebedeva MA, Eaton JS, Shadel GS. Loss of p53 causes mitochondrial DNA depletion and altered mitochondrial reactive oxygen species homeostasis. *Biochim Biophys Acta*. 2009;1787:328-334.

[39] Jiang P, Du W, Wu M. Regulation of the pentose phosphate pathway in cancer. *Protein Cell*. 2014;5:592-602.

[40] Manganelli G, Masullo U, Passarelli S, Filosa S. Glucose-6-phosphate dehydrogenase deficiency: disadvantages and possible benefits. *Cardiovasc Hematologic Disor Drug Targets*. 2013;13:73-82.

[41] Jiang P, Du W, Yang X. A critical role of glucose-6-phosphate dehydrogenase in TAp73-mediated cell proliferation. *Cell Cycle*. 2013;12:3720-3726.

[42] Langbein S, Frederiks WM, zur Hausen A, Popa J, Lehmann J, Weiss C, et al. Metastasis is promoted by a bioenergetic switch: new targets for progressive renal cell cancer. *Int J Cancer*. 2008;122:2422-2428.

[43] Varshney R, Dwarakanath B, Jain V. Radiosensitization by 6-aminonicotinamide and 2-deoxy-D-glucose in human cancer cells. *Int J Radiat Biol*. 2005;81:397-408.

[44] Santos CR, Schulze A. Lipid metabolism in cancer. *FEBS J*. 2012;279:2610-2623.

[45] Menendez JA, Vellon L, Mehmi I, Oza BP, Ropero S, Colomer R, et al. Inhibition of fatty acid synthase (FAS) suppresses HER2/neu (erbB-2) oncogene overexpression in cancer cells. *Proc Natl Acad Sci U S A*. 2004;101:10715-10720.

[46] De Schrijver E, Brusselmans K, Heyns W, Verhoeven G, Swinnen JV. RNA interference-mediated silencing of the fatty acid synthase gene attenuates growth and induces morphological changes and apoptosis of LNCaP prostate cancer cells. *Cancer Res*. 2003;63:3799-3804.

[47] Chajes V, Cambot M, Moreau K, Lenoir GM, Joulin V. Acetyl-CoA carboxylase alpha is essential to breast cancer cell survival. *Cancer Res*. 2006;66:5287-5294.

[48] Beckers A, Organe S, Timmermans L, Scheys K, Peeters A, Brusselmans K, et al. Chemical inhibition of acetyl-CoA carboxylase induces growth arrest and cytotoxicity selectively in cancer cells. *Cancer Res.* 2007;67:8180-8187.

[49] Li S, Qiu L, Wu B, Shen H, Zhu J, Zhou L, et al. TOFA suppresses ovarian cancer cell growth in vitro and in vivo. *Mol Med Rep.* 2013;8:373-378.

[50] Clendening JW, Pandyra A, Boutros PC, El Ghamrasni S, Khosravi F, Trentin GA, et al. Dysregulation of the mevalonate pathway promotes transformation. *Proc Natl Acad Sci U S A.* 2010;107:15051-15056.

[51] Jiang P, Mukthavaram R, Chao Y, Nomura N, Bharati IS, Fogal V, et al. In vitro and in vivo anticancer effects of mevalonate pathway modulation on human cancer cells. *Br J Canc.* 2014;111:1562-1571.

[52] Gopalan A, Yu W, Sanders BG, Kline K. Simvastatin inhibition of mevalonate pathway induces apoptosis in human breast cancer cells via activation of JNK/CHOP/DR5 signaling pathway. *Canc Lett.* 2013;329:9-16.

[53] Ahn KS, Sethi G, Aggarwal BB. Simvastatin potentiates TNF-alpha-induced apoptosis through the down-regulation of NF-kappaB-dependent antiapoptotic gene products: role of IkappaBalpha kinase and TGF-beta-activated kinase-1. *J Immunol.* 2007;178:2507-2516.

[54] Parks SK, Chiche J, Pouyssegur J. Disrupting proton dynamics and energy metabolism for cancer therapy. *Nat Rev Canc.* 2013;13:611-623.

[55] Withana NP, Blum G, Sameni M, Slaney C, Anbalagan A, Olive MB, et al. Cathepsin B inhibition limits bone metastasis in breast cancer. *Canc Res.* 2012;72:1199-1209.

[56] Chen Q, Fei J, Wu L, Jiang Z, Wu Y, Zheng Y, et al. Detection of cathepsin B, cathepsin L, cystatin C, urokinase plasminogen activator and urokinase plasminogen activator receptor in the sera of lung cancer patients. *Oncol Lett.* 2011;2:693-699.

[57] Roy R, Yang J, Moses MA. Matrix metalloproteinases as novel biomarkers and potential therapeutic targets in human cancer. *J Clin Oncol.* 2009;27:5287-5297.

[58] Webb BA, Chimenti M, Jacobson MP, Barber DL. Dysregulated pH: a perfect storm for cancer progression. *Nat Rev Canc.* 2011;11:671-677.

[59] Pinheiro C, Longatto-Filho A, Azevedo-Silva J, Casal M, Schmitt FC, Baltazar F. Role of monocarboxylate transporters in human cancers: state of the art. *J Bioenerg Biomembr.* 2012;44:127-139.

[60] Miranda-Goncalves V, Honavar M, Pinheiro C, Martinho O, Pires MM, Pinheiro C, et al. Monocarboxylate transporters (MCTs) in gliomas: expression and exploitation as therapeutic targets. *Neuro-oncology.* 2013;15:172-188.

[61] Wiedmann RM, von Schwarzenberg K, Palamidessi A, Schreiner L, Kubisch R, Liebl J, et al. The V-ATPase-Inhibitor Archazolid Abrogates Tumor Metastasis via Inhibition of Endocytic Activation of the Rho-GTPase Rac1. *Canc Res*. 2012;72:5976-5987.

[62] Hinton A, Sennoune SR, Bond S, Fang M, Reuveni M, Sahagian GG, et al. Function of a subunit isoforms of the V-ATPase in pH homeostasis and in vitro invasion of MDA-MB231 human breast cancer cells. *J Biol Chem*. 2009;284:16400-16408.

[63] Nishisho T, Hata K, Nakanishi M, Morita Y, Sun-Wada GH, Wada Y, et al. The a3 isoform vacuolar type H(+)-ATPase promotes distant metastasis in the mouse B16 melanoma cells. *Mol Cancer Res*. 2011;9:845-855.

[64] von Schwarzenberg K, Wiedmann RM, Oak P, Schulz S, Zischka H, Wanner G, et al. Mode of cell death induction by pharmacological vacuolar H+-ATPase (V-ATPase) inhibition. J Biol Chem. 2012.

Chapter 17

Modulation of Host Programmed Cell Death Pathways by the Intracellular Protozoan Parasite, *Toxoplasma gondii* — Implications for Maintenance of Chronic Infection and Potential Therapeutic Applications

Sandra K. Halonen

Additional information is available at the end of the chapter

http://dx.doi.org/10.5772/61529

Abstract

Programmed cell death (PCD) pathways are genetically programmed mechanisms that can trigger the cell to die or commit "cell suicide". There are three major forms of programmed cell death that are now recognized: apoptosis (type I), autophagy (type II) and necrotic cell death or necroptosis (type III). While these cell death processes were once thought to occupy discrete cell states, evidence suggests that apoptosis, autophagy and necrosis are often regulated by similar pathways and share initiator and effector molecules and some subcellular compartments indicating that crosstalk exists between these three main forms of cell death pathways, resulting in a balanced interplay by which the cell decides its fate. PCD pathways have important roles in many cellular processes such as development and oncogenic transformation, but PCD pathways also play important roles in host defense and elimination of pathogens. *Toxoplasma gondii* is a microbial pathogen for which programmed cell death pathways are a key part of the host defense. *T. gondii* is an obligate intracellular protozoan parasite that infects approximately one-third of the world's population. In most immunocompetent individuals, the chronic infection is asymptomatic due to an effective immune response that eliminates active parasite replication. The parasite has evolved immune evasion strategies that enable it to survive and persist long enough in the host however to establish a chronic infection in which the cyst stage persists within neurons in the brain and skeletal muscle in the periphery. *T. gondii* has evolved multiple mechanisms to resist killing by apoptotic, autophagic and necrotic cell death pathways, and the parasite's manipulation of host PCD pathways plays a crucial role in host–parasite interactions and maintenance of the chronic infection. While most individuals chronically infected with *T. gondii* are asymptomatic, severe disease can occur in immunocompromised individuals where the infection reactivates from the brain causing severe necrotizing encephalitis, and increasing evidence indicates chronic cerebral toxoplasmosis in some individuals may lead to neuropsychiatric disorders such as schizophrenia and suicidal behavior. This review will focus on the role of PCD pathways in host defense of *T. gondii* and the parasite manipulation of these PCD pathways. A bet-

ter understanding of the molecular components underlying the PCD pathways and the parasite manipulation of these pathways may yield new therapeutic targets for treatment of clinical sequelae of cerebral toxoplasmosis.

Keywords: Apoptosis, autophagy, necroptosis, necrosis, toxoplasmosis

---

1. Introduction

Overview of Programmed Cell Death (PCD) pathways and applications to *Toxoplasma gondii* infection

Programmed cell death (PCD) pathways are genetically programmed mechanisms that can trigger the cell to die or commit "cell suicide". There are three major forms of programmed cell death which are now recognized: apoptosis (type I), autophagic cell death (type II) and necrotic cell death or necroptosis (type III) [1–4]. Apoptosis, the best-characterized and dominant form of PCD, is a controlled physiological process of cellular self-destruction that removes dead cells without damage to the host. Apoptosis is critically important during development and morphogenesis and eliminates damaged cells such as cancerous cells or infected cells that may interfere with normal function. Necrosis, conversely, is a cell death process that is observed in response to severe stresses such that occurs after physical injury or prolonged ischemia and has been long thought to be an unregulated process. However, a programmed form of necrotic death, called necroptosis, is now recognized that can occur during physical traumas, in neurodegeneration, in cell death due to ischemia or infection and that, in contrast to unordered necrosis, has dedicated molecular pathways controlling necrotic cell death. Autophagy is predominantly a strategy for survival and not death, serving as a housekeeping mechanism of normal turnover of long-lived proteins and whole organelles and crucial to maintenance of healthy cells. However, autophagy can promote cell death during normal development and excessive autophagy, stimulated in times of stress such as nutrient deprivation and some diseases, can lead to what is now recognized as autophagic cell death [5–7]. While these cell death processes were once thought to occupy discrete cell states, evidence suggests apoptosis, autophagy and necroptosis are often regulated by similar pathways, share initiator and effector molecules and some subcellular compartments indicating a complex crosstalk exists between these three main forms of cell death pathways, resulting in a balanced interplay by which the cell decides its fate [8].

PCD pathways are also an important part of host defense against intracellular pathogens. The intracellular protozoal pathogen, *Toxoplasma gondii,* is a common infection that causes a chronic infection of central nervous system that persists for the lifetime of the individual. PCD pathways are a key part of the host defense against *T. gondii,* but the parasite has evolved multiple mechanisms to resist killing by host cell death pathways. Parasite manipulations of host cell death pathways are essential components leading to successful establishment of infection and for maintenance of the chronic infection in the host. In this review, the molecular components and signaling pathways of each of the three main types of PCD pathways will be discussed with the role of these PCDs in the pathogenesis of the intracellular pathogen, *T. gondii*, and potential therapeutic targets for toxoplasmosis and other diseases, the focus of this review.

2. Apoptosis, necroptosis and autophagic cell death pathways: Morphological features and molecular components

Apoptosis, necroptosis and autophagic cell death pathways operate via distinct genetically programmed molecular mechanisms containing dedicated molecular components and leading to characteristic morphological features (Figure 1). Apoptosis operates via a caspase-dependent mechanism while necroptosis and autophagic cell death are caspase-independent processes, with autophagic cell death being cathepsin-dependent and necroptosis mediated via the receptor interacting protein kinase 3 (RIP3) and its substrate, the mixed lineage kinase-like domain protein, MLKL [8]. Apoptosis results in mitochondrial membrane permeabilization, chromatin condensation and DNA fragmentation, resulting in cells become smaller (pknotic) and membrane blebbing into apoptotic bodies. Phagocytes, such as macrophages, subsequently engulf apoptotic cells and hence apoptosis is considered a noninflammatory form of programmed cell death. Necrotic cell death, in contrast to apoptosis, is marked by swelling of ER and mitochondrial organelles, increase in cell size, rapid rupture of the plasma membrane resulting in lysis of the cell and release of danger-associated molecular patterns (DAMPs) which stimulate inflammation. While necrosis has traditionally been considered to be "pathological cell death", necroptosis is now recognized to occur via a regulated mechanism that is inhibited by caspases [9, 10]. Autophagy is the process by which cells recycle cellular constituents and involves engulfment of cytoplasmic material and intracellular organelles within double-membrane vesicles called autophagosomes that fuse with lysosomes to make autophagolysosomes that degrade the cellular cargo via cathepsins located in the autolysosomes. The molecular components and signaling pathways of each cell death processes are reviewed below, followed by the current understanding of the crosstalk that exists between these PCD pathways.

2.1. Apoptosis

The term apoptosis was first used in 1972 by Kerr, Wyllie and Currie [11] to describe a form of cell death morphologically distinct from necrosis. Apoptosis has since been well studied and is now understood to be a regulated energy-dependent process mediated via cysteine-dependent aspartate-directed enzymes called caspases [3]. Apoptosis is a homeostatic process that balances cell numbers and plays a crucial role in several physiological processes including embryogenesis to shape morphological structures such as fingers, in the establishment of functional synaptic connections in the nervous system, in development of the immune response to remove self-reactive lymphocytes and at the termination of the immune response to remove antigen-specific lymphocytes. Apoptosis is also used to rid the body of cells in various pathological conditions such as removal of cancerous cells, infected cells or cells damaged by noxious agents [12].

Comparison of PCD Pathways

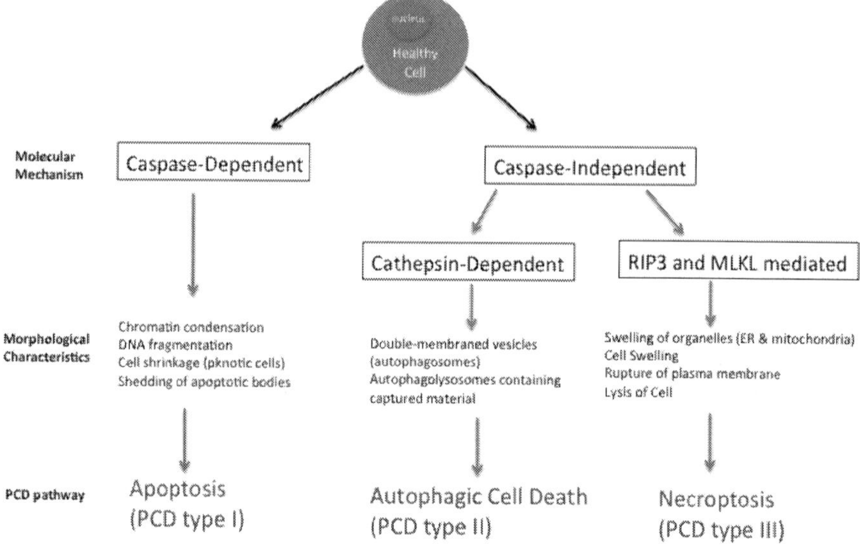

Figure 1. Comparison of three types of Programmed Cell (PCD) Pathways. The molecular mechanisms and morphological characteristics of Apoptosis, Autophagic Cell Death and Necroptosis cell death pathways, so-called PCD types I, II and III, respectively, are indicated. RIP3 = receptor-interacting protein kinase 3; MLKL = mixed lineage kinase-like domain protein

2.1.1. Signaling pathways and molecular components of apoptosis

Apoptosis is mediated via caspases which are present in the cytoplasm as pro-enzymes and when activated initiate a proteolytic cascade that results in apoptosis of the cell. There are three activation pathways that can initiate apoptosis: the extrinsic (death receptor) pathway, the intrinsic (mitochondrial) pathway and the granzyme pathway (Figure 2). There are about 14 known caspases which are divided into initiator and effector caspases. Initiator caspases, caspases 8, 9 and 10, are triggered by either the extrinsic, intrinsic or granzyme pathway, respectively. All three pathways converge on caspase 3, which activates effector caspases 6 and 7 that lead to apoptosis. A major target of effector caspases is poly-ADP-ribose polymerase (PARP) which is involved in DNA repair, cell survival, proliferation and differentiation.

The extrinsic pathway is initiated by binding of cell membrane receptors of Fas (CD95) or tumor necrosis factor (TNF) family receptor (TNFR), which after binding its relevant ligand of FasL or TNFα, respectively, trimerizes and attracts the docking proteins Fas-associated death domain (FADD) and TNF receptor-associated death domain (TRADD), respectively. FADD

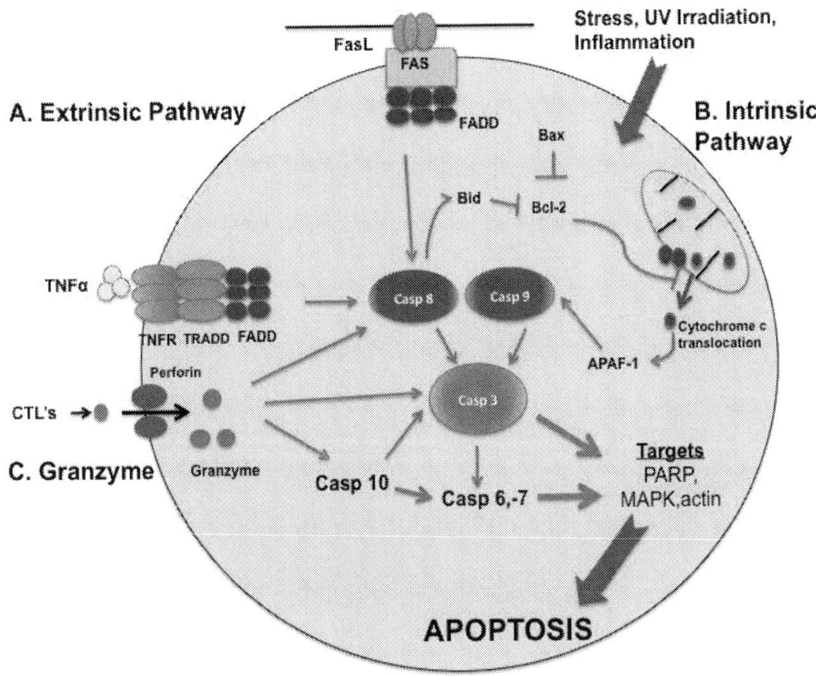

Figure 2. **Signaling Pathways of Apoptosis.** Apoptosis is induced through either the extrinsic pathway (A), intrinsic pathway (B) or the granzyme pathway (C). Activation of initiator caspases 8 and 9 (Casp 8,-Casp9), leads to apoptotic cell death through activation of Caspase 3 (Casp3) and subsequent activation of effector caspases 6 and 7 (Casp 6-Casp7). The extrinsic pathway (A) is initiated via extracellular molecules, FasL and TNFα, which bind to TNFR family members, Fas and TNFR respectively which activate Casp8 (green lines). The intrinsic pathway (B) is initiated by stress, UV irradiation and inflammation which act on the mitochondria through the pro-apoptotic Bcl-2 family members, such as Bax, resulting in the blockage of the anti-apoptotic activity of Bcl-2 (red lines). As a result, Cytochrome c is released into the cytoplasm and activates Casp9 through APAF-1 (green lines). Casp8 may also trigger the intrinsic pathway through activation of Bid, which inhibits the anti-apoptotic activity of Bcl-2. The granzyme pathway (C) is activated via cytotoxic T cells (CTLs) that introduce granzyme molecules into the target cells via secretion of perforin, which through a multimerization process forms a pore in the cell membrane allowing granzyme into the target cell. Granzymes cleave multiple caspases, including Caspase 10.

contains two death effector domains that can bind and cleave caspase 8. TRADD lacks death effector domains but binds FADD that then cleaves caspase 8. Activated caspase 8 then activates caspase 3. The intrinsic pathway is activated via changes in the mitochondria membrane potential which can be triggered via stress, toxic reagents, UV irradiation and inflammation. Changes in mitochondria membrane result in translocation of cytochrome c from the inner mitochondrial membrane into the cytoplasm. Cytochrome c binds to apoptosis protease activating factor 1 (APAF-1) which in the presence of ATP cleaves caspase 9. The intrinsic pathway is regulated by Bcl-2 which localizes to the outer mitochondria membrane. Bcl-2 members (Bcl-2 and Bcl-x) are anti-apoptotic while others such as Bax and Bad are pro-apoptotic. The granzyme activation pathway is stimulated by cytotoxic T lymphocytes (CTLs) or NK cell secretion of perforin, which creates pores in the target cell membrane allowing granzymes into the cytoplasm activating caspase 10 which can then activate caspase 3.

2.1.2. Apoptosis in physiological and pathological conditions and potential therapeutic application

Apoptosis is a tightly regulated process and is rarely observed in healthy animals because phagocytes rapidly remove apoptotic cells. Abnormalities in cell death regulation can be a significant component of diseases such as cancer, AIDS and neurodegenerative diseases such as Parkinson's disease, Alzheimer's disease, Huntington's disease and amyotrophic lateral sclerosis. Some of these conditions are characterized by insufficient apoptosis and others excessive apoptosis. Excessive apoptosis, for example, is a feature of neurodegenerative diseases while resistance to apoptosis characterizes many cancers. Understanding the mechanisms of apoptosis at the molecular level may provide therapeutic interventions in these diverse disease processes.

2.1.2.1. Autophagic cell death

Autophagy, which means literally to "eat oneself", is a physiological response to stress such as starvation and is considered the major cellular mechanism for generating the needed metabolic sources of energy and metabolites during times of nutrient deprivation. Autophagy can also be used to remove damaged or unwanted organelles, such as mitochondria and long-lived proteins. Autophagy is a homeostatic mechanism that is predominantly a cytoprotective process [7, 13]. Other substrates are also now recognized as potential cargo for autophagy including lipids, nucleic acids, reticulocytes and intracellular pathogens and viruses. The concept of autophagic cell death has been a matter of debate, but it is now recognized as a type of programmed cell death mechanism that can lead to both apoptotic and necrotic cell death in certain circumstances such as extreme stress conditions [14, 15]. Autophagic cell death or type II programmed cell death is now used to refer to cell death process distinct from apoptosis that is caspase-independent and occurs with accumulation of double-membrane organelles called autophagosomes.

2.1.3. Signaling pathways and molecular components of autophagy

Autophagy delivers cytosolic materials to the interior of the lysosomes for degradation via a process involving the *"de novo"* formation of cytoplasmic double-membrane vacuoles called autophagosomes which then fuse with lysosomes forming an autolysosome that degrades the cellular cargo (Figure 3). Autophagy is regulated by a genetic program with a number of autophagy-related genes (Atg) whose gene products (ATG proteins) regulate distinct steps in autophagy. The autophagic pathway proceeds through the following defined steps: (i) initiation phase involves the formation of an isolation membrane or phagophore, (ii) elongation of the phagophore, (iii) maturation of an autophagosome with accumulation of cytosolic cargo, (iv) fusion of the mature autophagosome with the lysosome and (v) degradation of the contents via lysosomal proteases (i.e., cathepsins) in the autolysosome. The initiation phase involves formation of an initiation complex comprised of the class III PI3 kinase, VSP34, which converts phosphatidylinositol to phosphatylinositol-3-phosphate (PI3P). The activity of VSP34, which binds to beclin 1, requires the activity of VSP15, the regulator beclin 1 and ATG14L (Figure 3a). Elongation of the autophagosome membrane requires action of two ubiquitin-like conjugation systems, the Atg5–Atg12 conjugation system and the microtubule-associated protein-1 light chain (LC3). The phosphatidylethanolamine-conjugated form of LC3, LC3-PE (also called LC-II) is generated by ATG4-dependent proteolytic cleavage of LC3 and the action of the E1 ligase, ATG7, the E2 ligase, ATG3 and the E3 ligase complex, ATG12/ATG5/ATG16L. LC3-PE stably associates with the autophagosome and is commonly used as a marker for autophagosomes.

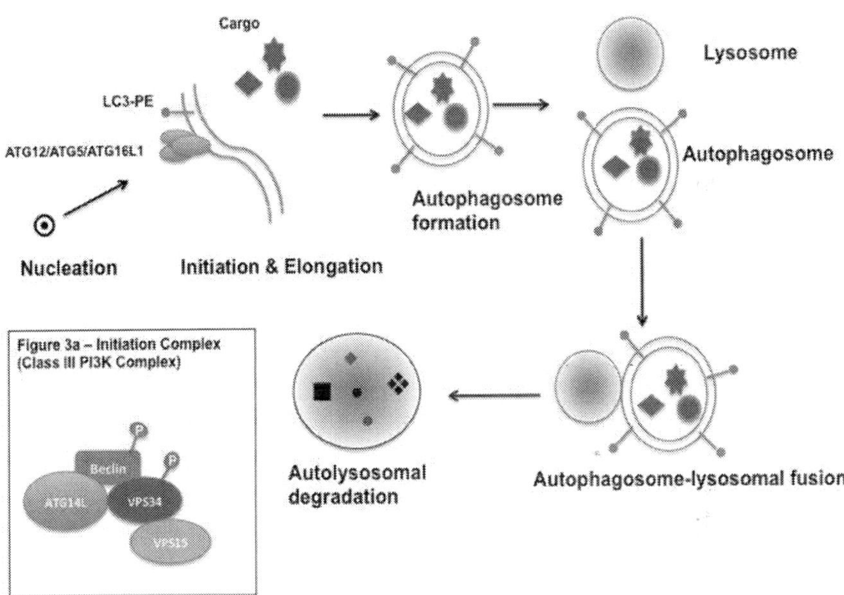

Figure 3. **Autophagy Pathway**. Autophagy is a membrane-dependent pathway that involves a series of defined steps beginning with a nucleation step which then leads to: 1). Initiation of the isolation membrane or phagophore, 2). Formation of the autophagosome around cellular cargo, 3). Fusion of autophagosome with lysosomes, and 4). Degradation of cargo by lysosomal proteases in autolysosomes. Formation of the initiation complex (also called Class III PI3K Complex) as shown in Figure 3a, consists of Beclin-1, ATG14L, VPS34 and VPS15. Activation of the initiation complex requires disruption of binding of Bcl-2 to Beclin-1. The Class III PI3K complex generates PI3P at the site of nucleation of the isolation membrane, which leads to binding of PI3P-binding proteins and the subsequent recruitment of proteins involved in the 'elongation reaction' of the isolation membrane. These proteins contribute to membrane expansion resulting in the formation of a double-membrane structure, the autophagosome that surrounds cellular cargo. The phosphatidylethanolamine-conjugated form of the LC3 (LC3-PE) is generated by the ATG4 dependent proteolytic cleavage of LC3 and the action of E1 ligase, E2 ligase and the E3 ligase ATG12/ATG5/ATG16L. LC3-PE stably associates with the autophagosome and is a marker of autophagosomes and therefore LC3 is commonly used as a marker of autophagy pathway.

While the autophagy pathway is primarily a homeostatic process promoting cell survival, increased autophagosomal formation can occur coincidentally with cell dying indicating autophagy may be involved in regulated cell death pathways, most notably apoptosis [7, 13]. For example, in apoptosis-comprised cells, cells die via a caspase-independent mechanism characterized by autophagosome accumulation, implicating a role for autophagy in the cell death process. Autophagy and apoptosis may also be regulated coordinately as anti-apoptotic proteins that downregulate apoptosis can also downregulate autophagy. For example, members of the Bcl-2 family can bind to beclin-1, thus inhibiting the formation of initiation complex that is necessary to stimulate autophagy. Other studies indicate once apoptosis is activated, apoptosis effector molecules may suppress autophagy as beclin-1 is cleaved and inactivated by caspases. Finally, studies also indicate autophagy proteins may play a dual role in regulation of apoptosis and autophagy. For example, Atg5 may affect the extrinsic apoptotic pathway through interaction with FADD proteins, while Atg12 is an effector of intrinsic apoptotic pathway with both Atg5 and Atg12 acting as pro-apoptotic regulators.

Autophagy pathway has also been implicated in necrosis and necroptosis cell death. Evidence indicates autophagy may be able to act as an inhibitor of necrosis/necroptosis by preserving cellular functions, removing toxic products and maintaining cellular energy. For example, knockdown studies of components of autophagy pathways, such as beclin-1, have found cytotoxicity is exacerbated suggesting that autophagy has a cytoprotective role. Autophagy may be able to act as a buffer to metabolic stress, providing a mechanism to generate ATP to maintain metabolic viability and thus prevent necrotic cell death. Conversely, studies using a specific inhibitor of necroptosis, necrostatin-1, found both necroptosis and autophagy were inhibited, indicating autophagy may be induced by necrosis. Molecular components underlying the relationship between autophagy and necroptosis/necrotic cell death at this point are poorly understood.

2.1.4. Physiological and pathological roles of autophagic cell death and therapeutic implications

Thus, while autophagy is now recognized to play a role in apoptotic and necrotic cell death programs, this process is complex and incompletely understood. However, increasing evidence indicates autophagic cell death may be necessary for cell death in certain circumstances. It has been suggested autophagic cell death represents a failed adaptive mechanism that may have prevented death under milder conditions. Understanding of the relationships of autophagic cell death with other programmed cell death modalities will require further study. The study of autophagy in disease is an emerging area of research. Autophagy pathway and/or autophagic cell death pathways which result in failure to remove damaged organelles and/or damaged cells are now thought to contribute to various diseases such as cancer, neurodegenerative diseases such as Parkinson's disease, aging and inflammation. Elucidation of the influence of autophagy with other cell death programs will be essential to the development of therapeutics targeting autophagy for treatment of these various diseases. In the context of infectious disease, autophagy also is now recognized to play an important role in intracellular pathogen defense, and it has been suggested that intracellular pathogen load could be a factor that disrupts the balance between cell survival and cell death [16].

2.2. Necroptosis

Pathologists first used the term "necrosis" in the early 19th century to describe tissue destruction. Necrosis is characterized by organelle swelling and cell lysis, releasing the cellular content of cells and resulting in inflammation. The first indication that necrosis could occur in a regulated manner arose from observation that tumor necrosis factor (TNFα) could trigger apoptosis and necrotic forms of cell death. Caspase inhibition strongly exacerbates necrotic cell death, indicating caspase activity negatively regulates necrosis. This accumulating evidence led to the concept of "programmed necrosis" and the coining of the term necroptosis in 2005 to designate regulated necrotic cell death [4, 9]. Receptor-interacting protein kinase 3 (RIP3) and its substrate, the mixed lineage kinase-like domain protein (MLKL), are now recognized as the key activation steps and the molecular hallmarks of regulated necrotic cell death or necroptosis.

2.2.1. Signaling and molecular components of necroptosis

Programmed necrotic cell death can be activated when cells are stimulated by ligation of death receptors including CD95 (also known as FAS), TNF receptors (TNFRs) and TNF-related apoptosis-inducing ligand (TRAIL). Ligation of cell death receptors leads to the formation of complex 1, comprised of TNFR-associated death domain protein (TRADD), TNFR-associated factor 2 (TRAF2), receptor-interacting kinase 1(RIP1) and cellular inhibitors of apoptosis (CIAPs) (Figure 4). Upon the release of second mitochondria-derived activator of caspases (Smac) from the mitochondria, RIP1 is de-ubiquinated and dissociates from the death receptor and induces the formation of FADD-RIP1-caspase 8 pro-apoptosis complex II. Conversion of complex 1 to complex II indicates a change in cell fate from pro-survival to cell death. When caspase 8 is inhibited, RIP1 and RIP3 form the necrotic death domain complex, the necrosome, and their kinase activities become activated. RIP3 recruits it substrate monomeric MLKL from the cytoplasm and phosphorylates it. The phosphorylation destabilizes MLKL and drives its oligomerization which enables it to bind to phosphatidylinositol phosphate lipids (PIPs) and cariolipin (CL). Different PIPs and CLs orchestrate the translocation of MLKL to different membrane compartments including the mitochondria, endoplasmic reticulum, the Golgi, lysosomes and plasma membranes. In addition to activation via death receptors, it is also now clear that necroptosis can also be initiated by pathogen recognition receptors (PRRs) such as toll-like receptors (TLRs) and cytokines via activation of RIP3 through distinct upstream mechanisms.

2.2.2. Physiological and pathological roles of necroptosis and potential therapeutic applications

Programmed necrosis has been found to be important in host antiviral responses and a variety of tissue-damage related diseases such as acute pancreatitis, ischemic reperfusion injury, retinal detachment, atherosclerosis, neuronal loss in Gaucher's disease, amongst others. It has been a paradigm that apoptotic cell death is anti-inflammatory with dead cells cleared by phagocytes while necrotic cell death leads to inflammation and tissue damage. Necrotic cells, however, can be internalized by macrophages although via macropinocytosis and thus in a manner distinct from phagocytosis of apoptotic cells, but they are nonetheless efficiently cleared by professional and nonprofessional phagocytes and hence rarely found in tissues. Large numbers of dying cells such as may occur in excessive tissue injury, autoimmune diseases such as systemic lupus erythematosus or infectious disease may lead to the defective clearance of necrotic cells and contribute to the persistent inflammation and tissue damage. However, smaller number of cells dying via necrosis or necroptosis may not necessarily lead to massive tissue damage and hence controlled necrotic cell death may not necessarily lead to massive tissue damage commonly associated with necrotic cell death.

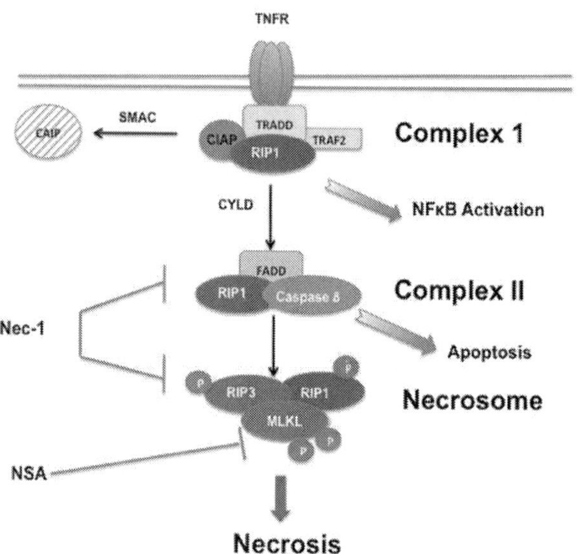

Figure 4. Necroptosis Signaling Pathway. Ligation of tumor-necrosis factor 1 (TNFR1) leads to the formation of Complex 1 that is comprised of TNFR-1, TFNR1-associated death domain (TRADD), TNFR-associated factor 2 (TRAF-2), Receptor-interacting kinase 1 (RIP1) and cellular inhibitors of apoptosis (CIAPs). RIP1 is initially modified by ubiquitin chains within Complex I after recruitment to the membrane and serves as the signaling platform for NF-κB activation. Upon release of the second mitochondria-derived activation of caspases (Smac) from the mitochondria, CIAPs are auto-degraded. RIP1 is subsequently de-ubiquinated by CYLD and dissociates from the death receptor, and associates with FADD-RIP1-Caspase8 forming Complex II, which is pro-apoptotic. Within Complex II, Caspase 8 activates Caspase 3/7 to induce apoptosis and blocks necrosis by cleaving and inactivating RIP1 and RIP3. When apoptosis is inhibited, as for example by Z-VAD, RIP1 and RIP3 form the necrotic death complex, the necrosome, and their kinase activities become activated. RIP3 recruits the mixed lineage kinase domain-like protein, MLKL, and phosphorylates MLKL, causing its activation. Activated MLKL then associates with membranes such as the plasma membrane, mitochondrial members and others, causing lysis of these membranes and death of the cell. Inhibitors of necroptosis include, Necrostatin 1 (Nec1) which inhibits RIP1 dependent necrosis by inhibiting the kinase activity of RIP1 to prevent necrosome formation and Necrosulfonamide (NSA), which inhibits MLKL-mediated necrosis.

Knockout of either RIP3 or MLKL genes in mice has been used to assess the physiological and pathological roles of programmed necrosis. These types of studies have confirmed that necroptosis is important in host antiviral responses and in some tissue-damage related diseases. Several inhibitors have been identified that can inhibit necroptosis. Necrostatins, for example, are small molecules that inhibit RIP1 kinase activity and thus inhibit necroptosis (Figure 4). RIP1 also contributes to other processes such as apoptosis and NK-kB activation, however, indicating other processes may also be affected by Nec-1. Targeting RIP3 or MLKL offer more specific inhibitors for study or treatment of necroptosis. A chemical inhibitor of MLKL called necrosulfonamide (NSA), for example, blocks necrosis by preventing oligomerization of MLKL and thus is a specific inhibitor of necroptosis (Figure 4). Use of these specific inhibitors of necroptosis will be essential to elucidate the role of necroptosis in cell death. As the role of programmed necrotic death is now known to occur in several human diseases, development of specific inhibitors of neroptosis could also have therapeutic uses.

2.3. Crosstalk between apoptosis, necroptosis and autophagy PCD pathways

As molecular components and regulation of each pathway has become better understood, it has become apparent that significant crosstalk exists between the three major pathways of programmed cell death, the focus of which has been the topic of several recent reviews [3, 8, 10, 13]. Extensive crosstalk exists between apoptosis and autophagy as discussed above in the section on autophagic cell death. The stimuli for apoptosis and autophagy are often the same and evidence indicates instances where they can cooperate, antagonize or assist each other. Caspases, Beclin-1 and/or the TOR kinase pathway have all been implicated in participating in the complex crosstalk between apoptosis and autophagy. Crosstalk between apoptosis and necrosis also appears to exist as receptors that stimulate apoptosis can also stimulate necroptosis. Energy is thought to play a central role in determining the interplay between these two pathways where instances of high ATP levels would enable a cell to undergo apoptosis while low ATP levels would favor necrosis. Other factors such as p53, Bcl-2 proteins and PARP1 are also suggested to play a role. Evidence also suggests a complex interplay between autophagy and necroptosis. Autophagy and necrosis can be activated in parallel or sequentially and can have either opposite or the same effects. Autophagy, for example, can protect against some types of necroptotic cell death with autophagy serving as the last resort before cell death via necrosis. The molecular basis for interlinked processes in autophagy and necroptosis and the impact of autophagy on necroptosis and other cell death pathways [13], however, remain unclear. While the molecular details are not fully understood, it is clear that a complex crosstalk and molecular interplay between apoptosis, autophagy and necroptosis occur and determine the ultimate fate of the cell to survive or die in a given situation or under a given stress signal. A better understanding of the mechanisms of apoptosis, autophagic cell death and necroptosis at the molecular level and the crosstalk between these cell death pathways would provide a deeper insight into different disease processes and may provide novel therapeutic strategies.

3. *Toxoplasma gondii:* Manipulation of programmed cell death pathways and impact on host–parasite interactions and maintenance of chronic infection

PCD can be important host cell defense mechanisms to eliminate intracellular pathogens. *Toxoplasma gondii* is an obligate intracellular protozoan parasite that causes a chronic infection with approximately one-third of the world's population chronically infected, making *T. gondii* one of the most prevalent human parasitic infections worldwide [17–19]. In humans, infection with *T. gondii* is characterized by rapid parasite replication and dissemination throughout the body with the parasite capable of infecting all types of nucleated cells. The acute infection is followed by a chronic infection that lasts for the lifetime of the individual with the parasite harbored within neurons in the brain and muscle tissue in intracellular cysts. In most immunocompetant individuals, the infection is asymptomatic due to an effective immune response that eliminates active parasite replication. In immunocompromised individuals such as AIDS patients that are chronically infected, the parasite reactivates from cysts in the brain leading to severe and potentially fatal encephalitis. The parasite can also be transmitted transplacentally and can cause severe neurological complications to the fetus including mental retardation, hydrocephaly and chorioretinitis which can reactivate through the first two decades of life potentially leading to blindness. Finally, increasing evidence indicates that in some immunocompetent individuals, chronic infection with *T. gondii* may lead to development of serious psychiatric disorders such as schizophrenia or suicidal tendencies [20–22].

While the host mounts an effective immune response against *T. gondii*, the parasite has evolved immune evasion strategies that enable it to survive and persist long enough in the host to establish the cyst stage of the infection that can persist for the lifetime of the host. PCD pathways are a key part of the host defense against *T. gondii*, but the parasite has evolved multiple mechanisms to manipulate these pathways. For example, the parasite induces apoptosis in selected immune cell lineages, thus suppressing the host immune response and helping to establish infection but the parasite also has multiple mechanisms to resist killing by apoptotic and, to lesser degree, autophagic and necrotic cell death pathways in different host cell types. Thus, parasite manipulations of host cell death pathways appear to be an essential component to the successful establishment of infection and maintenance of the chronic infection. A brief summary of the mechanisms by which the parasite manipulates these cell death pathways is reviewed below followed by a discussion of the potential therapeutic interventions.

3.1. Apoptosis

T. gondii has the ability to promote apoptosis of immune effector cells early after infection in selected cell types. In murine models of toxoplasmosis, apoptosis of CD4+, CD8+ T cells, B cells, NK cells and granulocytes is observed in the spleen early after infection [23]. Apoptosis in Peyer's patch T cells in perorally infected mice has also been observed [24]. Acute infection of *Toxoplasma* in both mice and humans induces a state of transient immunosuppression as determined by decreased antibody and T lymphocyte responses. Triggering of apoptosis of T cells and other immune effector cells by *T. gondii* may be one factor by which the parasite restricts the immune response, thus allowing the establishment of infection. Induction of apoptosis of immune effector cells may be somewhat parasite strain-dependent as, in murine models, high levels of apoptosis in T cells was induced when mice were infected with virulent RH strain while this phenomenon was not seen in mice infected with avirulent strains.

Inhibition of apoptosis by infection with *T. gondii* has been found to occur in a wide range of cell types including macrophages and a wide variety of non-immune effector cells such as fibroblasts, endothelial cells, muscle cells and astrocytes [23, 25, 26]. Infected host cells are resistant to induction of apoptosis to a wide range of stimuli including CTL-mediated cytotoxicity, irradiation, growth factor withdrawal, TNFα and several toxic reagents. Blockage of apoptosis in host cells establishes an anti-apoptotic condition of the host cell and favors parasite persistence. Neighboring uninfected cells are also rendered resistant to apoptosis. Inhibition of host cell apoptosis of non-immune effector cells likely ensures the host cell stays alive long enough to facilitate intracellular replication of the parasite in the tissues while suppression of apoptosis of uninfected neighboring cell may help create a microenvironment in different tissues in which the parasite can persist and replicate.

While blockage of apoptosis is known to occur in a wide variety of host cells, modulation of apoptosis in cells of the central nervous system (CNS) is less well studied and results are ambiguous. *T. gondii* can infect neurons, astrocytes and microglia supporting growth of the rapidly replicating tachyzoite form while bradyzoite stage and cysts develop only in neurons and astrocytes. In murine astrocytes, infection blocks apoptosis beginning at 6–24 h after infection, allowing time for the parasite to replicate, egress and initiate infection of a new host cell [26]. Thus, inhibition of apoptosis in astrocyte host cells may allow an increase in parasite numbers in the brain and help establish the chronic infection in the brain. Conversely, several studies indicate the parasite induced apoptosis in neurons. For example, tachyzoites induced apoptosis in mouse brain cells in adult mice including in neurons [27, 28]. Likewise, in a murine model of congenital toxoplasmosis, infection resulted in a decrease in neuron number and markers of apoptosis were found indicating infection induced apoptosis in CNS tissues [29]. In murine neural stem cells, infection was also found to induce apoptosis although only apoptosis of neighboring neurons were assessed in this study [30]. A study in murine brain indicated infected neurons were resistant to apoptosis and apoptosis was only induced in the uninfected neurons [31]. Microglia cells which are activated in the brains of mice infected with *T. gondii* with toxoplasmic encephalitis produced nitric oxide (NO), which induces neuronal apoptosis indicating activated microglia may be part of the mechanism leading to neuronal loss of uninfected neurons [32]. Collectively, the above results are suggestive of a mechanism in which *Toxoplasma* infection of neural tissues induces apoptosis of neurons, thus leading to loss of neurons which could lead to neurological abnormalities. However, it is unclear if apoptosis of infected neurons was induced or, alternatively, if infected neurons are resistant to apoptosis. Suppression of apoptosis of infected neuronal cells may facilitate development of the cysts in neurons and allow persistence of the parasite in the brain. A better understanding of parasite modulation of apoptosis in infected neurons is thus of importance. A better understanding of the modulation of apoptosis in infected neurons could lead to the development of antiparasitic drugs and other therapeutic interventions to control and manage the chronic phase of infection.

3.1.1. Parasite mechanisms modulating apoptosis

The parasite replicates within a membrane-bound compartment called the parasitophorous vacuole (PV), and hence it is not in direct contact with the host cell cytoplasm and regulators or effectors of apoptosis. However, parasite secretory molecules, which are either released into the host cell at invasion, secreted across the PV membrane (PVM) or act at the PVM cytosolic surface, have been identified which can modulate host cell apoptosis [25, 33]. These mechanisms are briefly summarized below.

3.1.2. Parasite mechanisms that stimulate apoptosis

Several mechanisms have been identified by which the parasite can stimulate apoptosis. In parasite-infected murine macrophages, supernatants can induce apoptosis of neighboring uninfected macrophages via nitric oxide [34]. Induction of apoptosis has also been found to occur via nitric oxide secretion from activated microglia. Nitric oxide secretion from activated microglia can lead to neuronal cell death and may be an important mechanism by which infection leads to neuronal loss in the brain. The parasite secretory molecule, GRA1, has been identified as capable of inducing apoptosis of infected monocytes and uninfected bystander cells [35]. Secreted GRA1 from sites of ongoing *T. gondii* replication could also induce apoptosis of monocytes recruited to the site of parasite replication. Given monocytes are essential to control parasite replication at the site of primary infection, inhibition of apoptosis in monocytes could serve to downregulate host responses early in infection, helping to establish the infection in the host.

3.1.3. Parasite mechanisms that inhibit apoptosis

Multiple mechanisms have been identified by which the parasite can inhibit apoptosis [25]. Activation of NF-κB is a mechanism that inhibits apoptosis in fibroblasts and some other host cells [36–38]. Early after infection, the parasite induces rapid translocation of host cell transcription factor, NF-kB, into the nucleus, which activates cell survival pathways and induction of an anti-apoptotic state of the host cell. Nuclear translocation of NF-κB and subsequent gene expression requires activity of the host IκB kinase (IKK). The activation of NF-κB was associated with localization of phosphorylated IκBα subunits to the PVM that was mediated by a parasite-derived IκB kinase (TgIKK). TgIKK is produced as the parasite replicates, and thus allows for a continued phosphorylation of IκB and sustained inhibition of apoptosis. Thus, *T. gondii* modulation of NF-κB gene expression and induction of anti-apoptotic state of the host cell involves both the host cell IKK and TgIKK activity at different phases of infection. NF-kB activation has not been found to occur in infected macrophages. Rather in macrophages, *T. gondii*, inhibits apoptosis by G_i-protein-mediated signaling, activating PI 3-kinase leading to phosphorylation of protein kinase B (PKB/AKt) as an inhibitor of apoptosis [39]. A micro-RNA-mediated mechanism has also recently been identified in macrophages that inhibited apoptosis via a reduction in Bim, a pro-apoptotic effector of the Bcl-2 family [40]. Other mechanisms have been identified in various cell types that can inhibit caspase cascade including blockage of caspases 8 and 9 and 3, increased expression of anti-apoptotic molecules of the Bcl-2 family, decreased activity of poly (ADP)-ribose polymerase and inhibition of granzymes [33, 41].

Only a few parasite proteins that modulate apoptosis have been identified. The parasite serine–threonine phosphatase, called TgPP2C, was recently identified which downregulates apoptosis in host cells [42]. TgPP2C is secreted into host cells from the PV and translocates to the host cell nucleus and has been shown to regulate growth and survival of the parasite. Using a yeast two-hybrid system, TgPP2C was found to interact with host cell protein, SSRP1 (structure-specific recognition protein 1), which binds to DNA and regulates DNA repair genes. It is speculated that SSRP1 might also be involved in expression of other genes involved in cell survival and apoptosis. Identification of other parasite molecules by which the parasite manipulates host cell apoptosis would facilitate a deeper understanding of host–parasite relationship and may lead to development of new therapeutic targets and antiparasitic drugs.

3.1.4. *Parasite modulation of apoptosis: significance to pathogenesis*

In conclusion, *T. gondii* has the ability to both promote and inhibit apoptosis of its own host cell and of uninfected neighboring cells by multiple mechanisms. The ability to modulate apoptosis varies by host cell type, parasite load and parasite virulence. Manipulation of apoptosis has been proposed to be crucial to promoting a stable parasite–host interaction and allowing establishment of persistent infection [23]. The multitude of anti-apoptosis mechanism that the parasite *T. gondii* employs is likely reflective of its status as an obligate intracellular parasite and the essentiality of maintenance of host cell viability to sustain parasite replication and survival. However, intracellular replication eventually leads to lysis of the host cell and typically occurs between 48 and 72 h post-infection, a temporal scale similar to apoptosis. Given the similar temporal scales of lytic death of the host cell and cell death by apoptosis, it can be argued inhibition of apoptosis of the host cell is not necessary to allow for parasite replication. However, as the parasite needs to obtain purine, cholesterol, tryptophan and other components from the host cell, the parasite inhibition of host cell apoptosis may also serve to maintain the host cell in a pro-survival state to enhance availability of host cell nutrients. Inhibition of apoptosis may also be critical for the chronic phase of infection allowing differentiation of bradyzoites and development of the cyst stage in neurons. A better understanding of manipulation of apoptosis by bradyzoites in host cells and, specifically in neurons, is of great importance as the CNS is the compartment in which the parasite predominantly persists in the chronic infection and the cause of serious clinical sequelae in the CNS of immunocompromised individuals causing reactivated toxoplasmosis, in immature immune system of the fetus and newborns causing congenital toxoplasmosis, and in immunocompetant individuals possibly contributing to the development of serious neuropsychiatric disorders.

3.2. Autophagy

Autophagy has been found to play a role in host defense against many intracellular pathogens as the autophagy pathway can degrade intracellular pathogens via autophagolysosomes including *T. gondii* [16]. *T. gondii* replicates within a membrane compartment in the host cell called the parasitophorous vacuole (PV) that is a non-fusogenic compartment and does not fuse with cell lysosomes. However, the PV can be delivered to the lysosomes via autophagy-mediated delivery in IFN-γ-activated macrophages resulting in killing of the parasite [43]. This autophagic-mediated parasite killing is mediated by the immunity-related GTPases (IRGs) which are stimulated by IFN-γ. Engagement of CD40 also has been found to induce killing of *T. gondii* via autophagy in many nonhematopoetic cells including endothelial cells lines, human and mouse retinal pigment epithelial cells as well as in hemopoietic cells such as macrophages [44, 45].

Interestingly, in non-immune-stimulated cells, evidence indicates the parasite can use autophagy to enhance its own survival. In the first 24 h after infection, host cell autophagy is upregulated and the parasite uses autophagy to acquire host cell nutrients, while at 24–36 h after infection when significant parasite replication has occurred, host cell autophagy is suppressed [46, 47]. This biphasic response may be due to the crosstalk that exists between apoptosis and autophagy. That is, inhibition of apoptosis occurs in infected cells during the first 24 h after infection and as a result of crosstalk, it has been suggested that autophagy is stimulated and then, vice versa, during the later stages of the intracellular infection cycle (> 24 h post-infection), as apoptosis inhibition wanes, suppression of autophagy increases. Thus, regulation of host cell autophagy by the parasite may be part of a cell survival mechanism in *T. gondii* infected host cells related to inhibition of apoptosis. Interestingly, parasite-mediated activation of AKt signaling has recently been found to prevent autophagy degradation of the parasite [48]. As AKt activation has also been linked to inhibition of apoptosis in *T. gondii* infected cells, this event suggests AKt signaling may be a pathway by which the parasite can regulate both host cell autophagy and apoptosis pathways [39, 48]. Dual manipulation of host cell apoptosis and autophagy pathways indicates a fine level of control of host PCD pathways by the parasite to promote parasite persistence in the host. Further work on the interplay between apoptosis and autophagy and the parasite modulation of these two processes is needed to further elucidate these mechanisms.

3.3. Necrotic cell death

A few *in vitro* studies have reported *T. gondii* inducing necrotic cell death of host cells [49–52]. In IFN-γ stimulated fibroblasts and astrocytes, infection with avirulent parasite strains of the parasite, rather than stimulating autophagic-mediated mechanism of parasite killing, induces disruption of the PV membrane via IRG proteins resulting in parasite death and subsequently triggering of necrotic host cell death [50, 53]. This type of necrotic death of the host cells involved killing of intracellular parasites and thus would promote host survival via elimination of the parasite. In these experiments, the parasite was found to egress within hours after infection, thus limiting replication in host cells which also would promote clearance of the parasite by immune cells. Experiments have similarly found early egress of the parasites in

IFN-γ-stimulated astrocytes, indicating this mechanism may also function in the brain to limit parasite numbers and aid in elimination of the parasite [54].

Conversely, *Toxoplasma*-specific primed T cells were found to be able to induce rapid egress of the parasite, causing necrotic cell death of host cells via ligation of the death receptor or perforin, but as egress was rapid the intracellular parasites escaped killing [55]. These egressed parasites were capable of infecting neighboring cells, thus suggesting that during *Toxoplasma* infection of T cells, death receptor-and perforin-mediated parasite egress may contribute to parasite dissemination of the parasite. Furthermore, as necrotic cell death induced by death receptor or perforin-dependent parasite egress contributes to inflammatory processes, it could also lead to further spread of the parasite due to infection of invading leukocytes. This process may also contribute to the establishment of infection in the brain as infected leukocytes can invade brain. Finally, death receptor-induced or perforin-mediated egress could be involved during the reactivation of *T. gondii* in chronically infected individuals, leading to reinfection of egressed parasites in the brain. Thus, parasite evasion of this cell death pathway may have significant consequences on the clinical sequelae of the chronic infection in the brain. Experiments in fibroblasts found necrotic host cell death was not dependent on the necroptosis mediator RIPK3 or caspases [52]. Further studies are needed to ascertain if necroptosis occurs in other cell types that have been associated with necrotic-type cell death in *T. gondii* infected cells.

4. Parasite manipulation of PCD pathways: Conclusions and potential therapeutic implications

A better understanding of the mechanisms by which the parasite modulates apoptosis, autophagic and necrotic host cell death pathways will enhance our understanding of the host/ parasite relationship in toxoplasmosis and may yield new therapeutic targets to treat pathological consequences of cerebral toxoplasmosis. For example, while the infection is asymptomatic in immunocompetant individuals, recent evidence indicates chronic infection may lead to psychiatric and other neurological disorders in some individuals. Some of these effects may be due to modifications of processes leading to cell death or resulting in dysregulated cell death, resulting in reduction in CNS cell numbers and specifically of neurons. Conversely, inhibition of apoptosis of astrocytes in the brain could be crucial for establishment of the parasite in this compartment, while inhibition of apoptosis of infected neurons may be essential for development of bradyzoite stage and thus crucial to long-term persistence of the parasite in the brain. The evidence indicates the parasite can manipulate both host cell apoptosis and autophagy pathways, indicating a sophisticated level of parasite control of host PCD pathways and elucidation of the molecular details of these mechanisms could lead to new therapeutic targets with which to control the infection. A better understanding of manipulation of apoptosis and other cell death pathways in the brain cells such as microglia, astrocytes and neurons is of particular importance. The identification of molecular components of PCD pathways crucial to parasite–host interactions in chronic toxoplasmosis could lead to development of new antiparasitic drugs and/or yield new potential therapeutic targets to treat pathological consequences of chronic cerebral toxoplasmosis in immunocompetant individuals and reactivated toxoplasmosis in immunosuppressed individuals. A similar harnessing of apoptotic, autophagic and programmed necrosis cell death pathways has been proposed to treat cancers, neurodegenerative disorders and other diseases, where PCD pathways have been identified as involved in pathogenesis [56]. The dissection of the molecular mechanisms by which *T. gondii* manipulates apoptosis and other cell death pathways thus could also serve as a benefit to probe pathways in normal cells and in diseases in which apoptosis and other cell death pathways play a central role.

Author details

Sandra K. Halonen*

Address all correspondence to: shalonen@montana.edu

Dept. of Microbiology and Immunology, Montana State University, Bozeman, MT, USA

References

[1] Tsujimoto Y, Shimizu S. Another way to die: autophagic programmed cell death. Cell Death Differ 2005;12(2):1528–34.

[2] Edinger AL, Thompson CB. Death by design: apoptosis, necrosis and autophagy. Curr Opin Cell Biol 2004;16(6):663–9.

[3] Elmore S. Apoptosis: a review of programmed cell death. Toxicol Pathol 2007;35(4): 495–516.

[4] Vandenabeele P, et al. Molecular mechanisms of necroptosis: an ordered cellular explosion. Nat Rev Mol Cell Biol 2010;11(10):700–14.

[5] Mizushima N, Levine B. Autophagy in mammalian development and differentiation. Nat Cell Biol 2010;12(9):823–30.

[6] Mizushima N, et al. Autophagy fights disease through cellular self-digestion. Nature 2008;451(7182):1069–75.

[7] Green DR, Levine B. To be or not to be? How selective autophagy and cell death govern cell fate. Cell 2014;157(1):65–75.

[8] Nikoletopoulou V, et al. Crosstalk between apoptosis, necrosis and autophagy. Biochim Biophys Acta 2013;1833(12):3448–59.

[9] Sun L, Wang X. A new kind of cell suicide: mechanisms and functions of programmed necrosis. Trends Biochem Sci 2014;39(12):587–93.

[10] Chan FK, Luz NF, Moriwaki K. Programmed Necrosis in the Cross Talk of Cell Death and Inflammation. Annu Rev Immunol 2014; 33:4.1-4.28.

[11] Kerr JF, Wyllie AH, Currie AR. Apoptosis: a basic biological phenomenon with wide-ranging implications in tissue kinetics. Br J Cancer 1972;26(4):239–57.

[12] Norbury CJ, Hickson ID. Cellular responses to DNA damage. Annu Rev Pharmacol Toxicol 2001;41:367–401.

[13] Ryter SW, Mizumura K, Choi AM. The impact of autophagy on cell death modalities. Int J Cell Biol 2014;2014:502676.

[14] Liu Y, Levine B. Autosis and autophagic cell death: the dark side of autophagy. Cell Death Differ 2015;22(3):367–76.

[15] Kroemer G, Levine B. Autophagic cell death: the story of a misnomer. Nat Rev Mol Cell Biol 2008;9(12):1004–10.

[16] Paulus GL, Xavier RJ. Autophagy and checkpoints for intracellular pathogen defense. Curr Opin Gastroenterol 2015;31(1):14–23.

[17] Nath A, Sinai AP. Cerebral Toxoplasmosis. Curr Treat Opt Neurol 2003;5(1):3–12.

[18] Halonen SK, Weiss LM. Toxoplasmosis. Handb Clin Neurol 2013;114:125–45.

[19] Tenter AM, Heckeroth AR, Weiss LM. Toxoplasma gondii: from animals to humans. Int J Parasitol 2000;30(12–13):1217–58.

[20] Okusaga O, Postolache TT. Toxoplasma gondii, the immune system, and suicidal behavior. In: Dwivedi Y. (Ed.) The Neurobiological Basis of Suicide. 2012: Boca Raton (FL).

[21] Niebuhr DW, et al. Selected infectious agents and risk of schizophrenia among U.S. military personnel. Am J Psychiatr 2008;165(1):99–106.

[22] Torrey EF, et al. Antibodies to Toxoplasma gondii in patients with schizophrenia: a meta-analysis. Schizophr Bull 2007;33(3):729–36.

[23] Luder CG, Gross U. Apoptosis and its modulation during infection with Toxoplasma gondii: molecular mechanisms and role in pathogenesis. Curr Top Microbiol Immunol 2005;289:219–37.

[24] Liesenfeld O, Kosek JC, Suzuki Y. Gamma interferon induces Fas-dependent apoptosis of Peyer's patch T cells in mice following peroral infection with Toxoplasma gondii. Infect Immun 1997;65(11):4682–9.

[25] Blader IJ, Koshy AA. Toxoplasma gondii development of its replicative niche: in its host cell and beyond. Eukaryot Cell 2014;13(8):965–76.

[26] Contreras-Ochoa CO, et al. Toxoplasma gondii invasion and replication within neonate mouse astrocytes and changes in apoptosis related molecules. Exp Parasitol 2013;134(2):256–65.

[27] Takahashi J, et al. Bax-induced apoptosis not demonstrated in the congenital toxoplasmosis in mice. Brain Dev 2001;23(1):50–3.

[28] el-Sagaff S, et al. Cell death pattern in cerebellum neurons infected with Toxoplasma gondii. J Egypt Soc Parasitol 2005;35(3):809–18.

[29] Shen DF, et al. Involvement of apoptosis and interferon-gamma in murine toxoplasmosis. Investig Ophthalmol Visual Sci 2001;42(9):2031–6.

[30] Wang T, et al. Toxoplasma gondii induce apoptosis of neural stem cells via endoplasmic reticulum stress pathway. Parasitology 2014;141(7):988–95.

[31] Handel U, et al. Neuronal gp130 expression is crucial to prevent neuronal loss, hyperinflammation, and lethal course of murine Toxoplasma encephalitis. Am J Pathol 2012;181(1):163–73.

[32] Zhang YH, et al. Activated microglia contribute to neuronal apoptosis in Toxoplasmic encephalitis. Parasit Vectors 2014;7:372.

[33] Sinai AP, et al. Mechanisms underlying the manipulation of host apoptotic pathways by Toxoplasma gondii. Int J Parasitol 2004;34(3):381–91.

[34] Nishikawa Y, et al. Toxoplasma gondii infection induces apoptosis in noninfected macrophages: role of nitric oxide and other soluble factors. Parasite Immunol 2007;29(7):375–85.

[35] D'Angelillo A, et al. Toxoplasma gondii dense granule antigen 1 stimulates apoptosis of monocytes through autocrine TGF-beta signaling. Apoptosis 2011;16(6):551–62.

[36] Molestina RE, et al. Activation of NF-kappaB by Toxoplasma gondii correlates with increased expression of antiapoptotic genes and localization of phosphorylated IkappaB to the parasitophorous vacuole membrane. J Cell Sci 2003;116(Pt 21):4359–71.

[37] Molestina RE, Sinai AP. Host and parasite-derived IKK activities direct distinct temporal phases of NF-kappaB activation and target gene expression following Toxoplasma gondii infection. J Cell Sci 2005;118(Pt 24):5785–96.

[38] Payne TM, Molestina RE, Sinai AP. Inhibition of caspase activation and a requirement for NF-kappaB function in the Toxoplasma gondii-mediated blockade of host apoptosis. J Cell Sci 2003;116(Pt 21):4345–58.

[39] Kim L, Denkers EY. Toxoplasma gondii triggers Gi-dependent PI 3-kinase signaling required for inhibition of host cell apoptosis. J Cell Sci 2006;119(Pt 10):2119–26.

[40] Cai Y, et al. Toxoplasma gondii inhibits apoptosis via a novel STAT3-miR-17-92-Bim pathway in macrophages. Cell Signal 2014;26(6):1204–12.

[41] Yamada T, et al. Toxoplasma gondii inhibits granzyme B-mediated apoptosis by the inhibition of granzyme B function in host cells. Int J Parasitol 2011;41(6):595–607.

[42] Gao XJ, et al. Protein phosphatase 2C of Toxoplasma gondii interacts with human SSRP1 and negatively regulates cell apoptosis. Biomed Environ Sci 2014;27(11):883–93.

[43] Martens S, et al. Disruption of Toxoplasma gondii parasitophorous vacuoles by the mouse p47-resistance GTPases. PLoS Pathogens 2005;1(3):e24.

[44] Van Grol J, et al. CD40 induces anti-Toxoplasma gondii activity in nonhematopoietic cells dependent on autophagy proteins. Infect Immun 2013;81(6):2002–11.

[45] Subauste CS. CD40, autophagy and Toxoplasma gondii. Memorias do Instituto Oswaldo Cruz, 2009;104(2):267–72.

[46] Orlofsky A. Toxoplasma-induced autophagy: a window into nutritional futile cycles in mammalian cells? Autophagy 2009;5(3):404–6.

[47] Lee YJ, et al. Proliferation of Toxoplasma gondii suppresses host cell autophagy. Korean J Parasitol 2013;51(3):279–87.

[48] Muniz-Feliciano L, et al. Toxoplasma gondii-induced activation of EGFR prevents autophagy protein-mediated killing of the parasite. PLoS Pathogens 2013;9(12):e1003809.

[49] Liesenfeld O, et al. Association of CD4+ T cell-dependent, interferon-gamma-mediated necrosis of the small intestine with genetic susceptibility of mice to peroral infection with Toxoplasma gondii. J Experiment Med 1996;184(2):597–607.

[50] Zhao YO, et al. Disruption of the Toxoplasma gondii parasitophorous vacuole by IFNgamma-inducible immunity-related GTPases (IRG proteins) triggers necrotic cell death. PLoS Pathogens 2009;5(2):e1000288.

[51] Persson EK, et al. Death receptor ligation or exposure to perforin trigger rapid egress of the intracellular parasite Toxoplasma gondii. J Immunol 2007;179(12):8357–65.

[52] Niedelman W, et al. Cell death of gamma interferon-stimulated human fibroblasts upon Toxoplasma gondii infection induces early parasite egress and limits parasite replication. Infect Immun 2013;81(12):4341–9.

[53] Halonen SK. Role of autophagy in the host defense against Toxoplasma gondii in astrocytes. Autophagy 2009;5(2):268–9.

[54] Melzer T, et al. The gamma interferon (IFN-gamma)-inducible GTP-binding protein IGTP is necessary for toxoplasma vacuolar disruption and induces parasite egression in IFN-gamma-stimulated astrocytes. Infect Immun 2008;76(11):4883–94.

[55] Persson EK, et al. Death receptor ligation or exposure to perforin trigger rapid egress of the intracellular parasite Toxoplasma gondii. J Immunol 2007;179(12):8357–65.

[56] Portt L, et al. Anti-apoptosis and cell survival: a review. Biochim Biophys Acta 2011;1813(1):238–59.

Necrosis

Chapter 18

HMGB1 in Cell Death

Daolin Tang and Rui Kang

Additional information is available at the end of the chapter

http://dx.doi.org/10.5772/61208

Abstract

High mobility group box 1 (HMGB1) is named for its electrophoretic mobility on pol-yacrylamide gels when it was first identified in calf thymus in 1973. HMGB1 plays a critical role in the stress response not only inside the cell as a DNA chaperone and cell death regulator, but also outside the cell as the prototypic damage-associated molecu-lar pattern molecule. The physiological and pathological role of HMGB1 in health and disease has been widely studied for years. In this chapter, we will focus on the release and function of HMGB1 in cell death types such as apoptosis, autophagy, and ne-crosis.

Keywords: hmgb1, autophagy, necrosis, apoptosis

1. Introduction

Cell death is the cell's process of losing its biological ability to carry out all the essential life processes. The Nomenclature Committee on Cell Death proposes several cell death classifica-tion criteria. According to morphological appearance, cell death is divided into apoptosis (type I), autophagy (type II), and necrosis (type III) [1, 2]. According to enzymological qualities, cell death is divided into several subtypes depending on the involvement or noninvolvement of nuclei or distinct protease classes such as caspases, calpains, cathepsins, and transglutami-nases. According to immunological characteristics, cell death is divided into immunogenic or tolerogenic cell death [3]. For example, apoptosis is generally considered nonimmunogenic cell death, whereas necrosis is considered immunogenic cell death. In addition, cell death can be classified into regulated or accidental cell death based on functional aspects [4]. Accidental cell death is caused by unexpected and accidental cell damage (e.g., ischemic and trauma), whereas regulated cell death is mediated by an expected program in response to different stimuli. The list of regulated cell death subtypes is rapidly increasing and includes anoikis, autophagic cell death, apoptosis, cornification, entosis, ferroptosis, mitotic catastrophe,

necroptosis, netosis, parthanatos, and pyroptosis [4]. Cell death is essential for a plethora of physiological processes, and its deregulation is implicated in several human diseases such as infections, neurodegeneration, cancer, autoimmunity, and ischemic disease [5-7]. During the past few decades, a number of important concepts regarding the regulation of cell death and its roles in human health and disease have arisen. Understanding the molecular mechanisms and signaling pathways of cell death is crucial for identifying new diagnostic and therapeutic targets.

Compared to pathogen-associated molecular pattern molecules (PAMPs), which are generated from the components of foreign pathogens such as bacteria and viruses, damage-associated molecular pattern molecules (DAMPs) are endogenous or self-molecules that are secreted, released, or undergo surface exposure by dead, dying, or injured cells [8-12]. Both PAMPs and DAMPs are mainly recognized by pattern recognition receptors such as receptor for advanced glycation end products (RAGE) and toll-like receptors (TLRs) to mediate the inflammatory, immunity, and metabolism response. The release and activity of DAMPs during cell death can determine whether cell death is immunogenic or tolerogenic [13]. Thus, DAMPs are suitable emergent targets for cell-death-associated immune therapy.

High mobility group box 1 (HMGB1) is named for its electrophoretic mobility on polyacrylamide gels when it was first identified in calf thymus in 1973 [14]. As an extremely conserved protein, HMGB1 originated before the divergence of the protostomes and deuterostomes, approximately 525 million years ago [15]. HMGB1 shares 100% amino acid sequence identity between mice and rats, and a 99% homology between rodents and humans [16-18]. The homolog of mammalian HMGB1 has been reported for several species such as *Nhp6A/B* in yeast and *HMG-D* and *DSP1* in *Drosophila* [19-21]. HMGB1 plays a critical role in the stress response not only inside the cell as a DNA chaperone and cell death regulator, but also outside the cell as the prototypic DAMP. The physiological and pathological role of HMGB1 in health and disease has been widely studied for years [22]. In this chapter, we will focus on the release and function of HMGB1 in cell death types such as apoptosis, autophagy, and necrosis.

2. HMGB1 structure and function

2.1. Structure

Human HMGB1 consists of 215 amino acid residues and has two L-shaped DNA-binding domains (HMG A box [9-79aa], HMG B box [95-163aa]) and a shorter C-terminal tail (186-215aa) [23]. Both A- and B-box domains are necessary for efficient DNA bending and flexure. HMGB1 binds to DNA without apparent sequence specificity. HMGB1 normally locates in the nucleus due to two nuclear-localization signals (NLS): NLS1 (28-44aa) and NLS2 (179-185aa) [24]. In contrast, HMGB1 contains nuclear-emigration signals in DNA-binding domains, which contributes to extranuclear HMGB1 during stress in a nuclear exportin chromosome-region maintenance 1-dependent manner. In addition to DNA, HMGB1 can bind a number of proteins involved in multiple biologic processes. For example, HMGB1 binds to RAGE, TLR4, and p53 by residues 150-183, 89-108, and 7-74, which mediates cell migration

[25], cytokine production [26], and gene transcription [27], respectively. The recombinant B box protein exhibits proinflammatory activity, whereas the recombinant A box protein displays anti-inflammatory activity [28], although the potential mechanism remains unknown. The C terminus is composed of 30 acidic amino acid residues and is able to regulate DNA binding/bending by intramolecular interaction with the A- and B- box [29, 30] or by intermolecular interaction with histones (e.g., H1 and H3) [31, 32]. Additionally, residues 201-205 in the C-terminal acidic tail region contribute to the antibacterial activity of recombinant HMGB1 [33]. Hence, the structural basis of HMGB1 determines its biological function.

2.2. Intracellular HMGB1

2.2.1. Nuclear HMGB1

HMGB1 translocates between the cytoplasm and the nucleus, but normally stays in the nucleus in most cells and tissues. Nuclear HMGB1 is the structural protein of chromatin and orchestrates a number of nuclear events by its DNA chaperone activity as follows: (1) Nucleosome stability and sliding. As basic unit of chromatin, nucleosome contains a short length of DNA wrapped around a core of histone proteins. HMGB1 and histone H1 can bind to linker DNA between successive nucleosomes in the chromatin fiber [34]. H1 stabilizes nucleosome with less mobility, whereas HMGB1 relaxes nucleosome and makes chromatin more accessible at the distorted site [35, 36]. (2) Nucleosome number and genome chromatinization. Loss of HMGB1 in mammalian and yeast cells leads to 20-30% less histones and nucleosomes and more RNA transcripts [37]. (3) Nuclear catastrophe and nucleosome release. Conditional knockout of HMGB1 in the pancreas causes nuclear oxidative injury and proinflammatory nucleosome release, which mediates sterile inflammation [38]. (4) DNA bending and binding. HMGB1 binds to DNA with structure-specificity, but not sequence-specificity [39]. After binding DNA, the major function of HMGB1 is to bend and change DNA conformation by unwinding [40], looping [41], or compacting DNA [42]. This DNA chaperone activity of HMGB1 is implicated in the regulation of gene transcription [43], DNA repair [44], DNA replication [45], V(D)J recombination [46], gene delivery [47], and gene transfer [48]. (5) Telomere homeostasis. Loss of HMGB1 in yeast and mammalian cells inhibits telomerase activity, decreases telomere length, and increases DNA damage and chromosomal instability [49].

2.2.2. Cytosolic HMGB1

Several cell types (e.g., fibroblasts [50], thymocytes [51]), and tissue types (e.g., liver, kidney, heart, and lung) [52] have normal cytosolic HMGB1 expression. The ratio of nuclear to cytoplasmic HMGB1 is about 30:1 [52]. Importantly, the translation of HMGB1 from the nucleus to the cytosol, including mitochondria and lysosomes, are observed in response to various stressors (e.g., cytokines, chemokines, heat, hypoxia, oxidative stress, and oncogenes). Although the function of cytosolic HMGB1 still remains poorly studied, HMGB1 may act as a positive regulator of mitochondrial quality in an autophagy-dependent and autophagy-independent manner [53, 54], which will be discussed later in the "Autophagy" section. In addition to autophagy, cytosolic HMGB1 is involved in the regulation of unconventional secretory pathways based on mass spectrometry-mediated binding partner analysis [55]. In one study, several HMGB1-binding partners in nuclear and cytosol fraction were identified in several cancer cells [55]. Among them, nine of the cytosolic HMGB1-binding proteins were related to protein translocation and secretion. In particular, immunoprecipitation analysis further confirmed four cytosolic HMGB1-binding proteins, including annexin A2, myosin IC isoform a, myosin-9, and Ras-related protein Rab10 [55]. These proteins are directly implicated in the process of unconventional protein secretion. Further studies are needed to define the function of cytosolic HMGB1 in unconventional protein secretion. In addition to nuclear and cytosolic HMGB1, intracellular HMGB1 presents on cell surface membranes and regulates neurite outgrowth [56], platelet activation [57, 58], cell differentiation [59], erythroid maturation [60], adhesion [61], and innate immunity [62].

2.3. Extracellular HMGB1

HMGB1 is released in two different ways. On the one hand, HMGB1 can be actively secreted by normal cells, especially immune and endothelial cells [63, 64]. On the other hand, HMGB1 can be passively released by dead, dying, or injured cells in response to autophagic cell death [65], apoptosis [66, 67], necrosis [68], necroptosis [69, 70], netosis [71], and pyroptosis [72]. Oxidative stress refers to elevated intracellular levels of reactive oxygen species (ROS) that play a central role in the regulation of HMGB1 secretion and release, although the actual mechanism of action remains ambiguous [73]. Once released, HMGB1 acts as a cytokine, chemokine, and growth factor that is implicated in multiple biological processes including inflammation, immunity, migration, invasion, metabolism, proliferation, differentiation, antimicrobial defense, angiogenesis, tissue regeneration, death, autophagy, senescence, and efferocytosis. Extracellular HMGB1 plays important roles in the pathogenesis of human disease and is a potential therapy target in infection and sterile inflammation [74-76]. Several factors can affect HMGB1 activity in different experimental settings. For example, RAGE [77] and TLRs [78, 79] are positive receptors in macrophages and fibroblasts, whereas CD24 [80] and T cell immunoglobulin mucin 3 [81, 82] are negative receptors of HMGB1-mediated signaling in macrophages and dendritic cells (DCs). In addition to receptors, HMGB1 can be directly taken up and mediate the inflammatory and metabolism response [83, 84]. Ultra-pure HMGB1 (free from contaminating bacterial proteins and nucleic acids) exhibits very low immune activity in macrophages. In contrast, extracellular HMGB1 is in fact a "sticky" protein and a synergistic immune effect is observed between HMGB1 and PAMPs (e.g., lipopolysaccharide), DAMPs (e.g., DNA), and other molecules (e.g., cytokines, chemokine, and IgG) in multiple cells [85]. Thus, serum and plasma components (e.g., immunoglobulins, phospholipids, thrombomodulin, and proteoglycans) can interfere with HMGB1 detection by enzyme-linked immunosorbent assay [86]. Another important factor affecting HMGB1 activity is its redox status [87]. HMGB1 contains three conserved redox-sensitive cysteine residues: C23, C45, and C106. Reduced all-thiol-HMGB1 only exhibits chemokine activity, whereas disulfide-HMGB1 displays only cytokine activity, and oxidized HMGB1 has neither in immune cells [88]. In addition, reduced HMGB1 induces autophagy, whereas oxidized HMGB1 triggers apoptosis in cancer cells [89]. This redox status of HMGB1 also affects the affinity between HMGB1 and its receptors [26]. A recent study demonstrates that HMGB1 is specifically cleaved by caspase-1 but not other caspases during inflammasome activation [90]. Collectively, the release and activity of HMGB1 is context-dependent.

3. HMGB1 regulates cell death

3.1. Mechanism of HMGB1-mediated autophagy regulation

Autophagy, including macroautophagy, microautophagy, and chaperone-mediated autophagy, is a highly conserved degradation process in organisms from yeasts to plants and animals [91]. The well-studied form of autophagy is macroautophagy (hereafter referred to as autophagy). As a complex dynamic process, autophagy is composed of the formation and maturation of three major membrane structures: the phagophore, autophagosome, and autolysosome [92]. Briefly, the phagophore originates from multiple membrane resources and engulfs the cytosolic materials, which leads to the formation of a closed autophagosome with a double membrane. Of note, microtubule-associated protein light chain 3 (LC3)-II is a widely used autophagosome marker [93]. Finally, autophagosomes fuse with lysosomes to form autolysosomes, which results in degradation of the engulfed material, including LC3-II, by lysosomal enzymes into elementary pieces that can be used for protein synthesis and energy production. Thus, autophagy is a programmed cell survival pathway in response to intracellular and extracellular stress [94]. However, excessive or impaired autophagy can cause cell death, indicating a dual role of autophagy in cell survival and cell death. In particular, autosis is an Na^+, K^+-ATPase-dependent form of cell death triggered by autophagy-inducing peptides, starvation, and hypoxia–ischemia [95]. The process of autophagy is controlled by multiple posttranslational modifications of the autophagy-related gene (Atg) family and shares regulators derived from other cell death pathways [96].

HMGB1 promotes autophagy in a location- and modification-dependent manner. Nuclear HMGB1 regulates heat shock protein β-1 (HSPB1) expression at a transactional level [54]. The protein expression of HSPB1, but not other heat shock proteins, is significantly inhibited in HMGB1 knockout or knockdown cells. Both HMGB1 and HSPB1 regulate mitochondrial selective autophagy, namely mitophagy, following mitochondrial injury [54]. Like other ATGs, it was recently suggested that HMGB1-independent autophagy exists in the regulation of mitochondrial quality, including the mitochondrial DNA damage response [53]. Cytosolic HMGB1 is a Beclin-1 binding protein [97]. HMGB1 C23S and C45S mutants lose their ability to bind Beclin-1 and therefore cannot promote autophagy [97]. The binding of HMGB1 with Beclin-1 is positively regulated by unc-51-like kinase 1 [98] mitogen-activated kinase-like protein [99], and nucleus accumbens-1 [100]. In contrast, p53 [101], SNCA/α-synuclein [102], lysosomal thiol reductase [103], miR34A [104], and miR22 [105] negatively regulate HMGB1-mediated autophagy by disrupting HMGB1-Beclin-1 complex formation. Moreover, activation of poly [ADP-ribose] polymerase 1 (PARP1) is required for tumor necrosis factor (TNF)-related apoptosis-inducing ligand (TRAIL)-induced ADP-ribosylation of HMGB1 and subsequent HMGB1-Beclin-1 complex formation in cancer cells [106]. Extracellular reduced HMGB1, but not oxidized HMGB1, significantly induced autophagy in cancer cells in a RAGE-dependent manner [89]. This process may sustain anaerobic energy production during tumor growth and development [107]. Collectively, these findings suggest an HMGB1-dependent autophagic pathway at multiple levels in response to stress. However, HMGB1-independent autophagy may exist in several organs, although the underlying mechanism of its action remains obscure [108].

3.2. Mechanism of HMGB1-mediated apoptosis regulation

Apoptosis is the process of programmed cell death and includes classical "extrinsic" and "intrinsic" pathways and nonclassical T and natural killer cell-mediated cytolytic pathways. The extrinsic pathway is primarily mediated by the binding of a ligand to a transmembrane death receptor (DR). DRs are members of the TNF receptor gene superfamily, including FasR, TNFR1, lymphotoxin receptor, DR3, and DR4/DR5 [109]. In addition to DRs, dependence receptors mediate apoptosis by monitoring the absence of certain trophic factors or the presence of anti-trophic factors [110]. The intrinsic pathway for apoptosis involves activation of a mitochondrial pathway including altering mitochondrial permeability and subsequent release of mitochondrial proteins such as cytochrome c and second mitochondrial-derived activator of caspases [111]. The process of apoptosis is tightly regulated by the Bcl-2, caspase, and nuclease families [112-114]. Caspases are a family of endoproteases linking inflammation and cell death. Initiator caspases (e.g., caspases-8 and -9) activate executioner caspases (e.g., caspases-3, -6, and -7) that mediate the cleavage of key structural proteins such as PARP1. However, caspase-independent apoptosis may exist by translocation of apoptosis-inducing factor [115, 116] and endonuclease G [117] from the mitochondria to the nucleus, or activation of Omi/HTRA2 (a mitochondrial serine protease) [118]. Remarkably, several caspases (e.g., caspase-1, -4, -5, and -12 in humans; caspase-1, -11, and -12 in mice) are critical mediators of innate immune responses partly by activation of inflammasome, but not activation of the apoptosis pathway.

Intracellular HMGB1 is generally an anti-apoptotic protein in response to several apoptotic stimuli such as ultraviolet radiation, CD95, TRAIL, caspase-8, and Bax [119]. Knockdown of HMGB1 increases drug sensitivity in cancer cells [120]. Mechanically, HMGB1 plays transcriptional-dependent (e.g., regulation of Bcl-2 family protein expression) and transcriptional-independent roles (e.g., regulation of autophagy and p53 location) in the regulation of apoptosis. For example, inhibition of HMGB1-mediated autophagy can increase caspase activity [121]. In addition to caspases, several non-caspase proteases such as calpain (Ca^{2+}-dependent proteases) may play a role in the execution of apoptosis. Interestingly, HMGB1 deletion can enhance calpain activity and trigger cleavage of Beclin-1 and ATG5 [122]. Thus, HMGB1 is an important regulator of the cross talk between apoptosis and autophagy. *In vivo*, conditional knockout of HMGB1 in pancreas, liver, intestinal epithelial and myeloid cells enhances sterile inflammation and infection partly through inhibition of autophagy and induction of apoptosis [38, 122-124]. In some cases, overexpression of HMGB1 renders cells sensitive to apoptosis in response to chemotherapy agents [125]. In addition, extracellular oxidized HMGB1 can induce caspase-dependent apoptosis in cancer cells [89]. These findings suggest that HMGB1 plays dual roles in the regulation of apoptosis.

3.3. Mechanism of HMGB1-mediated necrosis regulation

Necrosis includes accidental and regulated necrosis [2]. Partially, the term "necroptosis" has recently been used to describe regulated necrosis when cells lack the capacity to activate caspase [126]. Necroptosis is mediated by a signaling complex called necrosome, containing receptor-interacting protein (RIP)1, RIP3, and mixed-lineage kinase domain-like (MLKL) [127, 128], and can be inhibited by small molecule inhibitors necrostatin 1 and necrosulfonamide [129, 130] [126]. The fundamental causes of necrosis include calcium overload, ROS generation, cellular energy depletion, and membrane lipid injury [131]. PARP is a protein family involved in a number of cellular processes such as DNA repair and programmed cell death. Induced overactivation of PARP1 can lead to adenosine triphosphate (ATP) depletion and subsequent necrosis [132]. The process of necrosis ends with the leaking out of enzymes from lysosomes to digest cell components that are associated with HMGB1 release. *In vivo*, loss of HMGB1 in the pancreas increases L-arginine-induced apoptosis and necrosis due to oxidative injury [38]. However, the role of HMGB1 in necroptosis remains undefined.

4. HMGB1 release in cell death

4.1. Mechanism of HMGB1 release in autophagy

Autophagic cell death is not only a morphologic notion such as cell death associated with autophagosomes and autolysosomes, but also a functional description that excessive autophagy can cause cell death. Induction of autophagy facilitates both active secretion and passive release of HMGB1. For example, the release of HMGB1 is significantly increased in response to epidermal growth factor (EGF) receptor-targeted diphtheria toxin (DT-EGF)-induced autophagic cell death [65]. In contrast, suppression of ATG5, ATG7, or ATG12 expression by RNA interference (RNAi) inhibits autophagy and subsequent HMGB1 release after treatment with DT-EGF in cancer cells [65]. In addition, ATG5-dependent autophagy promotes HMGB1 secretion in fibroblasts and macrophages after treatment with Hank's balanced salt solution and lipopolysaccharide [97, 133]. Antioxidant (e.g., N-acetyl-L-cysteine) inhibits the cytosolic translocation and release of HMGB1 in starvation-induced autophagy [97]. In contrast, ROS and knockdown of superoxide dismutases (SOD)-1 and SOD2 by RNAi promotes cytosolic HMGB1 expression and extracellular release [134]. These findings suggest that oxidative stress is involved in autophagy-mediated HMGB1 release.

4.2. Mechanism of HMGB1 release in apoptosis

An early study indicated that HMGB1 is released only by necrotic cells, but not apoptotic cells [68]. However, recent studies demonstrated that activation of caspases and deoxyribonuclease (DNase) in apoptosis regulates HMGB1 release and activity in apoptosis. Caspase-3 and caspase-7 are important executioner caspases in apoptosis through amplified initiation signals from caspase-8 and caspase-9. Activation of caspase-3 and -7 induces mitochondrial complex 1 protein p75 NDUFS1 cleavage, which results in mitochondrial ROS production and subsequent HMGB1 release during apoptosis in DCs [135]. Interestingly, the activity of released HMGB1 in apoptosis is impaired, which promotes immunological resistance due to its oxidized form [135]. In addition to caspase-3 and -7, caspase-1 is responsible for HMGB1 cleavage and release in the response to pyroptosis in immune cells [72, 136, 137]. This caspase-1-mediated HMGB1 fragment can rescue apoptosis-induced immune tolerance in a RAGE-dependent manner [137]. Thus, different caspases can determine HMGB1 release and action in apoptosis and pyroptosis.

DNase is responsible for DNA fragmentation during cell death. Activation of DNA endonuclease (DNase-gamma) contributes to the degradation of DNA into nucleosomal units in apoptosis, whereas activation of DNase I and II facilitates degradation of DNA in necrosis [138]. The release of HMGB1 in apoptosis is triggered by DNase-gamma-mediated nucleosomal DNA fragmentation [139, 140]. Thus, inhibition of DNase gamma activity by small molecular compound DR396 can significantly diminish HMGB1 release in response to apoptotic stimuli [139, 140].

4.3. Mechanism of HMGB1 release in necrosis

The nuclear enzyme PARP1, which catalyzes the synthesis of the biopolymer poly(ADP-ribose), exhibits an essential role in the DNA damage response and genomic stability. However, overactivation of PARP1 may deplete the stores of cellular NAD+, which results in ATP depletion and subsequent necrosis [141]. In fact, HMGB1 release in necrosis is regulated by PARP1. Genetic and pharmacologic inhibition of PARP1 inhibits alkylating DNA damage agent-mediated necrosis as well as HMGB1 release [142]. In addition to necrosis, activation of PARP1 also contributes to HMGB1 translocation and release in autophagy and inflammation [106, 143]. Interestingly, loss of HMGB1 in tissue and cells accelerates DNA damage that results in PARP1 overactivation [144]. These findings suggest interplay between HMGB1 and PARP1 in response to cell death.

The RIP3-mediated signaling pathway is responsible for HMGB1 release in necroptosis. Upregulation of RIP3 expression *in vitro* triggers necroptosis, whereas suppression of RIP3 expression by RNAi *in vitro* or *in vivo* significantly inhibits inflammatory stimuli-induced necroptosis. RIP3-deficient mice exhibit resistance to sepsis and donor kidney inflammatory injury. This anti-inflammatory function of RIP3 is due partly to inhibition of HMGB1 and release of other DAMPs [145]. Additionally, RIP3-mediated necroptosis also contributes to dsRNA/poly (I:C)-induced HMGB1 release [146]. This process promotes retinal degeneration and triggers an inflammatory response in the mouse retina [146]. In addition to RIP3, interferon-β promoter stimulator 1 (an adaptor molecule for RIG-I-like receptors) may be critical for poly (I:C)-induced HMGB1 release in necroptosis in DCs.

Cysteine cathepsins are lysosomal proteases with housekeeping functions that also initiate a specific cell death pathway termed lysosomal cell death. This type of cell death includes morphological features of necrosis and apoptosis [147]. Cathepsin B, a critical lysosomal cysteine protease, mediates HMGB1 release following *L. pneumophila*-induced lysosomal cell death [148]. Mechanically, cathepsin B can translocate from the lysosome to the nucleus, where it interacts with HMGB1 and inhibits its cytosolic translocation. In addition to lysosomal cell death, cathepsin B is also important for HMGB1 release during inflammasome activation [149, 150]. In contrast, cathepsin D may facilitate HMGB1 release in necroptosis in DCs. The function of other cathepsins in the regulation of HMGB1 release remains unknown.

5. Concluding remarks

HMGB1 is a member of family containing the evolutionarily conserved HMG box domains. The function of HMGB1 depends on its cellular location. Besides its functions in the nucleus and cytosol, HMGB1 plays a critical role in extracellular signaling associated with multiple biological processes. Both intracellular and extracellular HMGB1 are involved in the regulation of types of cell death such as apoptosis, necrosis, and autophagy. Intracellular HMGB1 regulates cell death in both transactional-dependent or transactional-independent manners. In many cases, HMGB1 is a negative regulator of apoptosis and necrosis, but a positive regulator of autophagy. In addition, the release and activity of HMGB1 in cell death is context-dependent, which may cause immunogenic cell death or tolerogenic cell death. Future studies are needed to define the upstream and downstream signaling of HMGB1 in the regulation of cell death; clarify the interplay and cooperative role of HMGB1 and other DAMPs in the cell-death-associated microenvironment; and develop new therapeutic strategies for targeting HMGB1 in cell-death-associated disorders.

Acknowledgements

We apologize to the researchers who were not referenced due to space limitations. We thank Christine Heiner (Department of Surgery, University of Pittsburgh) for her critical reading of the manuscript. This work was supported by the USA National Institutes of Health (R01CA160417 and R01GM115366 to D.T.) and a 2013 Pancreatic Cancer Action Network-AACR Career Development Award (Grant Number 13-20-25-TANG). Work performed in support of findings reviewed in this manuscript was aided by core support of the University of Pittsburgh Cancer Institute (P30CA047904).

Author details

Daolin Tang* and Rui Kang*

*Address all correspondence to: tangd2@upmc.edu; kangr@upmc.edu

Department of Surgery, University of Pittsburgh Cancer Institute, University of Pittsburgh, USA

References

[1] Kroemer, G., Galluzzi, L., Vandenabeele, P., Abrams, J., Alnemri, E.S., Baehrecke, E.H., Blagosklonny, M.V., El-Deiry, W.S., Golstein, P., Green, D.R., Hengartner, M., Knight, R.A., Kumar, S., Lipton, S.A., Malorni, W., Nunez, G., Peter, M.E., Tschopp, J., Yuan, J., Piacentini, M., Zhivotovsky, B., Melino, G. (2009) Classification of cell death: recommendations of the Nomenclature Committee on Cell Death 2009. *Cell Death Differ* 16, 3-11.

[2] Galluzzi, L., Vitale, I., Abrams, J.M., Alnemri, E.S., Baehrecke, E.H., Blagosklonny, M.V., Dawson, T.M., Dawson, V.L., El-Deiry, W.S., Fulda, S., Gottlieb, E., Green, D.R., Hengartner, M.O., Kepp, O., Knight, R.A., Kumar, S., Lipton, S.A., Lu, X., Madeo, F., Malorni, W., Mehlen, P., Nunez, G., Peter, M.E., Piacentini, M., Rubinsztein, D.C., Shi, Y., Simon, H.U., Vandenabeele, P., White, E., Yuan, J., Zhivotovsky, B., Melino, G., Kroemer, G. (2012) Molecular definitions of cell death subroutines: recommendations of the Nomenclature Committee on Cell Death 2012. *Cell Death Differ* 19, 107-120.

[3] Green, D.R., Ferguson, T., Zitvogel, L., Kroemer, G. (2009) Immunogenic and tolerogenic cell death. *Nat Rev Immunol* 9, 353-363.

[4] Galluzzi, L., Bravo-San Pedro, J.M., Vitale, I., Aaronson, S.A., Abrams, J.M., Adam, D., Alnemri, E.S., Altucci, L., Andrews, D., Annicchiarico-Petruzzelli, M., Baehrecke, E.H., Bazan, N.G., Bertrand, M.J., Bianchi, K., Blagosklonny, M.V., Blomgren, K., Borner, C., Bredesen, D.E., Brenner, C., Campanella, M., Candi, E., Cecconi, F., Chan, F.K., Chandel, N.S., Cheng, E.H., Chipuk, J.E., Cidlowski, J.A., Ciechanover, A., Dawson, T.M., Dawson, V.L., De Laurenzi, V., De Maria, R., Debatin, K.M., Di Daniele, N., Dixit, V.M., Dynlacht, B.D., El-Deiry, W.S., Fimia, G.M., Flavell, R.A., Fulda, S., Garrido, C., Gougeon, M.L., Green, D.R., Gronemeyer, H., Hajnoczky, G., Hardwick, J.M., Hengartner, M.O., Ichijo, H., Joseph, B., Jost, P.J., Kaufmann, T., Kepp, O., Klionsky, D.J., Knight, R.A., Kumar, S., Lemasters, J.J., Levine, B., Linkermann, A., Lipton, S.A., Lockshin, R.A., Lopez-Otin, C., Lugli, E., Madeo, F., Malorni, W., Marine, J.C., Martin, S.J., Martinou, J.C., Medema, J.P., Meier, P., Melino, S., Mizushima, N., Moll, U., Munoz-Pinedo, C., Nunez, G., Oberst, A., Panaretakis, T., Penninger, J.M., Peter, M.E., Piacentini, M., Pinton, P., Prehn, J.H., Puthalakath, H., Rabinovich, G.A., Ravichandran, K.S., Rizzuto, R., Rodrigues, C.M., Rubinsztein, D.C., Rudel, T., Shi, Y., Simon, H.U., Stockwell, B.R., Szabadkai, G., Tait, S.W., Tang, H.L., Tavernarakis, N., Tsujimoto, Y., Vanden Berghe, T., Vandenabeele, P., Villunger, A., Wagner, E.F., et al. (2015) Essential versus accessory aspects of cell death: recommendations of the NCCD 2015. *Cell Death Differ* 22, 58-73.

[5] Ashkenazi, A., Salvesen, G. (2014) Regulated cell death: signaling and mechanisms. *Annu Rev Cell Dev Biol* 30, 337-356.

[6] Linkermann, A., Stockwell, B.R., Krautwald, S., Anders, H.J. (2014) Regulated cell death and inflammation: an auto-amplification loop causes organ failure. *Nat Rev Immunol* 14, 759-767.

[7] Vanden Berghe, T., Linkermann, A., Jouan-Lanhouet, S., Walczak, H., Vandenabeele, P. (2014) Regulated necrosis: the expanding network of non-apoptotic cell death pathways. *Nat Rev Mol Cell Biol* 15, 135-147.

[8] Zitvogel, L., Kepp, O., Kroemer, G. (2010) Decoding cell death signals in inflammation and immunity. *Cell* 140, 798-804.

[9] Bianchi, M.E. (2007) DAMPs, PAMPs and alarmins: all we need to know about danger. *J Leukoc Biol* 81, 1-5.

[10] Rubartelli, A., Lotze, M.T. (2007) Inside, outside, upside down: damage-associated molecular-pattern molecules (DAMPs) and redox. *Trends Immunol* 28, 429-436.

[11] Tang, D., Kang, R., Coyne, C.B., Zeh, H.J., Lotze, M.T. (2012) PAMPs and DAMPs: signal 0s that spur autophagy and immunity. *Immunol Rev* 249, 158-175.

[12] Zhang, Q., Kang, R., Zeh, H.J., 3rd, Lotze, M.T., Tang, D. (2013) DAMPs and autophagy: cellular adaptation to injury and unscheduled cell death. *Autophagy* 9, 451-458.

[13] Hou, W., Zhang, Q., Yan, Z., Chen, R., Zeh Iii, H.J., Kang, R., Lotze, M.T., Tang, D. (2013) Strange attractors: DAMPs and autophagy link tumor cell death and immunity. *Cell death & disease* 4, e966.

[14] Goodwin, G.H., Sanders, C., Johns, E.W. (1973) A new group of chromatin-associated proteins with a high content of acidic and basic amino acids. *Eur J Biochem* 38, 14-19.

[15] Sharman, A.C., Hay-Schmidt, A., Holland, P.W. (1997) Cloning and analysis of an HMG gene from the lamprey Lampetra fluviatilis: gene duplication in vertebrate evolution. *Gene* 184, 99-105.

[16] Gariboldi, M., De Gregorio, L., Ferrari, S., Manenti, G., Pierotti, M.A., Bianchi, M.E., Dragani, T.A. (1995) Mapping of the Hmg1 gene and of seven related sequences in the mouse. *Mamm Genome* 6, 581-585.

[17] Ferrari, S., Ronfani, L., Calogero, S., Bianchi, M.E. (1994) The mouse gene coding for high mobility group 1 protein (HMG1). *J Biol Chem* 269, 28803-28808.

[18] Wen, L., Huang, J.K., Johnson, B.H., Reeck, G.R. (1989) A human placental cDNA clone that encodes nonhistone chromosomal protein HMG-1. *Nucleic Acids Res* 17, 1197-1214.

[19] Bustin, M. (2001) Revised nomenclature for high mobility group (HMG) chromosomal proteins. *Trends Biochem Sci* 26, 152-153.

[20] Giavara, S., Kosmidou, E., Hande, M.P., Bianchi, M.E., Morgan, A., d'Adda di Fagagna, F., Jackson, S.P. (2005) Yeast Nhp6A/B and mammalian Hmgb1 facilitate the maintenance of genome stability. *Curr Biol* 15, 68-72.

[21] Wu, Q., Zhang, W., Pwee, K.H., Kumar, P.P. (2003) Cloning and characterization of rice HMGB1 gene. *Gene* 312, 103-109.

[22] Kang, R., Chen, R., Zhang, Q., Hou, W., Wu, S., Cao, L., Huang, J., Yu, Y., Fan, X.G., Yan, Z., Sun, X., Wang, H., Wang, Q., Tsung, A., Billiar, T.R., Zeh, H.J., 3rd, Lotze, M.T., Tang, D. (2014) HMGB1 in health and disease. *Mol Aspects Med* 40, 1-116.

[23] Bianchi, M.E., Falciola, L., Ferrari, S., Lilley, D.M. (1992) The DNA binding site of HMG1 protein is composed of two similar segments (HMG boxes), both of which have counterparts in other eukaryotic regulatory proteins. *Embo J* 11, 1055-1063.

[24] Bonaldi, T., Talamo, F., Scaffidi, P., Ferrera, D., Porto, A., Bachi, A., Rubartelli, A., Agresti, A., Bianchi, M.E. (2003) Monocytic cells hyperacetylate chromatin protein HMGB1 to redirect it towards secretion. *Embo J* 22, 5551-5560.

[25] Huttunen, H.J., Fages, C., Kuja-Panula, J., Ridley, A.J., Rauvala, H. (2002) Receptor for advanced glycation end products-binding COOH-terminal motif of amphoterin inhibits invasive migration and metastasis. *Cancer Res* 62, 4805-4811.

[26] Yang, H., Hreggvidsdottir, H.S., Palmblad, K., Wang, H., Ochani, M., Li, J., Lu, B., Chavan, S., Rosas-Ballina, M., Al-Abed, Y., Akira, S., Bierhaus, A., Erlandsson-Harris, H., Andersson, U., Tracey, K.J. (2010) A critical cysteine is required for HMGB1 binding to Toll-like receptor 4 and activation of macrophage cytokine release. *Proc Natl Acad Sci U S A* 107, 11942-11947.

[27] Rowell, J.P., Simpson, K.L., Stott, K., Watson, M., Thomas, J.O. (2012) HMGB1-facilitated p53 DNA binding occurs via HMG-Box/p53 transactivation domain interaction, regulated by the acidic tail. *Structure* 20, 2014-2024.

[28] Li, J., Kokkola, R., Tabibzadeh, S., Yang, R., Ochani, M., Qiang, X., Harris, H.E., Czura, C.J., Wang, H., Ulloa, L., Warren, H.S., Moldawer, L.L., Fink, M.P., Andersson, U., Tracey, K.J., Yang, H. (2003) Structural basis for the proinflammatory cytokine activity of high mobility group box 1. *Mol Med* 9, 37-45.

[29] Wang, Q., Zeng, M., Wang, W., Tang, J. (2007) The HMGB1 acidic tail regulates HMGB1 DNA binding specificity by a unique mechanism. *Biochem Biophys Res Commun* 360, 14-19.

[30] Stros, M. (1998) DNA bending by the chromosomal protein HMG1 and its high mobility group box domains. Effect of flanking sequences. *J Biol Chem* 273, 10355-10361.

[31] Ueda, T., Chou, H., Kawase, T., Shirakawa, H., Yoshida, M. (2004) Acidic C-tail of HMGB1 is required for its target binding to nucleosome linker DNA and transcription stimulation. *Biochemistry* 43, 9901-9908.

[32] Sheflin, L.G., Fucile, N.W., Spaulding, S.W. (1993) The specific interactions of HMG 1 and 2 with negatively supercoiled DNA are modulated by their acidic C-terminal domains and involve cysteine residues in their HMG 1/2 boxes. *Biochemistry* 32, 3238-3248.

[33] Gong, W., Li, Y., Chao, F., Huang, G., He, F. (2009) Amino acid residues 201-205 in C-terminal acidic tail region plays a crucial role in antibacterial activity of HMGB1. *J Biomed Sci* 16, 83.

[34] Carballo, M., Puigdomenech, P., Palau, J. (1983) DNA and histone H1 interact with different domains of HMG 1 and 2 proteins. *Embo J* 2, 1759-1764.

[35] Cato, L., Stott, K., Watson, M., Thomas, J.O. (2008) The interaction of HMGB1 and linker histones occurs through their acidic and basic tails. *J Mol Biol* 384, 1262-1272.

[36] Travers, A.A. (2003) Priming the nucleosome: a role for HMGB proteins? *EMBO Rep* 4, 131-136.

[37] Celona, B., Weiner, A., Di Felice, F., Mancuso, F.M., Cesarini, E., Rossi, R.L., Gregory, L., Baban, D., Rossetti, G., Grianti, P., Pagani, M., Bonaldi, T., Ragoussis, J., Friedman, N., Camilloni, G., Bianchi, M.E., Agresti, A. (2011) Substantial histone reduction modulates genomewide nucleosomal occupancy and global transcriptional output. *PLoS Biol* 9, e1001086.

[38] Kang, R., Zhang, Q., Hou, W., Yan, Z., Chen, R., Bonaroti, J., Bansal, P., Billiar, T.R., Tsung, A., Wang, Q., Bartlett, D.L., Whitcomb, D.C., Chang, E.B., Zhu, X., Wang, H., Lu, B., Tracey, K.J., Cao, L., Fan, X.G., Lotze, M.T., Zeh, H.J., 3rd, Tang, D. (2014) Intracellular Hmgb1 inhibits inflammatory nucleosome release and limits acute pancreatitis in mice. *Gastroenterology* 146, 1097-1107.

[39] Yu, S.S., Li, H.J., Goodwin, G.H., Johns, E.W. (1977) Interaction of non-histone chromosomal proteins HMG1 and HMG2 with DNA. *Eur J Biochem* 78, 497-502.

[40] Yoshida, M., Makiguchi, K., Chida, Y., Shimura, K. (1984) Unwinding of DNA by nonhistone protein HMG1 and HMG2. *Nucleic Acids Symp Ser*, 181-184.

[41] Paull, T.T., Haykinson, M.J., Johnson, R.C. (1993) The nonspecific DNA-binding and -bending proteins HMG1 and HMG2 promote the assembly of complex nucleoprotein structures. *Genes Dev* 7, 1521-1534.

[42] Javaherian, K., Liu, J.F., Wang, J.C. (1978) Nonhistone proteins HMG1 and HMG2 change the DNA helical structure. *Science* 199, 1345-1346.

[43] Stros, M., Ozaki, T., Bacikova, A., Kageyama, H., Nakagawara, A. (2002) HMGB1 and HMGB2 cell-specifically down-regulate the p53- and p73-dependent sequence-specific transactivation from the human Bax gene promoter. *J Biol Chem* 277, 7157-7164.

[44] Liu, Y., Prasad, R., Wilson, S.H. (2010) HMGB1: roles in base excision repair and related function. *Biochim Biophys Acta* 1799, 119-130.

[45] Song, M.J., Hwang, S., Wong, W., Round, J., Martinez-Guzman, D., Turpaz, Y., Liang, J., Wong, B., Johnson, R.C., Carey, M., Sun, R. (2004) The DNA architectural protein HMGB1 facilitates RTA-mediated viral gene expression in gamma-2 herpesviruses. *J Virol* 78, 12940-12950.

[46] Zhang, M., Swanson, P.C. (2009) HMGB1/2 can target DNA for illegitimate cleavage by the RAG1/2 complex. *BMC Mol Biol* 10, 24.

[47] Kim, I.D., Shin, J.H., Kim, S.W., Choi, S., Ahn, J., Han, P.L., Park, J.S., Lee, J.K. (2012) Intranasal delivery of HMGB1 siRNA confers target gene knockdown and robust neuroprotection in the postischemic brain. *Mol Ther* 20, 829-839.

[48] Ueda, T., Shirakawa, H., Yoshida, M. (2002) Involvement of HMGB1 and HMGB2 proteins in exogenous DNA integration reaction into the genome of HeLa S3 cells. *Biochim Biophys Acta* 1593, 77-84.

[49] Polanska, E., Dobsakova, Z., Dvorackova, M., Fajkus, J., Stros, M. (2012) HMGB1 gene knockout in mouse embryonic fibroblasts results in reduced telomerase activity and telomere dysfunction. *Chromosoma* 121, 419-431.

[50] Einck, L., Soares, N., Bustin, M. (1984) Localization of HMG chromosomal proteins in the nucleus and cytoplasm by microinjection of functional antibody fragments into living fibroblasts. *Exp Cell Res* 152, 287-301.

[51] Guillet, F., Tournefier, A., Denoulet, P., Capony, J.P., Kerfourn, F., Charlemagne, J. (1990) High levels of HMG1-2 protein expression in the cytoplasm and nucleus of hydrocortisone sensitive amphibian thymocytes. *Biol Cell* 69, 153-160.

[52] Kuehl, L., Salmond, B., Tran, L. (1984) Concentrations of high-mobility-group proteins in the nucleus and cytoplasm of several rat tissues. *J Cell Biol* 99, 648-654.

[53] Ito, H., Fujita, K., Tagawa, K., Chen, X., Homma, H., Sasabe, T., Shimizu, J., Shimizu, S., Tamura, T., Muramatsu, S., Okazawa, H. (2015) HMGB1 facilitates repair of mitochondrial DNA damage and extends the lifespan of mutant ataxin-1 knock-in mice. *EMBO Mol Med* 7, 78-101.

[54] Tang, D., Kang, R., Livesey, K.M., Kroemer, G., Billiar, T.R., Van Houten, B., Zeh, H.J., 3rd, Lotze, M.T. (2011) High-mobility group box 1 is essential for mitochondrial quality control. *Cell Metab* 13, 701-711.

[55] Lee, H., Shin, N., Song, M., Kang, U.B., Yeom, J., Lee, C., Ahn, Y.H., Yoo, J.S., Paik, Y.K., Kim, H. (2010) Analysis of nuclear high mobility group box 1 (HMGB1)-binding proteins in colon cancer cells: clustering with proteins involved in secretion and extranuclear function. *J Proteome Res* 9, 4661-4670.

[56] Merenmies, J., Pihlaskari, R., Laitinen, J., Wartiovaara, J., Rauvala, H. (1991) 30-kDa heparin-binding protein of brain (amphoterin) involved in neurite outgrowth. Amino acid sequence and localization in the filopodia of the advancing plasma membrane. *J Biol Chem* 266, 16722-16729.

[57] Maugeri, N., Franchini, S., Campana, L., Baldini, M., Ramirez, G.A., Sabbadini, M.G., Rovere-Querini, P., Manfredi, A.A. (2012) Circulating platelets as a source of the damage-associated molecular pattern HMGB1 in patients with systemic sclerosis. *Autoimmunity* 45:584-587.

[58] Fuentes, E., Rojas, A., Palomo, I. (2014) Role of multiligand/RAGE axis in platelet activation. *Thromb Res* 133, 308-314.

[59] Passalacqua, M., Zicca, A., Sparatore, B., Patrone, M., Melloni, E., Pontremoli, S. (1997) Secretion and binding of HMG1 protein to the external surface of the membrane are required for murine erythroleukemia cell differentiation. *FEBS Lett* 400, 275-279.

[60] Hanspal, M., Hanspal, J.S. (1994) The association of erythroblasts with macrophages promotes erythroid proliferation and maturation: a 30-kD heparin-binding protein is involved in this contact. *Blood* 84, 3494-3504.

[61] Parkkinen, J., Rauvala, H. (1991) Interactions of plasminogen and tissue plasminogen activator (t-PA) with amphoterin. Enhancement of t-PA-catalyzed plasminogen activation by amphoterin. *J Biol Chem* 266, 16730-16735.

[62] Ciucci, A., Gabriele, I., Percario, Z.A., Affabris, E., Colizzi, V., Mancino, G. (2011) HMGB1 and cord blood: its role as immuno-adjuvant factor in innate immunity. *PLoS ONE* 6, e23766.

[63] Wang, H., Bloom, O., Zhang, M., Vishnubhakat, J.M., Ombrellino, M., Che, J., Frazier, A., Yang, H., Ivanova, S., Borovikova, L., Manogue, K.R., Faist, E., Abraham, E., Andersson, J., Andersson, U., Molina, P.E., Abumrad, N.N., Sama, A., Tracey, K.J. (1999) HMG-1 as a late mediator of endotoxin lethality in mice. *Science* 285, 248-251.

[64] Yang, L., Xie, M., Yang, M., Yu, Y., Zhu, S., Hou, W., Kang, R., Lotze, M.T., Billiar, T.R., Wang, H., Cao, L., Tang, D. (2014) PKM2 regulates the Warburg effect and promotes HMGB1 release in sepsis. *Nat Commun* 5, 4436.

[65] Thorburn, J., Horita, H., Redzic, J., Hansen, K., Frankel, A.E., Thorburn, A. (2009) Autophagy regulates selective HMGB1 release in tumor cells that are destined to die. *Cell Death Differ* 16, 175-183.

[66] Bell, C.W., Jiang, W., Reich, C.F., 3rd, Pisetsky, D.S. (2006) The extracellular release of HMGB1 during apoptotic cell death. *Am J Physiol Cell Physiol* 291, C1318-1325.

[67] Jiang, W., Bell, C.W., Pisetsky, D.S. (2007) The relationship between apoptosis and high-mobility group protein 1 release from murine macrophages stimulated with lipopolysaccharide or polyinosinic-polycytidylic acid. *J Immunol* 178, 6495-6503.

[68] Scaffidi, P., Misteli, T., Bianchi, M.E. (2002) Release of chromatin protein HMGB1 by necrotic cells triggers inflammation. *Nature* 418, 191-195.

[69] Duprez, L., Takahashi, N., Van Hauwermeiren, F., Vandendriessche, B., Goossens, V., Vanden Berghe, T., Declercq, W., Libert, C., Cauwels, A., Vandenabeele, P. (2011)

RIP kinase-dependent necrosis drives lethal systemic inflammatory response syndrome. *Immunity* 35, 908-918.

[70] Zhang, A., Mao, X., Li, L., Tong, Y., Huang, Y., Lan, Y., Jiang, H. (2014) Necrostatin-1 inhibits Hmgb1-IL-23/IL-17 pathway and attenuates cardiac ischemia reperfusion injury. *Transpl Int* 27, 1077-1085.

[71] Mitroulis, I., Kambas, K., Chrysanthopoulou, A., Skendros, P., Apostolidou, E., Kourtzelis, I., Drosos, G.I., Boumpas, D.T., Ritis, K. (2011) Neutrophil extracellular trap formation is associated with IL-1beta and autophagy-related signaling in gout. *PLoS ONE* 6, e29318.

[72] Lu, B., Nakamura, T., Inouye, K., Li, J., Tang, Y., Lundback, P., Valdes-Ferrer, S.I., Olofsson, P.S., Kalb, T., Roth, J., Zou, Y., Erlandsson-Harris, H., Yang, H., Ting, J.P., Wang, H., Andersson, U., Antoine, D.J., Chavan, S.S., Hotamisligil, G.S., Tracey, K.J. (2012) Novel role of PKR in inflammasome activation and HMGB1 release. *Nature* 488, 670-674.

[73] Tang, D., Kang, R., Zeh, H.J., 3rd, Lotze, M.T. (2011) High-mobility group box 1, oxidative stress, and disease. *Antioxid Redox Signal* 14, 1315-1335.

[74] Andersson, U., Tracey, K.J. (2011) HMGB1 is a therapeutic target for sterile inflammation and infection. *Annu Rev Immunol* 29, 139-162.

[75] Wang, H., Zhu, S., Zhou, R., Li, W., Sama, A.E. (2008) Therapeutic potential of HMGB1-targeting agents in sepsis. *Expert Rev Mol Med* 10, e32.

[76] Wang, H., Yang, H., Czura, C.J., Sama, A.E., Tracey, K.J. (2001) HMGB1 as a late mediator of lethal systemic inflammation. *Am J Respir Crit Care Med* 164, 1768-1773.

[77] Hori, O., Brett, J., Slattery, T., Cao, R., Zhang, J., Chen, J.X., Nagashima, M., Lundh, E.R., Vijay, S., Nitecki, D., et al. (1995) The receptor for advanced glycation end products (RAGE) is a cellular binding site for amphoterin. Mediation of neurite outgrowth and co-expression of rage and amphoterin in the developing nervous system. *J Biol Chem* 270, 25752-25761.

[78] Park, J.S., Svetkauskaite, D., He, Q., Kim, J.Y., Strassheim, D., Ishizaka, A., Abraham, E. (2004) Involvement of toll-like receptors 2 and 4 in cellular activation by high mobility group box 1 protein. *J Biol Chem* 279, 7370-7377.

[79] Tian, J., Avalos, A.M., Mao, S.Y., Chen, B., Senthil, K., Wu, H., Parroche, P., Drabic, S., Golenbock, D., Sirois, C., Hua, J., An, L.L., Audoly, L., La Rosa, G., Bierhaus, A., Naworth, P., Marshak-Rothstein, A., Crow, M.K., Fitzgerald, K.A., Latz, E., Kiener, P.A., Coyle, A.J. (2007) Toll-like receptor 9-dependent activation by DNA-containing immune complexes is mediated by HMGB1 and RAGE. *Nat Immunol* 8, 487-496.

[80] Chen, G.Y., Tang, J., Zheng, P., Liu, Y. (2009) CD24 and Siglec-10 selectively repress tissue damage-induced immune responses. *Science* 323, 1722-1725.

[81]   Tang, D., Lotze, M.T. (2012) Tumor immunity times out: TIM-3 and HMGB1. *Nat Immunol* 13, 808-810.

[82]   Chiba, S., Baghdadi, M., Akiba, H., Yoshiyama, H., Kinoshita, I., Dosaka-Akita, H., Fujioka, Y., Ohba, Y., Gorman, J.V., Colgan, J.D., Hirashima, M., Uede, T., Takaoka, A., Yagita, H., Jinushi, M. (2012) Tumor-infiltrating DCs suppress nucleic acid-mediated innate immune responses through interactions between the receptor TIM-3 and the alarmin HMGB1. *Nat Immunol* 13, 832-842.

[83]   Kang, R., Tang, D., Schapiro, N.E., Loux, T., Livesey, K.M., Billiar, T.R., Wang, H., Van Houten, B., Lotze, M.T., Zeh, H.J. (2014) The HMGB1/RAGE inflammatory pathway promotes pancreatic tumor growth by regulating mitochondrial bioenergetics. *Oncogene* 33, 567-577.

[84]   Xu, J., Jiang, Y., Wang, J., Shi, X., Liu, Q., Liu, Z., Li, Y., Scott, M.J., Xiao, G., Li, S., Fan, L., Billiar, T.R., Wilson, M.A., Fan, J. (2014) Macrophage endocytosis of high-mobility group box 1 triggers pyroptosis. *Cell Death Differ* 21, 1229-1239.

[85]   Bianchi, M.E. (2009) HMGB1 loves company. *J Leukoc Biol* 86, 573-576.

[86]   Urbonaviciute, V., Furnrohr, B.G., Weber, C., Haslbeck, M., Wilhelm, S., Herrmann, M., Voll, R.E. (2007) Factors masking HMGB1 in human serum and plasma. *J Leukoc Biol* 81, 67-74.

[87]   Tang, D., Billiar, T.R., Lotze, M.T. (2012) A Janus tale of two active high mobility group box 1 (HMGB1) redox states. *Mol Med* 18, 1360-1362.

[88]   Venereau, E., Casalgrandi, M., Schiraldi, M., Antoine, D.J., Cattaneo, A., De Marchis, F., Liu, J., Antonelli, A., Preti, A., Raeli, L., Shams, S.S., Yang, H., Varani, L., Andersson, U., Tracey, K.J., Bachi, A., Uguccioni, M., Bianchi, M.E. (2012) Mutually exclusive redox forms of HMGB1 promote cell recruitment or proinflammatory cytokine release. *J Exp Med* 209, 1519-1528.

[89]   Tang, D., Kang, R., Cheh, C.W., Livesey, K.M., Liang, X., Schapiro, N.E., Benschop, R., Sparvero, L.J., Amoscato, A.A., Tracey, K.J., Zeh, H.J., Lotze, M.T. (2010) HMGB1 release and redox regulates autophagy and apoptosis in cancer cells. *Oncogene* 29, 5299-5310.

[90]   LeBlanc, P.M., Doggett, T.A., Choi, J., Hancock, M.A., Durocher, Y., Frank, F., Nagar, B., Ferguson, T.A., Saleh, M. (2014) An immunogenic peptide in the A-box of HMGB1 protein reverses apoptosis-induced tolerance through RAGE receptor. *J Biol Chem* 289, 7777-7786.

[91]   Mizushima, N., Levine, B. (2010) Autophagy in mammalian development and differentiation. *Nat Cell Biol* 12, 823-830.

[92]   Klionsky, D.J., Emr, S.D. (2000) Autophagy as a regulated pathway of cellular degradation. *Science* 290, 1717-1721.

[93] Mizushima, N., Yoshimori, T., Levine, B. (2010) Methods in mammalian autophagy research. *Cell* 140, 313-326.

[94] Kroemer, G., Marino, G., Levine, B. (2010) Autophagy and the integrated stress response. *Mol Cell* 40, 280-293.

[95] Liu, Y., Shoji-Kawata, S., Sumpter, R.M., Jr., Wei, Y., Ginet, V., Zhang, L., Posner, B., Tran, K.A., Green, D.R., Xavier, R.J., Shaw, S.Y., Clarke, P.G., Puyal, J., Levine, B. (2013) Autosis is a Na+,K+-ATPase-regulated form of cell death triggered by autophagy-inducing peptides, starvation, and hypoxia-ischemia. *Proc Natl Acad Sci U S A* 110, 20364-20371.

[96] Xie, Y., Kang, R., Sun, X., Zhong, M., Huang, J., Klionsky, D.J., Tang, D. (2015) Post-translational modification of autophagy-related proteins in macroautophagy. *Autophagy* 11, 28-45.

[97] Tang, D., Kang, R., Livesey, K.M., Cheh, C.W., Farkas, A., Loughran, P., Hoppe, G., Bianchi, M.E., Tracey, K.J., Zeh, H.J., 3rd, Lotze, M.T. (2010) Endogenous HMGB1 regulates autophagy. *J Cell Biol* 190, 881-892.

[98] Huang, J., Ni, J., Liu, K., Yu, Y., Xie, M., Kang, R., Vernon, P., Cao, L., Tang, D. (2012) HMGB1 promotes drug resistance in osteosarcoma. *Cancer Res* 72, 230-238.

[99] Tang, D., Kang, R., Livesey, K.M., Cheh, C.W., Farkas, A., Loughran, P., Hoppe, G., Bianchi, M.E., Tracey, K.J., Zeh, H.J., 3rd, Lotze, M.T. (2010) Endogenous HMGB1 regulates autophagy. *J Cell Biol* 190, 881-892.

[100] Zhang, Y., Cheng, Y., Ren, X., Zhang, L., Yap, K.L., Wu, H., Patel, R., Liu, D., Qin, Z.H., Shih, I.M., Yang, J.M. (2012) NAC1 modulates sensitivity of ovarian cancer cells to cisplatin by altering the HMGB1-mediated autophagic response. *Oncogene* 31, 1055-1064.

[101] Livesey, K., Kang, R., Vernon, P., Buchser, W., Loughran, P., Watkins, S.C., Zhang, L., Manfredi, J.J., Zeh, H.J., Li, L., Lotze, M., Tang, D. (2012) p53/HMGB1 complexes regulate autophagy and apoptosis. *Cancer Res* 72, 1996-2005.

[102] Song, J.X., Lu, J.H., Liu, L.F., Chen, L.L., Durairajan, S.S., Yue, Z., Zhang, H.Q., Li, M. (2014) HMGB1 is involved in autophagy inhibition caused by SNCA/alpha-synuclein overexpression: a process modulated by the natural autophagy inducer corynoxine B. *Autophagy* 10, 144-154.

[103] Chiang, H.S., Maric, M. (2011) Lysosomal thiol reductase negatively regulates autophagy by altering glutathione synthesis and oxidation. *Free Radic Biol Med* 51, 688-699.

[104] Liu, K., Huang, J., Xie, M., Yu, Y., Zhu, S., Kang, R., Cao, L., Tang, D., Duan, X. (2014) MIR34A regulates autophagy and apoptosis by targeting HMGB1 in the retinoblastoma cell. *Autophagy* 10, 442-452.

[105] Li, X., Wang, S., Chen, Y., Liu, G., Yang, X. (2014) miR-22 targets the 3' UTR of HMGB1 and inhibits the HMGB1-associated autophagy in osteosarcoma cells during chemotherapy. *Tumour Biol* 35, 6021-6028.

[106] Yang, M., Liu, L., Xie, M., Sun, X., Yu, Y., Kang, R., Yang, L., Zhu, S., Cao, L., Tang, D. (2015) Poly-ADP-ribosylation of HMGB1 regulates TNFSF10/TRAIL resistance through autophagy. *Autophagy*, 11, 214-224.

[107] Luo, Y., Yoneda, J., Ohmori, H., Sasaki, T., Shimbo, K., Eto, S., Kato, Y., Miyano, H., Kobayashi, T., Sasahira, T., Chihara, Y., Kuniyasu, H. (2014) Cancer usurps skeletal muscle as an energy repository. *Cancer Res* 74, 330-340.

[108] Sun, X., Tang, D. (2014) HMGB1-dependent and -independent autophagy. *Autophagy* 10, 1873-1876.

[109] Ashkenazi, A., Dixit, V.M. (1999) Apoptosis control by death and decoy receptors. *Curr Opin Cell Biol* 11, 255-260.

[110] Mehlen, P., Bredesen, D.E. (2011) Dependence receptors: from basic research to drug development. *Sci Signal* 4, mr2.

[111] Tait, S.W., Green, D.R. (2010) Mitochondria and cell death: outer membrane permeabilization and beyond. *Nat Rev Mol Cell Biol* 11, 621-632.

[112] Youle, R.J., Strasser, A. (2008) The BCL-2 protein family: opposing activities that mediate cell death. *Nature reviews. Molecular cell biology* 9, 47-59.

[113] Samejima, K., Earnshaw, W.C. (2005) Trashing the genome: the role of nucleases during apoptosis. *Nat Rev Mol Cell Biol* 6, 677-688.

[114] Riedl, S.J., Shi, Y. (2004) Molecular mechanisms of caspase regulation during apoptosis. *Nat Rev Mol Cell Biol* 5, 897-907.

[115] Susin, S.A., Lorenzo, H.K., Zamzami, N., Marzo, I., Snow, B.E., Brothers, G.M., Mangion, J., Jacotot, E., Costantini, P., Loeffler, M., Larochette, N., Goodlett, D.R., Aebersold, R., Siderovski, D.P., Penninger, J.M., Kroemer, G. (1999) Molecular characterization of mitochondrial apoptosis-inducing factor. *Nature* 397, 441-446.

[116] Daugas, E., Susin, S.A., Zamzami, N., Ferri, K.F., Irinopoulou, T., Larochette, N., Prevost, M.C., Leber, B., Andrews, D., Penninger, J., Kroemer, G. (2000) Mitochondrio-nuclear translocation of AIF in apoptosis and necrosis. *Faseb J* 14, 729-739.

[117] Li, L.Y., Luo, X., Wang, X. (2001) Endonuclease G is an apoptotic DNase when released from mitochondria. *Nature* 412, 95-99.

[118] Hegde, R., Srinivasula, S.M., Zhang, Z., Wassell, R., Mukattash, R., Cilenti, L., DuBois, G., Lazebnik, Y., Zervos, A.S., Fernandes-Alnemri, T., Alnemri, E.S. (2002) Identification of Omi/HtrA2 as a mitochondrial apoptotic serine protease that disrupts inhibitor of apoptosis protein-caspase interaction. *J Biol Chem* 277, 432-438.

[119] Brezniceanu, M.L., Volp, K., Bosser, S., Solbach, C., Lichter, P., Joos, S., Zornig, M. (2003) HMGB1 inhibits cell death in yeast and mammalian cells and is abundantly expressed in human breast carcinoma. *Faseb J* 17, 1295-1297.

[120] Kang, R., Zhang, Q., Zeh, H.J., 3rd, Lotze, M.T., Tang, D. (2013) HMGB1 in cancer: good, bad, or both? *Clin Cancer Res* 19, 4046-4057.

[121] Livesey, K.M., Kang, R., Vernon, P., Buchser, W., Loughran, P., Watkins, S.C., Zhang, L., Manfredi, J.J., Zeh, H.J., 3rd, Li, L., Lotze, M.T., Tang, D. (2012) p53/HMGB1 complexes regulate autophagy and apoptosis. *Cancer Res* 72, 1996-2005.

[122] Zhu, X., Messer, J.S., Wang, Y., Lin, F., Cham, C.M., Chang, J., Billiar, T.R., Lotze, M.T., Boone, D.L., Chang, E.B. (2015) Cytosolic HMGB1 controls the cellular autophagy/apoptosis checkpoint during inflammation. *J Clin Invest* 125, 1098-1110.

[123] Huang, H., Nace, G.W., McDonald, K.A., Tai, S., Klune, J.R., Rosborough, B.R., Ding, Q., Loughran, P., Zhu, X., Beer-Stolz, D., Chang, E.B., Billiar, T., Tsung, A. (2014) Hepatocyte specific HMGB1 deletion worsens the injury in liver ischemia/reperfusion: A role for intracellular HMGB1 in cellular protection. *Hepatology* 59, 1984-1997.

[124] Yanai, H., Matsuda, A., An, J., Koshiba, R., Nishio, J., Negishi, H., Ikushima, H., Onoe, T., Ohdan, H., Yoshida, N., Taniguchi, T. (2013) Conditional ablation of HMGB1 in mice reveals its protective function against endotoxemia and bacterial infection. *Proc Natl Acad Sci U S A* 110, 20699-20704.

[125] Guerin, R., Arseneault, G., Dumont, S., Rokeach, L.A. (2008) Calnexin is involved in apoptosis induced by endoplasmic reticulum stress in the fission yeast. *Mol Biol Cell* 19, 4404-4420.

[126] Linkermann, A., Green, D.R. (2014) Necroptosis. *N Engl J Med* 370, 455-465.

[127] Li, J., McQuade, T., Siemer, A.B., Napetschnig, J., Moriwaki, K., Hsiao, Y.S., Damko, E., Moquin, D., Walz, T., McDermott, A., Chan, F.K., Wu, H. (2012) The RIP1/RIP3 necrosome forms a functional amyloid signaling complex required for programmed necrosis. *Cell* 150, 339-350.

[128] Vandenabeele, P., Galluzzi, L., Vanden Berghe, T., Kroemer, G. (2010) Molecular mechanisms of necroptosis: an ordered cellular explosion. *Nat Rev Mol Cell Biol* 11, 700-714.

[129] Degterev, A., Huang, Z., Boyce, M., Li, Y., Jagtap, P., Mizushima, N., Cuny, G.D., Mitchison, T.J., Moskowitz, M.A., Yuan, J. (2005) Chemical inhibitor of nonapoptotic cell death with therapeutic potential for ischemic brain injury. *Nat Chem Biol* 1, 112-119.

[130] Degterev, A., Hitomi, J., Germscheid, M., Ch'en, I.L., Korkina, O., Teng, X., Abbott, D., Cuny, G.D., Yuan, C., Wagner, G., Hedrick, S.M., Gerber, S.A., Lugovskoy, A., Yuan, J. (2008) Identification of RIP1 kinase as a specific cellular target of necrostatins. *Nat Chem Biol* 4, 313-321.

[131] Zong, W.X., Thompson, C.B. (2006) Necrotic death as a cell fate. *Genes Dev* 20, 1-15.

[132] Ha, H.C., Snyder, S.H. (1999) Poly(ADP-ribose) polymerase is a mediator of necrotic cell death by ATP depletion. *Proc Natl Acad Sci U S A* 96, 13978-13982.

[133] Dupont, N., Jiang, S., Pilli, M., Ornatowski, W., Bhattacharya, D., Deretic, V. (2011) Autophagy-based unconventional secretory pathway for extracellular delivery of IL-1beta. *Embo J* 30, 4701-4711.

[134] Tang, D., Kang, R., Livesey, K.M., Zeh, H.J., 3rd, Lotze, M.T. (2011) High mobility group box 1 (HMGB1) activates an autophagic response to oxidative stress. *Antioxid Redox Signal* 15, 2185-2195.

[135] Kazama, H., Ricci, J.E., Herndon, J.M., Hoppe, G., Green, D.R., Ferguson, T.A. (2008) Induction of immunological tolerance by apoptotic cells requires caspase-dependent oxidation of high-mobility group box-1 protein. *Immunity* 29, 21-32.

[136] Kamo, N., Ke, B., Ghaffari, A.A., Shen, X.D., Busuttil, R.W., Cheng, G., Kupiec-Weglinski, J.W. (2013) ASC/caspase-1/IL-1beta signaling triggers inflammatory responses by promoting HMGB1 induction in liver ischemia/reperfusion injury. *Hepatology* 58, 351-362.

[137] Leblanc, P.M., Doggett, T.A., Choi, J., Hancock, M.A., Durocher, Y., Frank, F., Nagar, B., Ferguson, T.A., Saleh, M. (2014) An Immunogenic Peptide in the A-box of HMGB1 Reverses Apoptosis-Induced Tolerance Through RAGE. *J Biol Chem.* 289, 7777-7786.

[138] Nagata, S., Nagase, H., Kawane, K., Mukae, N., Fukuyama, H. (2003) Degradation of chromosomal DNA during apoptosis. *Cell Death Differ* 10, 108-116.

[139] Yamada, Y., Fujii, T., Ishijima, R., Tachibana, H., Yokoue, N., Takasawa, R., Tanuma, S. (2011) DR396, an apoptotic DNase gamma inhibitor, attenuates high mobility group box 1 release from apoptotic cells. *Bioorg Med Chem* 19, 168-171.

[140] Yamada, Y., Fujii, T., Ishijima, R., Tachibana, H., Yokoue, N., Takasawa, R., Tanuma, S. (2011) The release of high mobility group box 1 in apoptosis is triggered by nucleosomal DNA fragmentation. *Arch Biochem Biophys* 506, 188-193.

[141] Zong, W.X., Ditsworth, D., Bauer, D.E., Wang, Z.Q., Thompson, C.B. (2004) Alkylating DNA damage stimulates a regulated form of necrotic cell death. *Genes Dev* 18, 1272-1282.

[142] Ditsworth, D., Zong, W.X., Thompson, C.B. (2007) Activation of poly(ADP)-ribose polymerase (PARP-1) induces release of the pro-inflammatory mediator HMGB1 from the nucleus. *J Biol Chem* 282, 17845-17854.

[143] Yang, Z., Li, L., Chen, L., Yuan, W., Dong, L., Zhang, Y., Wu, H., Wang, C. (2014) PARP-1 mediates LPS-induced HMGB1 release by macrophages through regulation of HMGB1 acetylation. *J Immunol* 193, 6114-6123.

[144] Huang, H., Nace, G.W., McDonald, K.A., Tai, S., Klune, J.R., Rosborough, B.R., Ding, Q., Loughran, P., Zhu, X., Beer-Stolz, D., Chang, E.B., Billiar, T., Tsung, A. (2014) Hepatocyte specific HMGB1 deletion worsens the injury in liver ischemia/reperfusion: A role for intracellular HMGB1 in cellular protection. *Hepatology* 59, 1984-1997.

[145] Lau, A., Wang, S., Jiang, J., Haig, A., Pavlosky, A., Linkermann, A., Zhang, Z.X., Jevnikar, A.M. (2013) RIPK3-mediated necroptosis promotes donor kidney inflammatory injury and reduces allograft survival. *Am J Transplant* 13, 2805-2818.

[146] Murakami, Y., Matsumoto, H., Roh, M., Giani, A., Kataoka, K., Morizane, Y., Kayama, M., Thanos, A., Nakatake, S., Notomi, S., Hisatomi, T., Ikeda, Y., Ishibashi, T., Connor, K.M., Miller, J.W., Vavvas, D.G. (2014) Programmed necrosis, not apoptosis, is a key mediator of cell loss and DAMP-mediated inflammation in dsRNA-induced retinal degeneration. *Cell Death Differ* 21, 270-277.

[147] Aits, S., Jaattela, M. (2013) Lysosomal cell death at a glance. *J Cell Sci* 126, 1905-1912.

[148] Morinaga, Y., Yanagihara, K., Nakamura, S., Hasegawa, H., Seki, M., Izumikawa, K., Kakeya, H., Yamamoto, Y., Yamada, Y., Kohno, S., Kamihira, S. (2010) Legionella pneumophila induces cathepsin B-dependent necrotic cell death with releasing high mobility group box1 in macrophages. *Respir Res* 11, 158.

[149] Willingham, S.B., Bergstralh, D.T., O'Connor, W., Morrison, A.C., Taxman, D.J., Duncan, J.A., Barnoy, S., Venkatesan, M.M., Flavell, R.A., Deshmukh, M., Hoffman, H.M., Ting, J.P. (2007) Microbial pathogen-induced necrotic cell death mediated by the inflammasome components CIAS1/cryopyrin/NLRP3 and ASC. *Cell Host Microbe* 2, 147-159.

[150] Duncan, J.A., Gao, X., Huang, M.T., O'Connor, B.P., Thomas, C.E., Willingham, S.B., Bergstralh, D.T., Jarvis, G.A., Sparling, P.F., Ting, J.P. (2009) Neisseria gonorrhoeae activates the proteinase cathepsin B to mediate the signaling activities of the NLRP3 and ASC-containing inflammasome. J Immunol 182, 6460-6469.

Chapter 19

Necrosis as Programmed Cell Death

Ma. Luisa Escobar, Olga M. Echeverría and Gerardo H. Vázquez-Nin

Additional information is available at the end of the chapter

http://dx.doi.org/10.5772/61483

Abstract

The process of cell death is the mechanism through which organisms eliminate useless cells. Hence, it is a normal process that maintains homeostasis. Cell removal can be effectuated by several pathways that involve complex and regulated molecular events specific to each type of cell death. Diverse studies have evidenced different types of cell death: apoptosis, autophagy, and necrosis. This chapter presents a brief review of the apoptotic and autophagic cell death processes but focuses attention primarily on necrosis because it has previously been considered an accidental and uncontrolled form of cell death. More recent evidence, however, has shown that, under certain circumstances, necrosis is conducted by a controlled program called necroptosis, which is now included as a programmed cell death process.

Keywords: Apoptosis, autophagy, cell death, necrosis, necroptosis

1. Introduction

The tissular environment includes a series of signals that maintain the rates of cell proliferation and cell death so as to conserve structural integrity and functionality. Alterations in either one of these processes can cause certain pathologies, such as cancer. The cell death process is an ongoing event during the development of tissues and organs, one that is present right from embryonic development in the form of programmed cell death, which occurs under physiological conditions as a process that requires the active participation of highly regulated mechanisms. Traditionally, apoptosis was synonymous with programmed cell death; however, different routes of cell death, such as autophagy and, more recently, necroptosis, are now included as forms of programmed cell death. Morphologically, each one of these cell death processes has features that make it possible to distinguish among them. The different molecular mechanisms involved in the cell death pathways are responsible for the morphological changes that occur in the affected cells. However, each pathway has specific characteristics;

for instance, cellular shrinkage is a phenomenon that occurs in apoptosis [1], but is not present in other types of cell death, such as autophagy or necrosis. On the other hand, the extensive presence of vesicles evidences autophagy but does not appear to the same extent in the other types of cell elimination [2]. Necrosis, meanwhile, presents generalized swelling of membranous organelles that leads to cell rupture [3].

2. Brief description of two types of programmed cell death: apoptosis and autophagy

Apoptosis, or type I programmed cell death, is the most widely studied of the forms of cell death. Its morphological characteristics can be identified under light microscopy, and include cell shrinkage, compacting of the chromatin, blebbing of the cytoplasmic membrane, and, finally, the formation of apoptotic bodies [1] (Figure 1). Biochemically, apoptosis is characterized by the participation of proteases called caspases, orderly internucleosomal DNA fragmentation, phosphatidylserine externalization, changes in mitochondrial membrane permeability, and the participation of members of the Bcl-2 protein family.

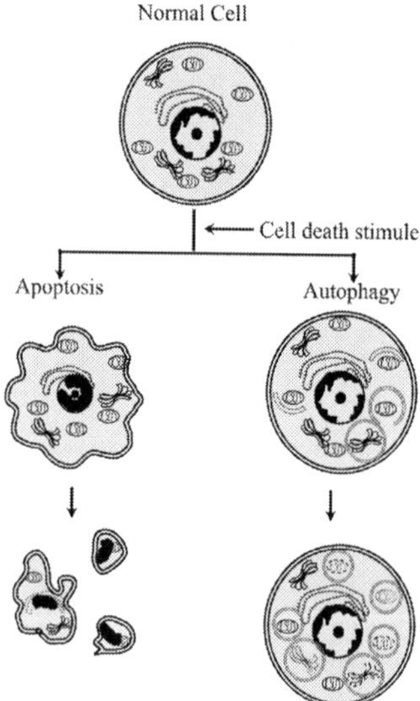

Figure 1. Schematic representation of the programmed cell death process type I (apoptosis) and type II (autophagy). Apoptosis is characterized by a cellular contraction, chromatin compaction, membranous blebs, and the formation of apoptotic bodies. Autophagy is characterized by the presence of a large number of autophagosomes with cytoplasmic content. Both types of cell death do not generate an inflammatory response since the cytoplasmic membrane is conserved until the cellular debris are eliminated by neighborhood or by specialized ones.

Caspases are cysteinyl-aspartate-specific proteases that are synthetized in an inactive form as zymogens called pro-caspases (Figure 2). It is this inactive form that allows the controlled execution of the cell death process. Caspases were first identified in the nematode *Ceanorhabditis elegans* [4], but homologous forms are present in mammals [5].

The hallmarks of apoptosis, such as DNA fragmentation and compacted chromatin, result from caspase activity. During apoptosis, DNA is fragmented into nucleosome size (200 bp) [6, 7]. The factor responsible for DNA fragmentation during apoptosis is a specific DNase (CAD, caspase-activated DNase) that is activated by active caspase-3 [8]. Active caspase-3, in turn, is involved in morphological cell changes during apoptosis, where it cleaves rho-associated kinase-1 (ROCK-1) in order to activate it and this, finally, affects the cytoskeletal arrangement causing the apoptotic shrinkage morphology [9].

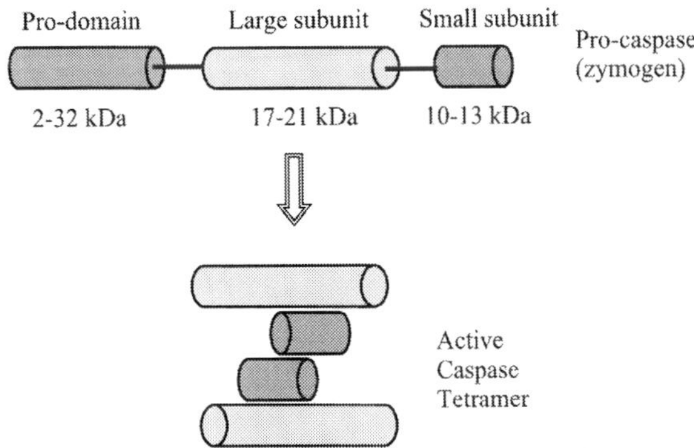

Figure 2. Caspases involved in the apoptotic process are synthetized in an inactive form as zymogens, are constituted by a pro-domain, a large subunit, and a small subunit. The zymogens are activated forming tetramers.

Apoptotic cell death is highly regulated by members of the B-cell lymphoma 2 (Bcl-2) family [10]. Bcl-2 family members have been classified as anti-apoptotic and pro-apoptotic proteins according to their Bcl-2 homology (BH) and domain organization (Figure 3). The presence of domains BH1, BH2, BH3, and BH4 corresponds to the group that inhibits apoptosis. The pro-apoptotic group, in contrast, is divided in two groups: those with domains BH1, BH2, and BH3, and those with only the BH3 domains (defined as BH3 only; see the review in [11]). This family of proteins performs its functions at the intracellular level inside the mitochondria, a key element in apoptosis.

Apoptosis can be initiated by two well-described routes: the extrinsic and intrinsic pathways (Figure 4). Extrinsic activation is conducted through the participation of death ligands (such as the tumor necrosis factor – TNF – superfamily, and TNF-related apoptosis-induced ligands, or TRAIL) with their cognate cell surface death receptors (such as TNF receptor 1, Fas, TRAIL receptor 1, or TRAIL receptor 2) (reviewed in [12]). Once the ligand recognizes and bonds to

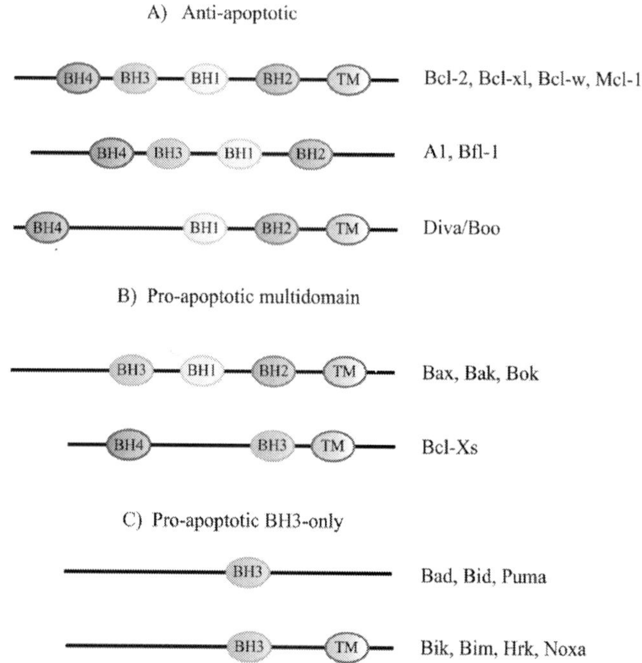

Figure 3. B-cell lymphoma 2 (Bcl-2) family proteins. A) General schematization of the structure of Bcl-2 proteins. B) The anti-apoptotic members – they possess all the four BH domains. C) The pro-apoptotic members which in turn are divided into two groups: multidomain and BH3 only.

its receptor, a series of intracellular complexes are formed to activate the initiator caspases (such as -8 and -10), which then activate the executioner caspases (such as -3, -6, and -7). In their activated form, these executioner caspases cleave multiple intracellular targets.

The intrinsic apoptotic pathway, in contrast, can be activated by various stimuli, including DNA damage, growth factor starvation, and oxidative stress [13]. During exposure of cells to these stimuli, the mitochondria are affected, since several members of the Bcl-2 family are activated and promote mitochondria outer membrane permeabilization (MOMP). The permeated external mitochondria membrane allows the release of cytochrome c (cyt c), which is associated with the Apaf-1 protein. The cyt c and Apaf-1 union then bonds to the initiator caspase-9 to form the complex that constitutes the apoptosome, which has the ability to activate the initiator caspases that perform their functions by cleaving specific cellular substrates.

The second process of cell death, autophagy, is a genetically programmed and evolutionarily conserved process that produces the degradation of obsolete organelles and proteins. It is activated by such extracellular stimuli as nutrient starvation, hypoxia, high temperature, and

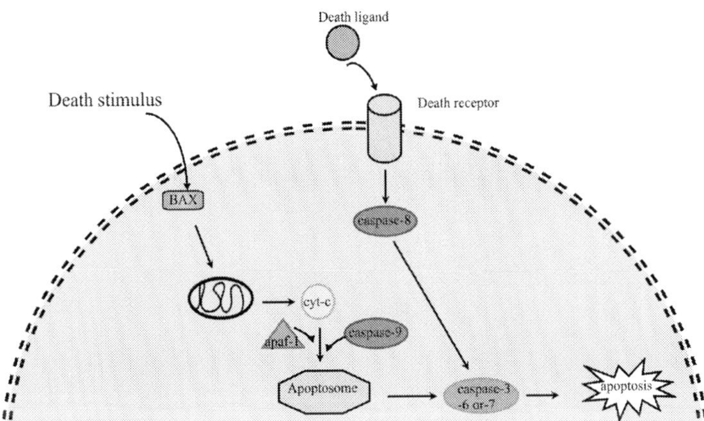

Figure 4. Routes of activation of apoptosis. The extrinsic route is mediated by external signals – a ligand – that activate to the membrane receptor. The ligand–receptor interaction induces the assembly of the death-inducing signaling complex (DISC) to promote the activation of caspase-8, which in turn is able to activate to the executor caspase -3, -6, or -7, conducting to the morphological changes of the apoptosis. The intrinsic route is directed by the mitochondrial outer membrane permeabilization, which allows the release of pro-apoptotic elements as cytochrome-C. Cytochrome-C induces the apoptosis protease-activating factor 1 (Apaf-1) to promote the activation of caspase-9 to assemble the apoptosome. The apoptosome is capable of activating to the executor caspases.

altered intracellular conditions, including the accumulation of damaged or superfluous organelles (reviewed in [2]).

In eukaryotic organisms, three types of autophagy have been described: microautophagy, macroautophagy (commonly called simply autophagy), and chaperone-mediated autophagy (Figure 5). Microautophagy involves the engulfing of cytoplasmic components directly at the level of the lysosome by means of an invagination process, while macroautophagy entails the formation of double-membrane vesicles that contain cellular components, which fuse with lysosomes to form an autophagolysosome. It is inside the autophagolysosome that the intra-vesicular components are degraded and, if possible, recycled by the cell (reviewed in [2 and 14]). Chaperone-mediated autophagy, finally, entails the participation of chaperones in recognizing the proteins designated for elimination by the lysosomes [14].

Autophagy is directed by *Atg* (AuTophaGy-related) genes, which are required to activate the signaling complex that triggers the formation of autophagosomes [15]. *Atg* genes were discovered in yeast, but many have orthologues in higher eukaryotes (Figure 6). Autophagosome formation entails the participation of the cytoplasmic protein LC3 (Atg8), which undergoes lipidation by phosphatidylethanolamine, and is then recruited to the nascent autophagosome membrane (Figure 7). Accumulation of lipidated LC3 protein (known as LC3-II) is used as a marker of autophagy [16].

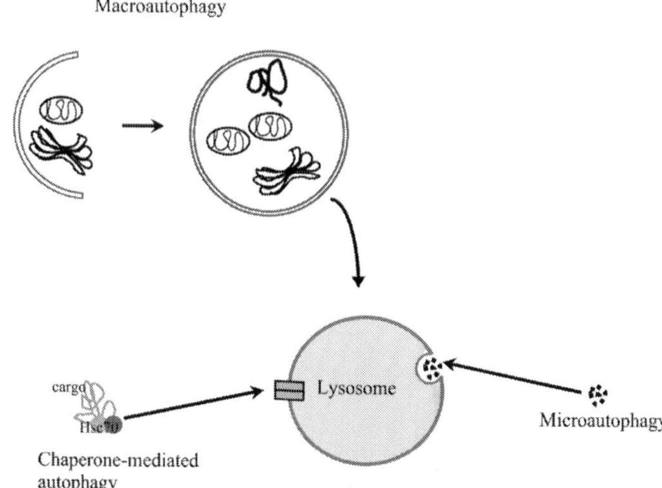

Figure 5. Schematic representations of the different mechanisms of autophagy. Macroautophagy – autophagy- implies the formation of a double-membrane vesicle, which engulfs cytoplasmic content that will be conducted to the lysosome to be degraded. Microautophagy is characterized by direct engulfing of cytoplasmic components by the lysosome. This process involves the remodeling of the membrane of the organelle by forming a lysosomal membrane invagination. During chaperone-mediated autophagy, the proteins to be degraded are targeted for an Hsp70, which in turn transport the target cargo to the lysosome.

Atg genes

Nucleation	Expansion	Fusion
Atg 6	Atg 1	
Atg 9	Atg 2	
Atg 11	Atg 3	Vam 3
Atg 13	Atg 4	Vam 7
Atg 14	Atg 5	VTI 1
Atg 17	Atg 8	YKT 6
Vps 15	Atg 10	
Vps 34	Atg 18	

Figure 6. Atg protein family includes more than 30 members that participate in the different events that constitute the autophagic process.

Autophagic cell death, or type II programmed cell death, is characterized by a massive engulfing of the cytoplasm by autophagic vesicles. This intense autophagic activity differs substantially from autophagy that occurs continuously at basal levels. Ultrastructural studies in Drosophila have revealed the accumulation of autophagic vacuoles in most larval tissues.

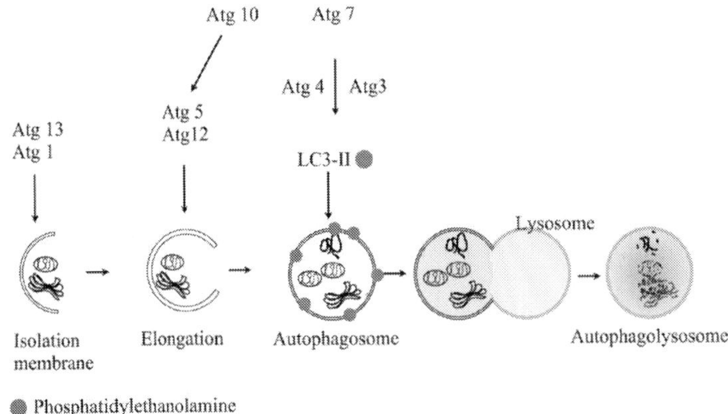

Figure 7. The formation of autophagic vacuoles involves the participation of different Atg proteins since the initial phases until the sequestration of cytoplasmic content. LC3 protein is lipidated by the phosphatidylethanolamine.

This type of programmed cell death begins with the degradation of cytoplasmic organelles by autophagy, though the cytoskeletal elements are conserved until the late stages of the process (reviewed in [17]).

3. Morphological characteristics of the necrosis process

Cell death caused by necrosis is considered an accidental, unprogrammed event that occurs under total ATP depletion [3], and that results from such external stimuli as extreme physical–chemical stress, heat, osmotic shock, mechanical stress, freezing, thawing, and high concentrations of hydrogen peroxide.

Necrotic cell death is characterized morphologically by generalized swelling of cell membranes, often accompanied by chromatin condensation and an irregular DNA degradation pattern [18]. The cytoplasmic membranes and membranous organelles dilate, and the increased cellular swelling causes the breakdown of the plasma membrane, which releases the cytoplasmic contents into the extracellular space (Figure 8). The release of the intracellular contents leads to massive cellular damage that affects neighboring cells, which explains why necrosis triggers inflammatory and autoimmune reactions. The necrosis process takes place in the absence of phagocytosis, and its final phase is characterized by the loss of the integrity of the cellular membrane. The release of the contents of necrotic cells includes molecules which act as signals that promote inflammation.

The most significant difference between programmed cell death (*i.e.*, apoptosis and autophagy) and necrosis is plasma membrane leakage and the consequent induction of inflammation in the affected tissue caused by the release of intracellular components [19].

because they included the multimerization of FADD in the absence of caspase activation. The morphology of apoptosis was not present, but ultrastructural analyses of those dying cells revealed necrotic morphological changes [29]. All these observations suggest that the receptor regulators of apoptosis were involved not only in that process, but also in the activation of a different signaling pathway that allows the formation of protein complexes which lead the cell toward a death process with necrotic features. These developments led to the emergence of a new concept of programmed cell death called necroptosis, whose morphological characteristics are similar to those of accidental necrosis, although the molecular events that occur indicate that it is a coordinated process. Necroptosis has been found under special conditions, where pro-apoptotic enzymes were absent or limited. In experimental embryo models, interdigital membrane regression in mouse embryos was effectuated by necrosis triggered by either caspase inhibition or drugs [30]. Necroptosis is thus a form of programmed cell death that has been demonstrated under experimental conditions, when apoptosis is inhibited.

6. Biochemical aspects of necroptosis

Several studies have succeeded in discerning the molecular events that occur during necroptosis, and it is those events that differentiate between necrosis (an accidental process) and necroptosis (a programmed process). Necroptosis has been observed in several pathological cell death events, such as ischemic brain injury, myocardial infarction, exotoxicity, and chemotherapy-induced cell death [31].

Necroptosis is morphologically characterized by several cytoplasmic changes. In fact, it is sometimes possible to distinguish the different degrees of advance of this process as the organelles swell, the cell membrane fragments, and cytoplasmic and nuclear disintegration become evident. During necroptosis, the nuclei remain intact and there is no massive caspase activation, chromatin condensation, spillage of cell contents, phagocytosis by macropinocytosis, lysosomal leakage, or oxidative bursts [32]. The term necroptosis has been introduced to identify a process of cell death with morphological characteristics distinct from those of apoptosis. Because there was no caspase activation during this process, it is called "caspase-independent".

Necroptosis is a programmed event that ends with the delivery of the cytoplasmic contents into the extracellular space. Membrane destabilization is a consequence of different intracellular mechanisms that generate osmotic changes by damaging the ion balance. When DNA damage occurred due to reactive oxygen species, the PARP protein was activated and began the reparation process; however, this process consumed abundant ATP and that reduction initiated a sequence of events that led to a deficient cellular efflux of calcium. The decreased ATP levels affected the activity of Na+-K+ ATPase, which requires a large amount of ATP in order to function correctly. This decreased Na+-K+ ATPase activity reduced calcium release and, as a result, increased intracellular calcium levels, leading to membrane destruction (reviewed in [33]). The breakdown of the cellular membrane, in turn, released several signals that activated the immune system. These soluble signals were proteins with pro-inflammatory properties that stimulate the recruitment of neutrophils to the site of cell death [34].

The mechanism proposed for the onset of necroptosis involves participation of the TNF-R (tumor necrosis factor-receptor), Fas, and TRAIL receptors, all of which belong to the tumor necrosis factor/nerve growth factor receptor superfamily and are involved in apoptotic cell death (Figure 10). Activation of TNF receptors by their ligands triggers different responses that involve pro-survival or pro-cell-death processes. Activated TNFR1 induces recruitment of TRADD, TRAF2/2, RIPK1, IAPs, and LUBAC to form a pro-survival complex that activates NF-kap-paB, JNK, and p38 MAPKs (reviewed in [35]). However, once this complex becomes establish-ed it is able to recruit FADD and procaspase-8, which produces a complex that could initiate either apoptosis or necroptosis. Under conditions of low levels of procaspase-8, a different complex is formed, – one that includes the receptor which interacts with protein 1 (RIP1 – a serine/threonine kinase activator -) and leads to the onset of necroptosis cell death. Biochemi-cally, necroptosis is defined as a form of cell death that is dependent on RIP1, which is the target protein in necrotic cell death induced by the TNFα, TRAIL, and CD95 receptors [36].

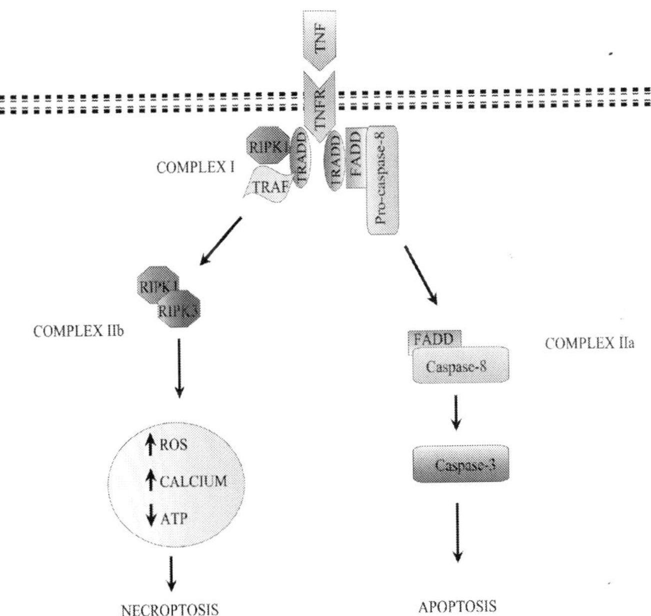

Figure 10. Biochemical aspects of necroptosis. TNFR recruits TRADD; this recruitment allows the formation of differ-ent complexes related to the RIPK1 protein or to the pro-caspase-8. The TRADD-FADD-pro-caspase-8 allows caspase-8 activation, which in turn activates the executer caspases, promoting the apoptosis process. Under conditions where caspase-8 is inhibited, the formation of the complex TRADD-FADD-RIPK1 initiates necroptosis.

The RIP3 is implicated in necroptosis during inflammatory responses to virus infections [37], and during cellular necrosis in response to the TNF-alpha family of death-inducing cytokines [38]. RIP3 mediates necroptosis induced by Smac mimetic and TNFα [38].

Activation of RIP can be directed not only by TNFR, but also by other death receptors, such as Fas. Activation of ligands associated with apoptotic cell death, such as Fas, in conditions that are unfavorable for apoptosis, that is, when caspases are absent or inhibited, allows RIP activation, which leads to death by necrosis [36, 39].

The biochemical process of necroptosis is a new and active field, so not all the routes of activation of this event have been determined. Diverse studies concur that kinase RIP is involved, as we have mentioned. Another protein, PARP-1 (poly (ADP-ribose) polymerase-1), has been shown to be involved in necrotic cell death by means of DNA-damaging agents [40], since PARP-1 is an abundant repair nuclear protein. PARP-1 is activated via TRAIL-induced necroptosis that induces ATP depletion [41]. DNA strand breaks promote the activation of PARP-1 (poly(ADP-ribose) polymerase-1) for DNA repair; PARP-1 binds to DNA strand breaks using NAD+ as substrate, generating a negatively charged PARP-1, which in turn is dissociated from DNA ends, allowing the DNA repair process [42, 43]. Intensive PARP-1 activation can generate increased NAD+ depletion and, as a result, an energy failure that leads to necrosis. In neuronal cells under severe oxidative stress, PARP-1 activation resulted in NAD + and ATP depletion that caused cell death [44].

7. Concluding remarks

Cell death is a normal event that controls tissue homeostasis. Today, we know that cells can be eliminated by means of different pathways that involve programmed or accidental mechanisms. Apoptotic cell death has been considered the major factor in physiological cell death, but recent evidence demonstrates that other routes of cell elimination – such as autophagy – also play important roles in maintaining homeostasis. A third route of cell death is necrosis, which was long considered an accidental form, characterized by general membrane swelling and ATP depletion. More recently, however, a new concept has been introduced: necroptosis. Necroptosis has been proposed as a kind of programmed cell death that is distinct from necrosis and apoptosis; one in which several signals involved in apoptosis participate significantly to initiate the process. It is important to note that necroptosis is an event that can be activated and regulated by such receptors as TNF or Fas – both of which are involved in the extrinsic route of apoptosis activation – when the pro-apoptotic signals are not available or are inhibited.

Activation of death receptors triggers a signaling cascade that includes activation of kinase RIP1, which in turn generates diverse intracellular reactions that lead the cell toward energy failure and conclude with the loss of intracellular homeostasis and the rupture of the membranes that, finally, generates an immunological response.

Acknowledgements

We thank CONACyT for grant 180526. The authors kindly thank Paul C. Kersey Johnson for reviewing the English word usage and grammar.

Author details

Ma. Luisa Escobar, Olga M. Echeverría and Gerardo H. Vázquez-Nin*

*Address all correspondence to: vazqueznin@ciencias.unam.mx

Laboratorio de Microscopía Electrónica, Depto. de Biología Celular, Facultad de Ciencias, Universidad Nacional Autónoma de México (UNAM), México

References

[1] Kerr JF, Wyllie AH, Currie AR. Apoptosis: a basic biological phenomenon with wide-ranging implications in tissue kinetics. *Br J Cancer* 1972; 26:239-257. http://www.ncbi.nlm.nih.gov/pubmed/4561027

[2] Levine B, Klionsky DJ. Development by self-digestion: molecular mechanisms and biological functions of autophagy. *Dev Cell* 2004; 6(4):463-477. DOI: 10.1016/S1534-5807(04)00099-1

[3] Nicotera P, Leist M, Ferrando-May E. Intracellular ATP, a switch in the decision between apoptosis and necrosis. *Toxicol Lett* 1998; 102-103: 139-142. PMID: 10022245

[4] Ellis RE, Jacobson DM, Horvitz HR. Genes required for the engulfment of cell corpses during programmed cell death in Caenorhabditis elegans. *Genetics* 1991; 129(1): 79-94. http://www.genetics.org/content/129/1/79.long

[5] Driscoll M. Cell death in C. elegans: molecular insights into mechanisms conserved between nematodes and mammals. *Brain Pathol* 1996; 6(4):411-425. PMID: 8944314

[6] Nagata S. Apoptotic DNA Fragmentation. *Exp Cell Res* 2000; 256:12-18. DOI: 10.1006/excr.2000.4834

[7] Blaisdell JO, Harrison L, Wallace SS. Base excision repair processing of radiation-induced clustered DNA lesions. *Radiat Prot Dosimetry* 2001, 97:25-31. PMID: 11763354

[8] McIlroy D, Sakahira H, Talanian RV, Nagata S. Involvement of caspase 3-activated DNase in internucleosomal DNA cleavage induced by diverse apoptotic stimuli. *Oncogene* 1999; 18: 4401-4408. http://www.stockton-press.co.uk/onc

[9]   Chang J, Xie M, Shah VR, Schneider MD, Entman ML, Wei L, Schwartz RJ. Activation of Rho-associated coiled-coil protein kinase 1 (ROCK-1) by caspase-3 cleavage plays an essential role in cardiac myocyte apoptosis. *Proc Natl Acad Sci U S A* 2006; 103:14495-14500. DOI: 10.1073/pnas.0601911103

[10]   Chipuk JE, Green DR. How do BCL-2 proteins induce mitochondrial outer membrane permeabilization? *Trends Cell Biol* 2008; 18:157-164. DOI: 10.1016/j.tcb. 2008.01.007

[11]   Tait SW, Green DR. Mitochondria and cell death: outer membrane permeabilization and beyond. *Nat Rev Mol Cell Biol* 2010; 11(9):621-632. DOI: 10.1038/nrm2952

[12]   Walczak H, Krammer PH. The CD95 (APO-1/Fas) and the TRAIL (APO-2L) apoptosis systems. *Exp Cell Res* 2000; 256: 58-66. DOI: 10.1006/excr.2000.4840

[13]   Schmitt CA, Lowe SW. Apoptosis and therapy. *J Pathol* 1999; 187:127-137. DOI: 10.1002/(SICI)1096-9896(199901)187:1<127::AID-PATH251>3.0.CO;2-T

[14]   Yang YP, Liang ZQ, Gu ZL, Qin ZH. Molecular mechanism and regulation of autophagy. *Acta Pharm Sinica* 2005; 26(12):1421-1434. DOI:10.1111/j. 1745-7254.2005.00235.x

[15]   Klionsky DJ, Emr SD. Autophagy as a regulated pathway of cellular degradation. *Science* 2000; 290(5497):1717-1721. DOI:10.1126/science.290.5497.1717

[16]   Kabeya Y, Mizushima N, Ueno T, Yamamoto A, Kirisako T, Noda T Kominami E, Ohsumi Y, Yoshimori T. LC3, a mammalian homologue of yeast Apg8p, is localized in autophagosome membranes after processing. *EMBO J* 2000; 19: 5720-5728. DOI: 10.1093/emboj/19.21.5720

[17]   Bursch W, Ellinger A, Gerner C, Schultze-Hermann R. Autophagocytosis and programmed cell death. In Klionsky D.J (Ed.), *Autophagy*, Landes Bioscience, Georgetown, TX, 2004.

[18]   Wyllie AH, Kerr JFR, Currie AR. Cell death: the significance of apoptosis. *Int Rev Cytol* 1980; 68:251-306. PMID: 7014501

[19]   Edinger AL, Thompson CB. Death by design: apoptosis, necrosis and autophagy. *Curr Opin Cell Biol* 2004; 16: 663-669. DOI:10.1016/j.ceb.2004.09.011

[20]   Festjens N, Vanden Berghe T, Vandenabeele P. Necrosis, a well-orchestrated form of cell demise: signalling cascades, important mediators and concomitant immune response. *Biochim Biophys Acta* 2006; 1757(9-10):1371-87. DOI:10.1016/j.bbabio. 2006.06.014

[21]   Griffiths EJ, Halestrap AP. Mitochondrial non-specific pores remain closed during cardiac ischaemia, but open upon reperfusion. *Biochem J* 1995; 307 (Pt. 1):93-98. http:// www.biochemj.org/bj/307/0093/3070093.pdf

[22] Wang L, Du F, Wang X. TNF-alpha induces two distinct caspase-8 activation pathways. *Cell* 2008; 133:693-703. DOI: 10.1016/j.cell.2008.03.036

[23] Bano D, Young KW, Guerin CJ, Lefeuvre R, Rothwell NJ, Naldini L, Rizzuto R, Carafoli E, Nicotera P. Cleavage of the plasma membrane Na+/Ca2+ exchanger in excitotoxicity. *Cell* 2005; 120:275-285. DOI:10.1016/j.cell.2004.11.049

[24] Yamashima T, Kohda Y, Tsuchiya K, Ueno T, Yamashita J, Yoshioka T, Kominami E. Inhibition of ischaemic hippocampal neuronal death in primates with cathepsin B inhibitor CA-074: a novel strategy for neuroprotection based on "calpain–cathepsin hypothesis." *Eur J Neurosci* 1998; 10:1723-1733. DOI: 10.1046/j.1460-9568.1998.00184.x

[25] Magno G, Joris I. Apoptosis, oncosis y necrosis. An overview of cell death. *Am J Pathol* 1995; 146:3-16. PMC1870771

[26] Roach HI, Clarke NM. Physiological cell death of chondrocytes in vivo is not confined to apoptosis. New observations on the mammalian growth plate. *J Bone Joint Surg Br* 2000; 82:601-613. PMID: 10855892

[27] Barkla DH, Gibson PR. The fate of epithelial cells in the human large intestine. *Pathology* 1999; 31(3):230-238. PMID: 10503269

[28] Kischkel FC, Hellbardt S, Behrmann I, Germer M, Pawlita M, Krammer PH, Peter ME. Cytotoxicity-dependent APO-1 (Fas/CD95)-associated proteins from a death-inducing signaling complex (DISC) with the receptor. *EMBO J* 1995; 14(22):5579-5588. http://www.ncbi.nlm.nih.gov/pmc/articles/PMC394672/

[29] Kawahara A, Ohsawa Y, Matsumura H, Uchiyama Y, Nagata S. Caspase-independent cell killing by Fas-associated protein with death domain. *JCB* 1998; 143(5): 1353-1360. DOI: 10.1083/jcb.143.5.1353

[30] Chautan M, Chazal G, Cecconi F, Gruss P, Golstein P. Interdigital cell death can occur through a necrotic and caspase-independent pathway. *Curr Biol* 1999; 9:967-970. DOI:10.1016/S0960-9822(99)80425-4

[31] Vandenabeele P, Declercq W, Vanden Berghe T. Necrotic cell death and 'necrostatins': now we can control cellular explosion. *TIBS* 2008; 33: 352-355. DOI: 10.1016/j.tibs.2008.05.007

[32] Vanden Berghe T, Vanlangenakker N, Parthoens E, Deckers W, Devos M, Festjens N, Guerin CJ, Brunk UT, Declercq W, Vandenabeele P. Necroptosis, necrosis and secondary necrosis converge on similar cellular disintegration features. *Cell Death Differ* 2010; 17:922-930. DOI: 10.1038/cdd.2009

[33] Ueda H, Fujita R. Cell death mode switch from necrosis to apoptosis in brain. *Biol Pharm Bull* 2004 Jul; 27(7):950-955. DOI.org/10.1248/bpb.27.950

[34] Yamasaki S, Ishikawa E, Sakuma M, Hara H, Ogata K, Saito T. Mincle is an ITAM-coupled activating receptor that senses damaged cells. *Nat Immunol* 2008; 9(10): 1179-88. DOI: 10.1038/ni.1651

[35] Kaczmarek A, Vandenabeele P, Krysko DV. Necroptosis: the release of damage-associated molecular patterns and its physiological relevance. *Immunity* 2013; 38(2): 209-23. DOI: 10.1016/j.immuni.2013.02.003

[36] Holler N, Zaru R, Micheau O, Thome M, Attinger A, Valitutti S, Bodmer JL, Schneider P, Seed B, Tschopp J. Fas triggers an alternative, caspase-8-independent cell death pathway using the kinase RIP as effector molecule. *Nat Immunol* 2000; 1(6):489-95. DOI:10.1038/82732

[37] Cho YS, Challa S, Moquin D, Genga R, Ray TD, Guildford M, Chan FK. Phosphorylation-driven assembly of the RIP1–RIP3 complex regulates programmed necrosis and virus-induced inflammation. *Cell* 2009; 137:1112-1123. DOI: 10.1016/j.cell.2009.05.037

[38] He S, Wang L, Miao L, Wang T, Du F, Zhao L, Wang X. Receptor interacting protein kinase-3 determines cellular necrotic response to TNF-alpha. *Cell* 2009; 137:1100-1111. DOI: 10.1016/j.cell.2009.05.021

[39] Lin Y, Choksi S, Shen HM, Yang QF, Hur GM, Kim YS, Tran JH, Nedospasov SA, Liu ZG. Tumor necrosis factor-induced nonapoptotic cell death requires receptor-interacting protein-mediated cellular reactive oxygen species accumulation. *J Biol Chem* 2004; 279:10822-10828. DOI: 10.1074/jbc.M313141200

[40] Andrabi SA, Dawson TM, Dawson VL. Mitochondrial and nuclear cross talk in cell death: parthanatos. *Ann NY Acad Sci* 2008; 1147:233-241. DOI: 10.1196/annals. 1427.014

[41] Jouan-Lanhouet S, Arshad MI, Piquet-Pellorce C, Martin-Chouly C, Le Moigne-Muller G, Van Herreweghe F, Takahashi N, Sergent O, Lagadic-Gossmann D, Vandenabeele P, Samson M, Dimanche-Boitrel MT. TRAIL induces necroptosis involving RIPK1/RIPK3-dependent PARP-1 activation. *Cell Death Differ* 2012; 19(12):2003-2014. DOI:10.1038/cdd.2012.90

[42] Lindahl T, Satoh MS, Poirier GG, Klungland A. Post-translational modification of poly(ADP-ribose) polymerase induced by DNA strand breaks. *Trends Biochem Sci* 1995; 20:405-411. DOI:10.1016/S0968-0004(00)89089-1

[43] Kim MY, Zhang T, Kraus WL. Poly(ADP-ribosyl)ation by PARP-1: 'PAR-laying' NAD+ into a nuclear signal. *Genes Dev* 2005; 19: 1951-1967.8-10 DOI:10.1101/gad. 1331805

[44] Diaz-Hernandez JI, Moncada S, Bolaños JP, Almeida A. Poly(ADP-ribose) polymerase-1 protects neurons against apoptosis induced by oxidative stress. Cell Death Differ 2007; 14:1211-1221. DOI:10.1038/sj.cdd.4402117